INTRODUCTION TO DISCRETE STRUCTURES

FOR COMPUTER SCIENCE AND ENGINEERING

INTRODUCTION
TO DISCRETE STRUCTURES
FOR COMPUTER SCIENCE
AND ENGINEERING

Franco P. Preparata *University of Illinois at Urbana-Champaign*
Raymond T. Yeh *University of Texas at Austin*

ADDISON-WESLEY PUBLISHING COMPANY
Reading, Massachusetts
Menlo Park, California · London · Amsterdam · Don Mills, Ontario · Sydney

This book is in the
Addison-Wesley Series in
Computer Science and Information Processing

Consulting Editors
Michael A. Harrison
Richard S. Varga

ISBN 0-201-05968-1
JKLMNOPQRS-MA-8987654321

PREFACE

TO THE STUDENT

It is almost superfluous to point to the pervasive expansion of the field of digital computers and systems in order to set the stage for the presentation of this book. That expansion has naturally determined the need for trained professionals with high levels of education. Thus college curricula in computer science and engineering have been established. These curricula—as one would expect in a rapidly growing field—represented at their inception a compromise between tradition and innovation, especially in the area of mathematical training. In fact, the required mathematical instruction was largely confined to infinitesimal analysis and calculus, and each computer-oriented course was supposed to develop its own formal prerequisites (usually algebraic notions). This policy clearly has two weaknesses. It is inefficient, because duplication of efforts is inevitable and precious time is withdrawn from specific subjects, and it is ineffective, because the presentation of background material for each individual course is understandably kept to an indispensable minimum. Although the exceptional student does not suffer from the lack of coordination, the background of the majority of students shows undesirable holes—a clear effect of piecemeal exposure to the pertinent mathematical material.

The need for complementing the studies in beginning analysis with an introduction to formal topics indispensable to students in the computer field was clearly recognized by distinguished groups of educators, formed in the last decade to formulate curricular recommendations. The most prominent such groups were the ACM Curriculum Committee on Computer Science and the COSINE Committee of the National Academy of Engineering. Specifically, the *Curriculum 68* report of the ACM committee (*Communications of the ACM*, Vol. 11, No. 3, March 1968) explicitly recommends course B3, "Introduction to Discrete Structures," which has strongly influenced this book, not only in title but also in selection of topics. In accordance with the recommendations of the committee, the purpose of this book is to serve as a text for an introductory course in discrete mathematics, with special emphasis on topics relevant both to the hardware and to the software of digital computers. It is addressed to upper-division undergraduates in computer science, computer engineering, and electrical engineering, and it may also be quite useful to students in business and economics with a strong computer orientation.

The purpose of Chapter 0 is to familiarize the reader with the terminology and techniques of mathematical discourse. Although we are perfectly aware that one cannot formally teach logical deduction, we see definite merit in presenting and discussing the meaning of various terms and phrases, such as axiom, lemma, theorem, proof by contradiction, inductive definitions, etc. The student is urged to read this chapter carefully, either to refresh his knowledge or to acquire familiarity with the type of language used. The reader will occasionally find sections or exercises marked with an asterisk (*). We have adopted the common convention of using this signal to indicate particular difficulty. The topic of the section may be highly specialized (and can therefore be skipped), or the section or exercise is especially difficult. In accordance with this convention, we have marked the next part of the preface with an asterisk to suggest that the student now turn to the text proper. The remainder of the preface is in fact intended as a guide for the instructor. To avoid being frustrated, therefore, the student should not read what follows until the completion of the course.

*TO THE INSTRUCTOR

The philosophy of our presentation is to proceed from general to specific. We start with the simplest structures, sets, and by adding properties, we reach more complicated structures. In this framework, graphs are viewed as relations on a set; lattices as a class of partially ordered sets; boolean algebras as a class of lattices; and so on. We feel that this approach may best illustrate to the reader how new properties add structure to formal systems.

All chapters have essentially the same format. Normally we approach new concepts by means of informal examples, highlighting the important features in simple language. In this manner the motivation is built up, and formal statements and definitions come much more easily. Examples—some with a computer flavor—are interwoven with the theoretical presentation in order to maintain a high degree of motivation. Each section is followed by a set of exercises, aimed at applying the concepts presented. Finally, each chapter concludes with a rather long section called "Notes and Applications," in which some side topics are presented, and important applications drawn from the computer field are discussed somewhat informally. Thus we have tried to maintain a close tie between theory and applications.

In Chapter 1, we introduce sets and their main properties. Some organization is added to sets by the introduction of binary relations, whose properties are analyzed in detail. Binary relations are then specialized to functions and relations on a set. The latter lead to the important classes of compatibility and equivalence relations, with their interesting characterization theorems. Note that no mention is made in this chapter of set algebra. Although set union and intersection are defined here for didactic purposes, the full appreciation of the structural richness of set algebra can be reached only at a much later stage (Chapter 5).

The graphical representations of binary relations are studied in Chapter 2, with the objective of investigating their topological properties. Directed and un-directed graphs are discussed, and concepts relating to connectedness—such as paths, chains, cut-sets, and articulation points—are studied. Graphs are then specialized to trees, and further specializations, such as rooted trees and oriented rooted trees, are investigated for their great usefulness as models. The section on path problems concerns itself with eulerian and hamiltonian circuits; planarity and map coloration are also discussed.

Chapter 3 presents algebraic systems. First we define binary operations, and by means of the concepts of associativity and inverses, the reader is led to appreciate the structural enrichment in passing from semigroups to groups. We then cover rings and their substructures in some detail. The important notions of iso-morphism and homomorphism are analyzed—in that order for clear pedagogical reasons. The study of congruence relations on semigroups leads to the fundamental theorem of homomorphisms. Finally, the various concepts are synthesized in a coherent and compact view through the unifying notion of universal algebra.

The concept of partial ordering opens Chapter 4, and its properties are brought to evidence. Next comes the study of posets, from which the notion of lattice emerges naturally. The general structure of lattices is investigated, and special attention is given to distributivity. Subsequently, the representation of lattices in terms of join-irreducible elements is presented very simply. Finally, partition lattices, an important class of lattices, are considered in connection with their relevance to the structural description of finite state machines.

Chapter 5 is devoted to boolean algebras. We first discuss set algebra and then view the power set of a set as a special case of a finite distributive lattice. The intuitive notion of complement in power sets is made abstract in the context of boolean algebras. We next show that boolean algebras are isomorphic to power sets (Stone theorem) and then tackle the general representation problem. Sub-sequently, the structural and the manipulative aspects of an algebraic system are clearly separated, and the calculus of boolean algebras is introduced as the application of a set of identities on free boolean algebras. The study of boolean functions precedes the discussion of important boolean algebras, such as propo-sitional calculus and switching algebra.

The first five chapters cover the main body of the course. A line had to be drawn somewhere between the topics to be presented and those to be excluded. Most of our decisions to exclude topics were based either on the fact that there were well-established courses covering them or on their advanced character. Therefore we have not discussed, for example, linear algebra, Galois fields, and predicate calculus. But there are two subjects which we felt had to be presented at least at a very introductory level: combinatorics and algorithms. We discuss these topics in Chapters 6 and 7, which have the same format as the other chapters but are somewhat shorter and less detailed. We suggest that they be viewed as long appendixes rather than as full-fledged chapters.

In Chapter 6, we present some elements of combinatorics that should enable the reader to solve simple counting problems. Permutations, combinations, distributions, the principle of inclusion and exclusion, and enumeration by recursion are presented, along with many simple, interesting applications. In the interest of completeness, we have included a simple introduction to Pólya's theory.

Chapter 7 begins with an informal discussion of the concept of algorithm, and through the notion of effective procedure, we arrive at Turing machines. We discuss the latter in some detail and use their halting problem as an eye-opener into the far-reaching field of algorithmic insolvability.

There are several possible teaching plans for a course in discrete structures, depending on the available time (semester, quarter, etc.) and the preference of the instructor. The instructor can therefore tailor his own course. To aid him in this endeavor we provide a detailed illustration of the prerequisite structure of the book as an appendix at the end of the text. Here, however, we wish to suggest what we believe is the essential core of a course in discrete structures: sets, relations, graphs, lattices, an introduction to boolean algebras, semigroups, and groups. After covering these core topics, the instructor may use any remaining time to cover supplementary topics, such as algebraic structures, combinatorics, and algorithms.

ACKNOWLEDGMENTS

The idea of writing this book grew out of the encouragement of our friends at the University of Texas, C. V. Ramamoorthy and C. L. Coates. Several persons have contributed significantly to the selection of some topics, the organization of the presentation, and the overall quality of the text. Some have remained unknown through the inscrutable reviewing process of publishers; we thank them all collectively. Others, whom we would like to thank individually for their extremely valuable recommendations, are D. E. Muller, C. L. Liu, and M. A. Harrison. Finally, we cannot omit expressing our gratitude to our editors at Addison-Wesley for their attention, which we are convinced was very special.

Urbana, Illinois   F.P.P.
Austin, Texas   R.T.Y.
March, 1973

CONTENTS

PRELIMINARY

0.1 INTRODUCTION

The purpose of this chapter is to familiarize the reader with the terminology and techniques of mathematical discourse. Essentially, mathematical discourse is logical deduction. There is substantial agreement on the fact that one develops an ability for rigorous deduction, not because one has been formally taught what deduction is, but rather because one is using an innate skill and has been slowly assimilating the "techniques" through exposure to a large number of examples. We concur entirely with this view. However, like many other well-developed disciplines, mathematical discourse has its own "jargon." Meanings of words and phrases are much sharper than the meanings of the same words and phrases in everyday usage. Therefore, we feel there is definite merit in discussing this jargon, so that the reader will clearly perceive the significance and the function of definitions, axioms, theorems, etc.

Another point needs some clarification. We shall frequently use such phrases as "true proposition," "valid statement," and the like. The adjectives "true" and "valid" (to be used synonymously), when applied to a statement, express the fact that we "accept" that statement or, equivalently, that we are convinced of its truth. Similarly, we shall use the phrases "logical argument," "valid proof," and the like; they mean that we are convinced of the truth of the conclusion reached through those arguments. In other words, we shall not discuss the philosophical grounds of truth and logic; we shall simply appeal to the reader as a "native user" of logic and try to guide him through the most common patterns of logical deduction. Later in this text we shall present a formal model for logic (Section 5.7 on propositional calculus).

We conclude this section by challenging the layman's view regarding the absolute "truth" of mathematical statements. In this regard we quote a modern paraphrase of Euclid's celebrated fifth postulate:

There is at most one line parallel to a given line through a given point not belonging to that line.[1]

Although this statement certainly holds in euclidean spaces, such as the plane, it fails on a spherical surface, since two lines whose interior angles on the same side sum exactly to two right angles meet on such a surface (for example, think of the earth as a sphere and of two meridians that intersect a parallel at right angles and yet meet at the poles).

This example indicates that the truth of a statement depends on what other statements are assumed to be true. A collection of statements held simultaneously true is usually referred to as a mathematical or formal system. We shall explore the nature of formal systems in the following sections.

0.2 THE NATURE OF FORMAL SYSTEMS

Roughly speaking, a formal system consists of notions, or concepts, and of assertions about their properties. It would be ideal to have a precise method that established the absolute nature both of the definitions of the concepts and of the validity of the assertions. However, a moment's reflection shows that such an ideal is unattainable since making use of other concepts is inherent in the nature of explanation. And in order to explain the meaning of these new concepts, one has to resort to different concepts (in order to avoid a vicious circle), and so on. Hence one finds oneself in a process which apparently cannot be brought to an end. To see how common this process is, one can look up almost any word in a dictionary, then look up the words used for its explanation, and so on. It usually takes only a few steps for the original word to reappear in an explanation. For example, starting from the word "dimension," we find in a standard dictionary the sequence of explanations shown in figure 0.2.1 (where the arrows mean "is explained as"). To avoid this circularity, certain principles regarding the construction of formal systems have emerged which are now illustrated.

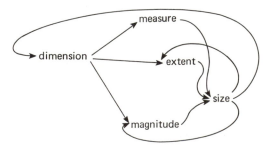

Figure 0.2.1. A sequence of explanations for the word "dimension"

---

1. For a discussion on Euclid's parallel axiom, the reader is referred to *Foundations of Geometry* by C. R. Wylie, Jr. (McGraw-Hill).

In constructing a formal or mathematical system, we first select a collection of concepts that seem to us immediately understandable (usually because of their closeness to immediate experience). These concepts are called *primitives* or *undefined terms*, and we use them without explaining their meanings. For example, the notions of "point" and "line" and the relationship [a point] "is on" [a line] are primitives in euclidean geometry. Thus the primitives are entities which enable us to begin the construction of a formal system. In other words, we must not employ any new concept in the system under consideration without first explaining its meaning with the help of primitives and of other concepts of the system whose meanings have been explained previously. The sentence that explains the meaning of a concept in this fashion is called a *definition*, and the concept whose meaning has been thereby explained is called a *defined* term. For example, using the primitives "point," "line," and [a line] "passes through" [a point], we define the concept of "collinearity" as follows:

Three points are *collinear* if and only if there is a line passing through them.

Note that the phrase "if and only if" in the preceding definition has the function of stipulating what meaning is to be attributed to a concept which has not occurred previously. The same phrase stipulates the *condition* that an object, denoted by the defined term, satisfies. Moreover, the "only if" part—referred to as *restriction* —says that no objects other than those covered by the definition receive that denomination. Usually the restriction is omitted for brevity, although it is understood that, *in definitions*, "if" has the same meanings as "if and only if."

As another example, consider a system whose primitives are positive integers, and we want to introduce the concept of "greater than." Again, it is necessary to explain the exact meaning of this concept in terms of notions already known. Let us assume that addition of positive integers (symbolized by "+") has been defined. We give the following definition of "greater than":

A positive integer x is *greater than* another positive integer y (denoted by the symbols $x > y$) if and only if there is a positive integer z such that $x = y + z$.

That definition states the equivalence of the two statements

x is greater than y

and

$x = y + z$ for some positive integer z.

In other words, the definition permits the transformation of the statement "x is greater than y" into an equivalent statement, which no longer contains the words "greater than" but is formulated entirely in terms of concepts we already understand.

We now turn our attention to the statements of the system. Again, we select a collection of statements, of whose self-evidence we are reasonably convinced;

these statements are called *axioms*[2] (or postulates) and are accepted as true. At the same time, we will not accept any other statements as true unless we have established their validity through logical arguments, using primitive terms, definitions, and other statements whose validity has been established previously. Statements in the system whose validity is established in this way are called *theorems*, and the process of establishing their validity is called a *proof*. It is appropriate to reflect briefly on the nature of statements. Consider the statement "Parallel lines meet." That statement has a *content* (parallel lines meet) and a *truth value* (*false* in euclidean geometry, *true* in elliptic geometry). It is convenient, for economy of writing, to symbolically designate statements in the system by capital letters, sometimes subscripted. Specifically, we shall use letters such as P and Q to denote statements with no reference to their truth values, and we shall denote *valid statements* with the letters $S_1, S_2, S_3, \ldots$. Note that a valid statement may be of the type "P is true" or of the type "If P is true, then Q is true." In other words, it is in general a statement about statements. A theorem S_k is a statement of the type "If P is true, then Q is true"; "P is true" is called the *hypothesis* or *premise*, and "Q is true" is called the *thesis* or *conclusion*. With this notation, a proof of a theorem S_k consists of a sequence of statements $S_1, S_2, \ldots, S_k$ such that for each $1 \leqslant i \leqslant k$, S_i either is an axiom or is deducible from statements $S_1, S_2, \ldots, S_{i-1}$ by logical arguments. The format of theorem-proof presentations is pretty standardized, with only minor variations. In this book we shall adhere to the following format:

Theorem. S_k.

Proof. $S_1, S_2, \ldots, S_k$.$\|$

where the symbol "$\|$" denotes completion of the proof. (In many texts $\|$ is replaced by Q.E.D., the abbreviation of the Latin sentence "quod erat demonstrandum," meaning "which was to be proved.")

Example 0.2.1. We shall now illustrate the concepts introduced. The undefined terms here are "points," denoted by P's, "lines," denoted by L's, and the relationship [point] "is on" [line]. The following axioms hold for the given primitives.

Axiom 1. If P_1 and P_2 are two distinct points, there is at least one line L such that both P_1 and P_2 are on it.

Axiom 2. If P_1 and P_2 are two distinct points, there is at most one line L such that both P_1 and P_2 are on it.

Axiom 3. If L_1 and L_2 are two lines, there is at least one point P on both L_1 and L_2.

---

2. The term axiom comes from the Greek ἄξιος (áxios) meaning "worthy"; therefore, an axiom is a fact worthy of general acceptance.

Axiom 4. There exists at least one line.

Axiom 5. Every line L must have three distinct points on it.

Axiom 6. Not all points are on the same line.

The reader can verify for himself that the system given in figure 0.2.2, which consists of seven points $P_1, P_2, \ldots, P_7$ and seven lines $L_1, L_2, \ldots, L_7$, does satisfy the six axioms given.

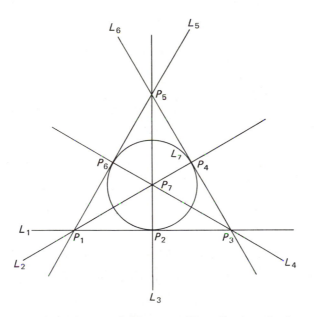

Figure 0.2.2. A system satisfying a specific collection of axioms

Two consequences of those six axioms will now be given as theo of the system under consideration.

Theorem 1. Any two distinct points are on a unique line.

Proof. Let P_1 and P_2 be two distinct points. By axiom 1, there is a line L such that both P_1 and P_2 are on L. However, by axiom 2, L must be the only line passing through P_1 and P_2. Hence the statement of the given theorem is true.‖

Theorem 2. There exist three points which are not on the same line.

Proof. By axiom 4, there exists a line L. By axiom 5, L must contain three points P_1, P_2, P_3. By axiom 6, there must exist a point P not on L. Finally, by theorem 1, P_1, P_2 cannot be simultaneously on a line L' different from L. Hence the three points P_1, P_2 and P cannot be on the same line.‖

There are two important properties concerning a collection of axioms, namely, independence and consistency.

A collection of axioms is said to be *independent* if it does not contain any superfluous statement. That is, no axiom is a logical consequence of the others. For example, axiom 2 in example 0.2.1 is independent of the other five axioms since it does not hold for the system shown in figure 0.2.3, which satisfies only axioms 1, 3, 4, 5, and 6.

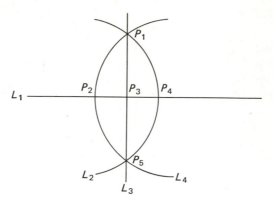

Figure 0.2.3. A geometry satisfying axioms 1, 3, 4, 5, and 6 in example 0.2.1

A formal system is said to be *consistent* or noncontradictory if no two axioms in the system contradict each other. In other words, if there are two contradictory statements, at least one cannot be proved, or a statement cannot be simultaneously true and false within the system. For example, let us assume that axiom 1 in example 0.2.1 is replaced by axiom 1′: If P_1 and P_2 are two distinct points, then there exist at least two lines L_1 and L_2 such that both P_1 and P_2 are on L_1 and on L_2. Then the resulting system is clearly not consistent, since axioms 1′ and 2 contradict each other.

0.3 ELEMENTS OF LOGICAL DEDUCTION

Usually in formal systems there are elements which are different descriptions of the same entity. For example, $(a + b)^2$ and $(a^2 + 2ab + b^2)$ are different expressions describing the same entity in the algebra of real numbers. It is natural to say that these two expressions are "equivalent" in the sense that they can be used interchangeably. We shall discuss the notion of "equivalence" at some length in Section 1.5. The equivalence we are considering here, however, is of a special type, because it enjoys what is called the *substitution property*. A general formulation of equivalence with the substitution property will be given in Section 3.4, but for the present the reader may think of it as "equality" in the algebra of real numbers. The following rule may then be taken as the statement of the characteristic property of an equivalence with the substitution property.

Substitution principle[3]. If a and b are two elements in a formal system that are equivalent with the substitution property, then b can be replaced by a in any statement or formula without affecting the validity of that statement or formula.

Consider the following example.

Example 0.3.1. The algebraic inequality a, $3x - 3y \geqslant 1$, with x and y positive real numbers, describes a region in the plane (x, y). Let P be the statement "If $3x - 3y \geqslant 1$, then $x > y$" (P is true). We know that b, $x \geqslant (1 + 3y)/3$, is equivalent to a with the substitution property, because they both describe the same region of the plane. Therefore we may substitute b for a in P, thereby obtaining Q, "If $x \geqslant (1 + 3y)/3$, then $x > y$," which is also true.

The substitution principle is a very useful rule in proving assertions.

As noted, earlier, a proof is a sequence of true statements $S_1, S_2, \ldots, S_k$. If the sequence is too long, there is a definite risk that the reader may lose the "thread" of the arguments. This may be a reason for breaking up the sequence into shorter sequences; the technique corresponds to "pausing," so to speak, on some intermediate statements $S_{i_1}, S_{i_2}, \ldots,$ of $S_1, S_2, \ldots, S_k$. This segmentation may also be motivated by aesthetic rather than utilitarian reasons; in fact, the "prover" may feel that some result S_i is so interesting and elegant that it is worth stating as a self-standing theorem. The segmentation therefore leads to a sequence of proofs of theorems. The writer can express the role and the importance he attributes to the theorems of the sequence by identifying them with different terms. A *theorem* is normally a main result; a preparatory result is called a *lemma*. On the other hand, a theorem which follows so obviously from the proof of another theorem that almost no proof is necessary is usually referred to as a *corollary*.[4]

Two theorems may be so related that one can be obtained from the other by interchanging the hypothesis and the thesis. If so, each theorem is called the *converse* of the other. For example, the converse of "In an isosceles triangle, two angles are equal" is "If two angles of a triangle are equal, the triangle is isosceles." If the converse of a theorem in a given system is also a theorem in the same system, then it is desirable to express the two as a single theorem. For example, combining

---

3. Principles, as a category of statements, have a rather anomalous status since sometimes they have the character of axioms, sometimes that of theorems. The term "principle," a product of tradition, is commonly used to emphasize the great importance of a statement.

4. The word *theorem* comes from the Greek θεορεῖν (theorein) meaning "to contemplate." *Lemma* comes from the Greek λαμβάνειν (lambanein), "to accept, to take"; that is, a lemma is a statement which is accepted as a premise. *Corollary* comes from the Late Latin *corollarium*, "gratuity"; that is, a corollary is a statement derived with no effort.

the two previous statements, we have "A triangle is isosceles if and only if two angles are equal."

The statement "A is true if and only if B is true" can be broken down into two separate statements: (1) A is true *if* B is true; and (2) A is true *only if* B is true. The first statement means that if B is true, then so is A. Alternatively, this statement may also be interpreted as "A sufficient condition for A to be true is that B is true" or, equivalently, "B is a sufficient condition for A." The second statement means that "the only way that A can be true is to have B true also." In other words, the statement "A is true only if B is true" is equivalent to "A necessary condition for A to be true is that B is true" or "B is a necessary condition for A." Theorems involving the phase "if and only if" or, equivalently, "a necessary and sufficient condition" are usually called *characterization* theorems. Note that there is no assurance that a necessary condition will also be a sufficient condition, or that a sufficient condition will also be a necessary condition. The customary notation for "B is a sufficient condition for A" is $B \Rightarrow A$ (read "B implies A") and the customary notation for "B is a necessary and sufficient condition for A" is $B \Leftrightarrow A$ (read "B implies and is implied by A" or "B means A" or "A and B are *logically equivalent*").

Example 0.3.1. In euclidean geometry, a necessary condition that a given quadrilateral $ABCD$ is a rectangle is that it must be a parallelogram. This condition is clearly not sufficient, since a parallelogram need not be a rectangle. On the other hand, that $ABCD$ is a square is clearly a sufficient condition for it to be a rectangle. However, this is not a necessary condition, since not all rectangles are squares. Finally, a rectangle is characterized by a necessary and sufficient condition given in the form of a theorem:

A parallelogram $ABCD$ is a rectangle if and only if one of its angles is a right angle.

We will now briefly discuss two commonly used techniques for proving a theorem in the form "If P is true, then Q is true." In the first type of proof, called *direct proof*, we derive a sequence of deductions (implications) of the form

If P is true, then Q_1 is true; if Q_1 is true, then Q_2 is true; ...; if Q_n is true, then Q is true.

In this manner, from the truth of P we deduce the truth of Q_1, from the truth of Q_1 the truth of Q_2, and so on. The proofs of both theorems in example 0.2.1 are direct.

When a direct proof is difficult to obtain, we may resort to a demonstration technique called *proof by contradiction*, which consists of the following four steps (for notational convenience we denote by NOT(P) the *negation* of a statement P).

1. We have as hypothesis a collection of statements, "P_1 is true," "P_2 is true," ..., "P_m is true," and as thesis "Q is true."

2. We make the arbitrary hypothesis "Q is false" (that is, "NOT(Q) is true").

3. From our collection of hypotheses, those given in (1) and the arbitrary one in (2), we deduce two simultaneous statements, "P is false" and "P is true"— a contradiction.

4. A contradiction is unacceptable, because our system is consistent. Since the only uncertified element in our line of reasoning was the arbitrary hypothesis "Q is false," we conclude that "Q is false" is false; that is, Q is true. This completes the proof of our thesis.

Example 0.3.3. The following is Euclid's classical proof that there is no largest prime integer. We adhere to the format given above.

1. We assume as hypothesis all the commonly known properties of the integers. Specifically, we assume the so-called fundamental theorem of arithmetic, which states that every integer can be written as the product of powers of primes (for example, $168 = 2^3 \cdot 3 \cdot 7$ is true).

2. We make the arbitrary hypothesis "NOT(Q) is true"; that is, "There is a largest prime."

3. Since there is a largest prime, there is only a finite number of primes. Let them be $p_1, p_2, \ldots, p_n$. Consider the number $N = p_1 p_2 \cdots p_n + 1$. Since N is an integer, it is a product of powers of primes. Let $P =$ "N is divisible by some prime." But when N is divided by any prime p_i (for $i = 1, 2, \ldots, n$), there is a remainder of 1, so we have NOT(P) = "N is not divisible by any prime"—a contradiction.

4. We must reject our arbitrary hypothesis; therefore we have proved the statement "There is no largest prime."‖

A similar technique one may resort to when a direct proof is hard to obtain is a *proof by contraposition* or *contrapositive proof*. It is based on the rule of logic (to be discussed formally in Section 5.7) that "P implies Q" if and only if "NOT(Q) implies NOT(P)." Each statement is the *contraposition* of the other, and they are logically equivalent.

We shall return to the topic of logical deduction later in the book, after developing some background for handling logic as a formal system (Section 5.7).

0.4 INDUCTION

Suppose that we want to determine the sum $S(n)$ of the positive integers not greater than n, for $n = 1, 2, 3, 4$. We may perform this task, by carrying out the actual calculation for each of the four integers, and we find: $S(1) = 1$; $S(2) = 1 + 2 = 3$; $S(3) = 1 + 2 + 3 = 6$; $S(4) = 1 + 2 + 3 + 4 = 10$. Let us now consider the theorem "The sum $S(n)$ of the consecutive positive integers not greater than n for $n = 1, 2, 3, 4$ is given by $n(n + 1)/2$". By noting that $1 \times 2/2 = 1$, $2 \times 3/2 = 3$,

$3 \times 4/2 = 6$, $4 \times 5/2 = 10$ and comparing the theorem's statement with these results, we would have proved the theorem. In its extreme simplicity, the theorem has been proved by *exhaustive enumeration*, or *perfect induction*. In other words, we have shown that the theorem's thesis holds true for *each* of the cases indicated by the theorem's hypothesis.

Exhaustive enumeration is clearly possible when the collection of cases is reasonably small. Such proof, however, may be very tedious for a large collection, and it is impossible for an infinite collection. In fact, consider the extension of the preceding theorem to an arbitrary positive integer n. Then, relying on accumulated experience, we conjecture that

$$S(n) = \frac{n(n+1)}{2} \, . \tag{0.4.1}$$

For any n greater than or equal to 1, we immediately see that

$$S(n+1) = S(n) + (n+1) = \frac{n(n+1)}{2} + n + 1 = \frac{(n+1)(n+2)}{2}. \tag{0.4.2}$$

Indeed, since we know the value of $S(1)$, we can prove that for any $n \geqslant 1$, $S(n) = n(n+1)/2$ by constructing $S(2)$ from $S(1)$, $S(3)$ from $S(2)$, ..., $S(n)$ from $S(n-1)$. (Of course, we assume here the usual arithmetic rules governing addition and multiplication of numbers.) We observe that this proof is for a particular n, and we cannot prove that, for all n, $S(n) = n(n+1)/2$, because the proof for all n must involve an infinite number of steps, i.e., constructing the infinite sequence $S(1)$, $S(2)$, $S(3)$, ...

Taking a closer look at the problem just described, we see that the statement "$S(n) = \frac{1}{2}n(n+1)$" (briefly denoted as P_n) is a statement dependent on the integer n (in other words, for any integer j we can formulate such a statement). However, we do not know whether the statement P_n is true. Assume now that $k \geqslant 1$ is the smallest integer for which P_n is true (in the above example, $k = 1$, but this is not always true), and assume that we have an argument that shows that if P_{n-1} is true, we can derive that P_n is true (for $n-1 \geqslant k$). In other words, we have a mechanism by which to assert the validity of P_n from the validity of P_{n-1}. Thus we imagine using this mechanism in going from "P_j is true" to "P_{j+1} is true," and so on for each pair of consecutive integers $(j, j+1)$ until we reach "P_n is true"; we said *we imagine*, because in reality we do not have to perform the sequence of derivations, since we know that for given n we *can* perform such a sequence of derivations. This is the spirit of a method of reasoning known as *finite* or *mathematical induction*, expressed by the following axiom.

Principle of finite induction. Let there be associated with each positive integer $n \geqslant k$ a statement P_n which is either true or false. If, "P_k is true", and for each positive integer j greater than or equal to k, "if P_j is true, then so is P_{j+1}," then "P_n is true" for all integers $n \geqslant k$. Formally, from the two statements

$$P_k \text{ is true, some } k \geqslant 1 \tag{0.4.3}$$

$$\text{if } P_j \text{ is true, then } P_{j+1} \text{ is true, } j \geqslant k \tag{0.4.4}$$

we derive

$$P_n \text{ is true, for every } n \geqslant k. \tag{0.4.5}$$

Statement (0.4.3) is referred to as the *base* (or basic step); statement (0.4.4) is referred to as *induction*, or the *inductive step*; the statement "P_j is true" is referred to as the *inductive hypothesis*. The formulation expressed by (0.4.3), (0.4.4), and (0.4.5) is referred to as the principle of (finite) *weak induction* because *only* P_j is assumed true in order to prove the truth of P_{j+1} (a weak assumption, therefore). In some problems, however, in order to prove the truth of P_{j+1}, we may have to assume the truth of all P_i for all i such that $k \leqslant i \leqslant j$ (apparently a stronger assumption). This procedure is referred to as *strong induction*, and its formulation follows.

From the conditions

$$P_k \text{ is true, some } k \geqslant 1 \tag{0.4.6}$$

and

$$\text{if } P_k, P_{k+1}, \ldots, P_j \text{ are true, then } P_{j+1} \text{ is true, } j \geqslant k \tag{0.4.7}$$

we derive

$$P_n \text{ is true, for every } n \geqslant k. \tag{0.4.8}$$

Despite appearances, strong finite induction is no stronger than weak finite induction. We use an inductive argument to prove their equivalence. Let "U_j is true" denote the statement "$P_k, P_{k+1}, \ldots, P_j$ are true." Then clearly "P_k is true" is equivalent to "U_k is true" (base). The statement "if $P_k, \ldots, P_j$ are true, then P_{j+1} is true" can be rephrased as "if $P_k, \ldots, P_j$ are true, then $P_k, \ldots, P_j, P_{j+1}$ are true"; that is, "if U_j is true, then U_{j+1} is true" (induction). It follows by the principle of induction that "U_n is true" for $n \geqslant k$, which of course implies that P_n is true.

The principle of finite induction can be used in two main applications: to prove theorems and to formulate *inductive definitions*. In the proof of theorems, the base and the inductive step are statements to be proved true; in the formulation of definitions, the base and the inductive step are stated as true.

We conclude this section with examples of a theorem proved by induction and of inductive definitions.

Example 0.4.1. We will use the principle of finite induction to prove that $S(n) = 1 + 2 + \cdots + n = n(n+1)/2$.

Proof. 1) basic step: $S(1) = 1 \times 2/2 = 1$ is clearly valid.

2) inductive step: Assume that $S(k) = k(k+1)/2$ for some $k \geqslant 1$. Then
$S(k+1) = S(k) + k + 1 = k(k+1)/2 + k + 1 = (k+1)(k+2)/2$.

Hence, by the principle of finite induction, we conclude that $S(n) = n(n + 1)/2$ for all positive integers n.∥

Example 0.4.2. Consider the sequence of integers $i_1, i_2, \ldots$ defined by the following conditions.

i) base: $i_1 = 1$.

ii) inductive step: If i_k is defined, then $i_{k+1} = i_k + 2$. We see that the given sequence is $1,3,5,7,\ldots$, namely, the sequence of odd positive integers.

The following example defines a notion that is useful in the formal systems considered in this book.

Example 0.4.3. Consider the collection of real numbers $\{c_1, c_2, \ldots\}$ referred to here as *constants* and the collection of *variables* $\{x_1, x_2, \ldots\}$. Constants and variables are referred to as *literals*. *Algebraic expressions* with respect to the algebraic operations $+$ (addition) and $\cdot$ (multiplication) are defined inductively, as follows:

i) base: A literal, α, is an algebraic expression.

ii) inductive step: If $\alpha_1, \alpha_2, \ldots, \alpha_n$ are algebraic expressions, so are $(\alpha_1 + \alpha_2 + \cdots + \alpha_n)$ and $(\alpha_1 \cdot \alpha_2 \cdot \cdots \cdot \alpha_n)$.

Note that in the preceding example we are able to write an algebraic expression as a string of symbols including—besides literals and the operation symbols $+$ and $\cdot$—nested symmetric pairs of parentheses enclosing the expressions obtained according to part (ii) of the preceding definition. Therefore, for some literals $x_1, x_2, x_3, x_4, c_1,$ and c_2,

$$((x_1 + x_2 + ((x_3 + x_4) \cdot (x_2 \cdot c_2))) \cdot (x_2 + (x_3 \cdot x_4) + (c_1 \cdot (x_1 + x_3)))) \qquad (0.4.9)$$

is an expression written according to the preceding rule. We refer to an algebraic expression containing literals $a_1, \ldots, a_n$ as "an expression in the n literals $a_1, a_2, \ldots, a_n$," and we denote it by $E(a_1, a_2, \ldots, a_n)$ or, for short, E. Since it may be difficult, especially in complicated cases, to spot which parenthesis symbols form a symmetric pair, not only parentheses, but various types of bracket pairs, such as [ ] and { }, are sometimes nested in expressions to simplify their reading, for example, {[( )]( )}. Other simplification rules are also common in the use of parentheses if no ambiguity results. For example, an expression of the type $(E_1 + (E_2 \cdot E_3))$ is often written simply as $(E_1 + E_2 \cdot E_3)$. (That is, multiplication is performed *before* addition in accordance with the so-called *precedence rule*.) Taking advantage of these rules for simplifying the notation, we may write the expression in (0.4.9) as

$$[x_1 + x_2 + (x_3 + x_4) \cdot x_2 \cdot c_2] \cdot [x_2 + x_3 \cdot x_4 + c_1 \cdot (x_1 + x_3)].$$

   The concept of expression as explained in the preceding paragraphs need not be restricted to the two operations of multiplication and addition. For example, arithmetic expressions use also the operations of subtraction and division.

   We want to stress that an expression, in general, is a special string of symbols of the following types: *literals*, *parentheses*, and *operation symbols*; and it is generated according to an inductive definition. Parentheses in expressions can be omitted according to a *precedence rule* defined on the operations. We shall return to these notational conventions in subsequent chapters, whenever the need arises for defining expressions within various formal systems.

CHAPTER 1

SETS AND BINARY RELATIONS

1.1 THE CONCEPT OF SET

The concept of *set* plays a fundamental role in the development of the structures whose investigation is the subject of this book. Indeed, as we shall subsequently see, all these structures are sets if their additional *characterizing* properties are temporarily ignored. Loosely phrased, since properties usually confer structure, a set is the most general unstructured structure. The word "set," in fact, is synonymous with *collection*.

The motivation for the concept of set has its deep roots in experience. Indeed, intuition points to the ability of intelligent beings to discriminate objects on the basis of the recognition of uniform properties (such as color and texture), and usually to associate abrupt discontinuities with object boundaries. We note, then, that a primary experiential activity is the *identification* of "things" in the environment, and that such identification is based on the distinctness of these things, which are usually referred to as "objects." Once this primary step has been accomplished, objects—or rather, their concepts—are associated with properties (or attributes), as suggested by sentences like "bananas are yellow and oblong." A successive natural step in the manipulation of abstractions is the association of objects on the basis of common properties. Thus we have, imperceptibly, landed on the idea of collection, that is, an aggregate of items whose association rests exclusively on the possession of some particular properties.

The preceding paragraph contains, in colloquial terms, the essentials of the concept of "set." The sentence

A set is a collection of distinct objects (characterized by common properties)

should be viewed more as a description than as a definition, since the notion of set is taken as primitive.

In the following discussions, sets will always be denoted by capital letters, and elements of sets by lower-case letters. The symbolic notation for indicating that a is an element of the set A, is $a \in A$. The symbol "$\in$" denotes set membership and is read "belongs to." The symbol "$\notin$" means "does not belong to."

At this point we may start thinking of "objects" as having a somewhat more abstract connotation than suggested by the experiential context. That is, an object is whatever can be distinguished (from other objects) and, within the current contextual circumstances, is not worth analyzing into simpler constituents. For this reason the objects of a set are referred to as *elements*.

For example, some sets are "the letters of the English alphabet," "the males in the United States," etc. Note that each set is defined by means of a sentence, expressing the properties that an element must possess in order to belong to a set. This is not accidental; in fact, we are exhibiting the so-called *axiom of specification*, which states that any (meaningful) property specifies a set. We have purposely placed the adjective "meaningful" within parentheses, because it suggests an assignment of importance to properties; yet a discussion of that importance is not pertinent to this exposition. The defining, or characteristic, property of a set is the criterion by which elements of the set are associated. In the set defined as "the males in the United States," an element—in this case, a human being— qualifies for membership if it satisfies the properties of being male and of being physically located in the United States.

As an alternative to specification by properties, a set may be described by listing its elements. This can usually be done when the number of elements is of manageable size or, as is often true, when the set has no other defining property than that membership—which is itself a property! This should not seem far-fetched; the set defined as "the letters of the English alphabet" is a clear example of this situation. The defining property here is, in fact, membership in a list of symbols used for writing words of the English language. We realize that this is equivalent to listing the letters A, B, C, ..., Z.

As the preceding discussion shows, the number of elements of a set may vary widely. The number of elements a set contains is an important characteristic of the set, and by long-established convention, it is taken for granted that one can determine the number of elements of a set by counting them. However, it is worth noting that the concept of number has its roots in the concept of set. Two sets have the same number of elements (or are of the same "size") if each and every element of one set can be paired with exactly one element of the other, and the converse (that is, the pairing leaves no leftover elements in either set). Technically, this pairing is referred to as a *one-to-one correspondence*. Accordingly, a number is the abstraction common to all the sets such that a one-to-one correspondence can be established between the elements of any two of these sets. With this premise, the set size is obtained by arbitrarily labeling each and every element of a set with distinct successive natural numbers (starting with 1), and then by determining the largest natural number assigned. Since natural numbers are called "cardinals," the technical term *cardinality* denotes the set size. The cardinality of a set A is denoted by the symbol $\#(A)$.

We see that any set which contains a finite number of elements can always be counted in this way, and most sets encountered in our experience are of this type. However, we need not restrict ourselves to finite sets, since we easily realize that there is no largest number which can be selected as a bound to set cardinalities. We are therefore led to the concept of infinite sets. An important class of infinite sets is given by the following definition.

Definition 1.1.1. A set A which is either finite or such that its elements can be placed in a one-to-one correspondence with the positive integers is called a *countable*[1] set. If A is countable and infinite, it is called a *countably infinite* set.

For example, the set of all positive even integers is countably infinite since each element of the form $2n$ in the set can be assigned a distinct integer n, with the initial assignment of 1 to the element "2" in the set.

It should not be construed, however, that all infinite sets are countable. Consider the set E of the real numbers between 0 and 1. We shall show, by contradiction, that E is not countably infinite (the proof is the celebrated Cantor's diagonal argument). Let us now assume that E is countable. Thus, there is an assignment of successive positive integers, starting with 1, to elements of E. Let r_i denote the element in E with integer i assigned to it. Then elements of E may be listed in an infinite array, where each element is expressed by its decimal expansion, as shown in figure 1.1.1 (a_{ij} being decimal digits). We note that not all numbers

$$r_1 = 0.a_{11}a_{12}a_{13}\ldots$$
$$r_2 = 0.a_{21}a_{22}a_{23}\ldots$$
$$\vdots \qquad \vdots$$
$$r_k = 0.a_{k1}a_{k2}a_{k3}\ldots$$

Figure 1.1.1. Illustration of Cantor's diagonal argument

have a unique decimal expansion, because there are numbers, called dyadic fractions (i.e., rational numbers such that the periodic part of their expansion is 0), which admit of two different decimal expansions. For example, the number 0.232400 ... is a dyadic fraction which also admits of the decimal expansion of 0.2323999.... To enforce uniqueness, we will always use the former expansion for dyadic fractions, i.e., the representation whose periodic part is 0.

Consider the real number $r' = 0.a'_{11}a'_{22}a'_{33}\ldots$, where $9 \neq a'_{ii} \neq a_{ii}$. That is, the number r' differs from r_1 in figure 1.1.1 in the first decimal place, from r_2 in the second decimal place, and in general, from r_k in the kth decimal place. Therefore, the real number r' is equal to no r_k, contradicting our assumption that the array contained all the real numbers in the closed interval $[0, 1]$. The reached contradic-

---

1. Many authors use the word "denumerable" instead of "countable."

tion shows that the set E is not countable: in fact, E is said to have the *cardinality of the continuum*. In this book, only countable sets will be considered.

Each finite set A may be described as the list of the symbols denoting its elements, separated by commas, within a pair of braces. For example, the set A, consisting of the symbols $*, \Delta, +, -$, can be represented by $A \equiv \{*, \Delta, +, -\}$. Since the set is simply a collection of distinct objects, the order of appearance of symbols within the braces is irrelevant. Thus A can also be represented by $\{\Delta, -, *, +\}$. Moreover, each symbol appears once and only once in the given list; so one never uses the notation $A \equiv \{\Delta, -, \Delta, *, +\}$, in which elements are repeated. Alternatively, as we noted earlier, a set A may be specified by its characteristic property. A very common notation is

$$A \equiv \{x | x \text{ has the property } P\},$$

which is read: "A is the set of all elements x such that x has the property P." The latter notation is the only one strictly applicable to the countably infinite sets, although for some sets the listing of just a few elements makes clear what the elements of the set are. For example, the set of the positive integers is adequately denoted by $\{1, 2, 3, \ldots\}$.

Although manipulations of sets will be the subject of extensive study in the following chapters, we now need two important definitions concerning the comparison of two sets.

Definition 1.1.2. A set A is said to be *equal* (or *identical*) to a set B (denoted by $A = B$) if every element of A is also an element of B, and vice versa.

When the "vice versa" aspect of that definition does not necessarily hold we have a relationship covered by the next definition.

Definition 1.1.3. A set A is said to be *contained* within a set B or is said to be a *subset* of B (denoted by $A \subseteq B$) if every element of A is also an element of B. If every element of A is an element of B but some element of B is not an element of A, then A is said to be *properly contained* within B or is said to be a *proper subset* of B (denoted by $A \subset B$). We will also use the notation $A \nsubseteq B$ to denote that A is not a subset of B.

For example, the integral multiples of 12 are a proper subset of the integral multiples of 3, and the inhabitants of Chicago are a proper subset of the inhabitants of the state of Illinois.

We conclude this section with additional useful terminology. Because of their closeness to the fundamentals of precise thinking, the concepts involved are quite intuitive; yet they may be viewed in considerably deeper contexts, and we shall have several opportunities to return to them in later sections and chapters.

Definition 1.1.4. The *union* of two sets A and B, denoted by $A \cup B$, is the set which contains all the elements of A and all the elements of B. Using the notation previously introduced,

$$A \cup B \equiv \{x | x \in A \text{ or } x \in B \text{ or both}\}.$$

For example, if $A \equiv \{a_1, a_2, a_3, a_4\}$ and $B \equiv \{a_2, a_3, a_7\}$, the union of A and B is the set $U \equiv \{a_1, a_2, a_3, a_4, a_7\}$. Obviously, since every element of A or of B is also an element of U, by definition 1.1.4 we have $A \subseteq U$ and $B \subseteq U$.

Definition 1.1.5. The *intersection* of two sets A and B, denoted by $A \cap B$, is the set which contains all the elements common to both A and B. In set notation,

$$A \cap B \equiv \{a \mid a \in A \text{ and } a \in B\}.$$

For example, the intersection of the two sets A and B described above is the set $C \equiv \{a_2, a_3\}$. Obviously, since each element of C is also an element of A and of B, by definition 1.1.3 we have $C \subseteq A$ and $C \subseteq B$.

Bear in mind that definition 1.1.5 is not always applicable in its present formulation. Indeed, the intersection of two sets A and B is said to be a set in all cases. What happens when A and B have no common elements? There are two ways out of this impasse: amend definition 1.1.5 by adding a clause which restricts its applicability, or "invent" a *conventional* set, the empty set, which will make the definition uniformly valid. This latter alternative makes a more general definition possible and hence is more desirable.

Definition 1.1.6. The *empty set*, denoted by $\varnothing$, is the set which contains no elements. Two sets whose intersection is the empty set are said to be *disjoint*.

Theorem 1.1.1. The empty set is a subset of every set.

Proof. We use contradiction. Suppose that the statement of the theorem is false. Then there exists a set, say A, such that $\varnothing$ is not a subset of A. Therefore, there must be an element of $\varnothing$ which is not an element of A. However, since $\varnothing$ has no elements, by definition, we have reached a contradiction. Hence the theorem holds.‖

Sets, their union, and their intersection are very effectively represented in a pictorial form known as the *Venn diagram*. A generic set A is represented by a connected (usually circular) domain of the plane, that is, by a set of points, as shown in figure 1.1.2(a). For two given sets A and B, the union of two sets, A and B and their intersection are represented by the crosshatched domains in figure 1.1.2(b) and (c), respectively.

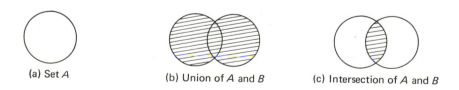

(a) Set A (b) Union of A and B (c) Intersection of A and B

Figure 1.1.2. The Venn diagram

EXERCISES 1.1

1. Give an example of a set whose elements are sets.

2. If $A \equiv \{a,b\}$, determine the truth or falsehood of each of the following statements.

   a) $\{b\} \in A$ b) $\varnothing \in A$
   c) $\{a\} \subseteq A$ d) $a \subset A$

3. Let $A \equiv \{a,b,c\}$, $B \equiv \{a,b\}$, $C \equiv \{b,c,d\}$, $D \equiv \{b\}$, and $E \equiv \{c,d\}$. State whether each of the following statements is valid or not.

   a) $B \subseteq A$ b) $D \neq C$
   c) E and D are disjoint   d) $A = B$
   e) $B \cap C = D$

4. Let $A \equiv \{1,2,3,4,5,6\}$, $B \equiv \{4,5,6,7,8,9\}$, $C \equiv \{2,4,6,8\}$, $D \equiv \{4,5\}$, $E = \{5,6\}$, $F \equiv \{4,6\}$. Let G be a set which satisfies the conditions

   $$G \subseteq A, \, G \subseteq B, \text{ and } G \subseteq C.$$

   Determine which of the sets A, B, C, D, E, F can be identified with G.

5. Determine which of the following sets is the empty set.

   a) $\{a | a \text{ is an odd integer and } a^2 = 4\}$ b) $\{a | a \text{ is an integer and } a + 9 = 9\}$
   c) $\{a | a \text{ is a positive integer and } a^2 < 1\}$ d) $\{a | a \text{ is an integer and } a < 1\}$

6. Let A, B, and C be countable sets; prove the following identities.

   a) $A \cup B = B \cup A$ b) $A \cap B = B \cap A$
   c) $(A \cup B) \cup C = A \cup (B \cup C)$ d) $(A \cap B) \cap C = A \cap (B \cap C)$
   e) $A \cap (B \cup C) = (A \cap B) \cup (A \cap C)$ f) $A \cup A = A \cap A = A$
   g) $A \cup (B \cap C) = (A \cup B) \cap (A \cup C)$

   (*Hint:* The set equality $A = B$ is usually established by showing that $A \subseteq B$ and $B \subseteq A$ hold simultaneously, that is, that each element of A is also an element of B and *vice versa*. For example, $A \cup B \subseteq B \cup A$ can be proved as follows: Let $a \in A \cup B$; then either $a \in A$ or $a \in B$ (or both). Suppose that $a \in A$; then we conclude that $a \in B \cup A$; that is, $A \cup B \subseteq B \cup A$, etc.)

7. Give an example for which $A \cup B = A \cup C$ and $B \neq C$.

8. Draw Venn diagrams illustrating the following conditions.

   a) $A \cup B \subset A \cup C$, but $B \nsubseteq C$ b) $A \cap B \subset A \cap C$, but $B \nsubseteq C$
   c) $A \cup B = C \cup B$, but $A \neq C$ d) $A \cap B = C \cap B$, but $A \neq C$

9. Give an example of an infinite set A such that there is a one-to-one correspondence between A and a proper subset of A.

10. What is the union of the set of positive integers divisible by 2 and the set of positive integers divisible by 3? What is their intersection?

11. Determine whether or not each of the following sets is countable.
   a) The set of positive integers divisible by 5
   b) The set of all integers
   *c) The set of all English sentences
   *d) The set of rational numbers
   *e) The set of all points in the plane whose coordinates are both rational numbers

*12. Show that $A \cup (B_1 \cap \cdots \cap B_n) = (A \cup B_1) \cap \cdots \cap (A \cup B_n)$.

*13. Show that every subset of a countable set is either finite or countable.

*14. Show that the union of a countable collection of countable sets is again countable.

1.2 BINARY RELATIONS BETWEEN SETS

A rather common experience is to associate spare parts and items of machinery. In particular, a given piece of equipment is related to all of its component parts, and each individual part is related to the various pieces of equipment of which it is a component. This "relatedness" is embodied in familiar "parts catalogues."

This real-world situation lends itself to the following mathematical description. Two finite sets $A \equiv \{a_1, a_2, \ldots, a_n\}$ (spare parts) and $B \equiv \{b_1, b_2, \ldots, b_m\}$ (machines) are given. The statement that "part a_i is used in machine b_j" may be represented by an *ordered* pair (a_i, b_j). The order of the elements in the pair is quite important. The pair (b_j, a_i), involving the same elements, a_i and b_j, but in reverse order, would represent the verbal statement "part b_j is used in machine a_i," which is clearly meaningless because b_j is not a part nor is a_i a machine, by assumption. Therefore, the association between parts and machines is adequately described by a set of ordered pairs of the type (a_i, b_j) where $a_i \in A$ and $b_j \in B$.

We immediately realize, however, that not every such pair is legitimate, since not every part is a component of every machine. Therefore, if we consider the set P of the nm possible pairs (a_i, b_j)—nm because each of the n elements $a_1, \ldots, a_n$ can be paired with each of the m elements $b_1, \ldots, b_m$—the relation "part is used in machine" is described by a subset of P. We note that the cardinality of set P is equal to the product of the cardinalities of the sets A and B; that is $\#(P) = \#(A) \cdot \#(B)$. Now we are ready for the following definitions.

Definition 1.2.1. The *cartesian product* $A \times B$ of the sets $A \equiv \{a_1, \ldots, a_n\}$ and $B \equiv \{b_1, \ldots, b_m\}$ is the set consisting of the ordered pairs (a_i, b_j) for every $a_i \in A$ and $b_j \in B$.

A convenient way to represent a cartesian product is a rectangular array having n rows and m columns, such that rows and columns are labeled in order $a_1, \ldots, a_n$ and $b_1, \ldots, b_m$, respectively. This array, conventionally referred to as an $n \times m$ *array*,[2] read "n by m array," contains nm intersections of rows with columns,

---

2. Usually such an array is called a *matrix*. Since, however, we shall very seldom refer to matrix calculus in this book, the term "matrix" will not be used hereafter.

and there is a one-to-one correspondence between such intersections and the ordered pairs of $A \times B$. Each such intersection will be referred to as a *position*.

For example, the cartesian product of sets $A \equiv \{a\}$ and $B \equiv \{b\}$ is the set $\{(a,b)\}$. The cartesian product of the empty set $\varnothing$ and any set A contains no element and hence must be the empty set; that is, $\varnothing \times A = \varnothing$. Note also that $A \times \varnothing = \varnothing$.

Definition 1.2.2. A *binary relation* α from a set A to a set B (or between two sets A and B) is a subset R_α of the cartesian product $A \times B (R_\alpha \subseteq A \times B)$.[3] The notation $a_i \, \alpha \, b_j$ indicates that $(a_i, b_j) \in R_\alpha$.

The adjective "binary" used in the definition implies that the relation defined exists between *two* sets. Indeed, it is possible to construct n-ary relations, which relate n sets $A_1, A_2, \ldots, A_n$ by defining subsets of n-term cartesian products of $A_1, A_2, \ldots, A_n$ in a manner analogous to the definition of a binary relation. However, n-ary relations will not be explicitly treated in this book.

A binary relation α from a set A with n elements to a set B with m elements is conveniently represented on an $n \times m$ array M_α by marking the positions in M_α which correspond to the pairs belonging to R_α with some special symbol, say "$\sqrt{}$," and leaving the other positions in M_α blank. (Note that "blank" itself becomes a symbol, so that "$\sqrt{}$" and "blank" have opposite meanings.) However, the choice of the two symbols $\{\sqrt{}, \text{blank}\}$ is entirely arbitrary, and any two other symbols represent a valid choice. Therefore, we may choose "1" and "0" in place of "$\sqrt{}$" and "blank," respectively. Valuable dividends will accrue from this choice in further developments, but as of now 1 and 0 are to be viewed exclusively as two different symbols, with no other significance, numerical or otherwise.

Example 1.2.1. Let $A \equiv \{a_1, a_2, a_3, a_4\}$, let $B \equiv \{b_1, b_2\}$, and let the relation α from A to B be defined by $R_\alpha \equiv \{(a_1, b_2), (a_2, b_1), (a_2, b_2)(a_4, b_2)\}$. The corresponding array representation is shown in figure 1.2.1.

$$M_\alpha = \quad \begin{array}{c|c|c|} & b_1 & b_2 \\ \hline a_1 & 0 & 1 \\ \hline a_2 & 1 & 1 \\ \hline a_3 & 0 & 0 \\ \hline a_4 & 0 & 1 \\ \hline \end{array}$$

Figure 1.2.1. Array (tabular) representation of the relation α

3. Note that when the phrase "relation between two sets A and B" is used, the order in which A and B appear is essential. Fortunately, the phrase "relation from A to B" does not suffer from the same mild ambiguity and is therefore preferred in this book. Furthermore, one must always keep in mind that R_α is a subset of $A \times B$; that is, it is not specified independently of $A \times B$. Therefore, a relation α from A to B is always to be intended as a triplet (A, B, R_α).

The given array representation of a relation, hereafter referred to as "tabular representation," is equivalent to a graphical representation, which is particularly attractive because of its pictorial merits. Elements of sets are represented by circles, technically referred to as *vertices*. An ordered pair is represented by an edge connecting the vertices that correspond to the pair elements, with an arrow-head pointing to the second element of the pair. This representation is usually referred to as a *directed edge*, or an *arc*. For example, an ordered pair (a_i, b_j) is graphically represented by

$$a_i \qquad\qquad b_j$$
$$\text{O}\!\!-\!\!\!-\!\!\!-\!\!\!-\!\!\!-\!\!\!\rightarrow\!\!\text{O}$$

and a relation simply becomes a collection of arcs. We immediately realize that each arc of the graphical representation corresponds to exactly one symbol "1" of the tabular representation; the converse is also true. Hence, given either representation, the other is readily obtainable.

Example 1.2.2. The relation given by R_α in example 1.2.1 has the graphical representation shown in figure 1.2.2.

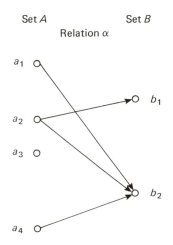

Figure 1.2.2. Graphical representation of the relation α

Consider now the following case. John is a son of Paul, and Paul is a brother of Peter, and we therefore conclude that John is a nephew of Peter. We have established a new relation, "nephew of," by combining two given relations, "son of" and "brother of," in the appropriate order. Combining two relations to obtain a new one is an instance of an important operation on relations, which we shall now formalize. Incidentally, each of the three mentioned relations ("son of," "brother of," and "nephew of") is from a set A (the human beings) to itself. However, this restriction is inessential, as expressed by the following general definition.

Definition 1.2.3. If α is a binary relation from the set A to the set B and β is a binary relation from the set B to the set C, then the ordered pair (α, β) is said to be *composable*. If (α, β) is a composable pair of binary relations, the *composite* of α and β, denoted by $\alpha\beta$, is a binary relation from the set A to the set C such that, for $a \in A$ and $c \in C$,

$$a(\alpha\beta)c \text{ if, for some } b \in B, \text{ both } a\alpha b \text{ and } b\beta c.$$

The application of this rule to α and β to obtain $\alpha\beta$ is called *composition of binary relations*. If (α, β) is a composable pair, we say that $\alpha\beta$ is defined.

Given a binary relation α from A to B, we refer to A as the "origin set" and to B as the "destination set" of the relation α. Then definition 1.2.3 indicates that, unless there is a set B which acts at the same time as the destination set of α and as the origin set of β, the composite of two arbitrary binary relations is not defined. We avoided this difficulty in the preceding example because any pair of binary relations from a set A to the same set A is automatically composable.

It is now interesting to examine how we can construct $\alpha\beta$, given α and β, that is, how we can construct a representation of $\alpha\beta$ (tabular or graphical), given the homologous representations of α and β.

Let us first consider the graphical representation. Here the three sets $A \equiv \{a_1, \ldots, a_n\}$, $B \equiv \{b_1, \ldots, b_m\}$, and $C \equiv \{c_1, \ldots, c_p\}$ can be represented as three linear arrays of vertices (see figure 1.2.3) vertically displayed. Arcs from A to B describe α, and arcs from B to C describe β. Now, the definition of composition states that an arc of $\alpha\beta$ exists if and only if there is at least one "path" from a vertex

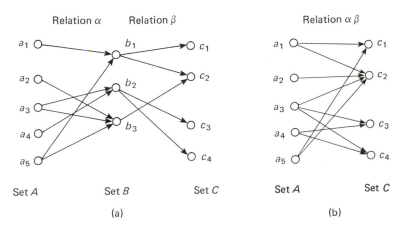

Figure 1.2.3. Graphical representation of the composition of binary relations

of A to a vertex of C, passing through some intermediate vertex of B. For example, in figure 1.2.3(a), there is a path from a_5 to c_2, since there is an arc from a_5 to b_3 and an arc from b_3 to c_2. Note that there are two paths from a_5 to c_2, but one is sufficient to establish $a_5(\alpha\beta)c_2$. The resulting composite relation $\alpha\beta$ is represented graphically in figure 1.2.3(b); an arc from a_i to c_k in figure 1.2.3(b) is present if and only if there is a corresponding path in figure 1.2.3(a).

Suppose now that for given sets A, B, C, and D we have three binary relations: α from A to B, β from B to C, γ from C to D. Then the composite $(\alpha\beta)\gamma$ is defined; indeed, we form first the composite $(\alpha\beta)$ since (α,β) is a composable pair, and then the composite $(\alpha\beta)\gamma$ since $((\alpha\beta),\gamma)$ is a composable pair.

Theorem 1.2.1. For binary relations α, β, and γ, such that $(\alpha\beta)\gamma$ is defined, the following equality holds.

$$(\alpha\beta)\gamma = \alpha(\beta\gamma) \tag{1.2.1}$$

Proof. We must show that every pair $(a,d)(a \in A,\ d \in D)$ in $R_{(\alpha\beta)\gamma}$ is also in $R_{\alpha(\beta\gamma)}$, and the converse. The proof is illustrated in figure 1.2.4. In order for $a[(\alpha\beta)\gamma]d$ to hold, there must be a vertex $c \in C$ such that $a(\alpha\beta)c$ and $c\gamma d$. But if $a(\alpha\beta)c$, there is a vertex $b \in B$ such that $a\alpha b$ and $b\beta c$. From $b\beta c$ and $c\gamma d$ it follows that $b(\beta\gamma)d$, which, together with $a\alpha b$, yields $a[\alpha(\beta\gamma)]d$ (see figure 1.2.4). The converse is proved in exactly the same way.$\|$

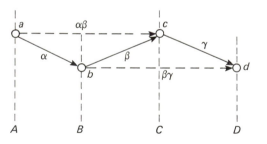

Figure 1.2.4. Illustrations of the proof of the associative property

Equality (1.2.1) is referred to as the *associative property*. By the associative property, we may omit parentheses when writing the composite $(\alpha\beta)\gamma$ or $\alpha(\beta\gamma)$; that is, $(\alpha\beta)\gamma = \alpha(\beta\gamma) = \alpha\beta\gamma$.

The procedure given for constructing the graphical representation of $\alpha\beta$ leads readily to a procedure for constructing its tabular representation. The tabular representations of α and β are an $n \times m$ array M_α and an $m \times p$ array M_β, respectively. Consider now the generic row (or column) of M_α (or M_β). Each row (or column) is a sequence of the symbols 1 and 0, ordered from left to right for rows (from top to bottom for columns). A path from a_i to c_k exists if and only if simultaneously, for some b_j,

1. there exists an arc from a_i to b_j, and
2. there exists an arc from b_j to c_k.

First we note that any row of M_α contains as many symbols (1 and 0) as any column of M_β; indeed, the number of these symbols is the cardinality of the set B. Then we observe that condition (1) is illustrated by a 1 in the jth position (see figure 1.2.5) of the row of M_α pertaining to a_i; condition (2) is illustrated by a

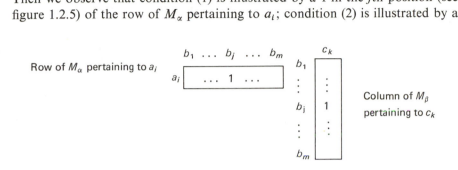

Figure 1.2.5. Illustration in M_α and M_β of a path from a_i to c_k

1 in the jth position of the column of M_β pertaining to c_k. The array $M_{\alpha\beta}$ can be related to the arrays M_α and M_β with the aid of the following definition.

Definition 1.2.4 (Array multiplication). The *product* $M_\alpha M_\beta$ of an $n \times m$ array M_α and an $m \times p$ array M_β is defined by the following rule: The position at the intersection of the ith row and kth column of $M_\alpha M_\beta$ contains a 1 if and only if there is at least one $j (1 \leqslant j \leqslant m)$ such that the jth positions of the ith row in M_α and of the kth column in M_β are simultaneously 1.

The following example should clarify this apparently complicated definition.

Example 1.2.3. From the relations α and β graphically illustrated in figure 1.2.3, the array $M_{\alpha\beta}$ is constructed (figure 1.2.6) by applying the stated rule on array multiplication. The encircled entry in $M_\alpha M_\beta$ results from the encircled row and

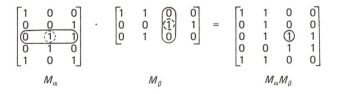

Figure 1.2.6. Illustration of array multiplication

column in M_α and M_β, respectively. Each of the latter has a 1 in the same position, shown by a broken-line circle.

It is worth pointing out that the product $M_\alpha M_\beta$ is *defined* if and only if M_α is an $n \times m$ array and M_β is an $m \times p$ array for arbitrary n, m, p (in other words, the number of columns of M_α must equal the number of rows of M_β).

The preceding discussion is summarized by the following theorem.

Theorem 1.2.2. The array $M_{\alpha\beta}$ of the composite $\alpha\beta$ is the product $M_\alpha M_\beta$ of the arrays M_α, M_β. Formally,

$$M_{\alpha\beta} = M_\alpha M_\beta. \tag{1.2.2}$$

Theorem 1.2.2, combined with theorem 1.2.1, yields the following straightforward corollary.

Corollary 1.2.3. The associative property holds for the multiplication of arrays, i.e., if $(M_\alpha M_\beta) M_\gamma$ is defined, then

$$(M_\alpha M_\beta) M_\gamma = M_\alpha (M_\beta M_\gamma). \tag{1.2.3}$$

Given a relation, we can construct a new relation, called its *inverse* (or *converse*).

Definition 1.2.5. The *inverse* of a binary relation α from a set A to a set B is a binary relation α^{-1} from B to A specified by

for $a \in A$ and $b \in B$, $b \, \alpha^{-1} \, a$ if and only if $a\alpha b$.

Let us obtain the two familiar representations of α^{-1}. The graphical representation of α^{-1} differs from that of α in that the direction of the arrows has been reversed on each arc. As for the tabular representation, if M_α is an $n \times m$ array, $M_{\alpha^{-1}}$ will be an $m \times n$ array, since α^{-1} is a relation from B to A. Furthermore, since in α and in α^{-1} the generic $b_j \in B$ is related ("connected") to the same subset of elements of A (this subset being indicated by the 1's in the column of b_j in M_α), we conclude that the column of M_α and the row of $M_{\alpha^{-1}}$ corresponding to b_j have 1's in exactly the same positions (see figure 1.2.7).

$$
M_\alpha =
\begin{array}{c|ccc}
 & b_1 & b_2 & b_3 \\
\hline
a_1 & 1 & 0 & 0 \\
a_2 & 0 & 0 & 1 \\
a_3 & 0 & 1 & 1 \\
a_4 & 0 & 1 & 0 \\
a_5 & 1 & 0 & 1 \\
\end{array}
\qquad
M_{\alpha^{-1}} =
\begin{array}{c|ccccc}
 & a_1 & a_2 & a_3 & a_4 & a_5 \\
\hline
b_1 & 1 & 0 & 0 & 0 & 1 \\
b_2 & 0 & 0 & 1 & 1 & 0 \\
b_3 & 0 & 1 & 1 & 0 & 1 \\
\end{array}
$$

Figure 1.2.7. Construction of the array of the inverse relation

The *transpose array* M' of M is the array whose rows are the columns of M—and clearly, whose columns are the rows of M. The array of the inverse relation is the transpose of the array of the (direct) relation, which is summarized formally as

$$M_{\alpha^{-1}} = M'_\alpha. \tag{1.2.4}$$

To combine the concepts of composite and inverse, consider the composite $\alpha\beta$ of the relation α from A to B and of the relation β from B to C. We now want to describe the inverse relation $(\alpha\beta)^{-1}$ of $\alpha\beta$. The meaning of $(\alpha\beta)^{-1}$ is evident in the graphical representation. Indeed $c \in C$ is related to $a \in A$ in $(\alpha\beta)^{-1}$ if there is at least one path from c to a passing through some intermediate vertex $b \in B$. But any such path coincides with a path of the direct composite relation $\alpha\beta$, the only difference being that it is traversed in the opposite direction. Specifically,

$$c(\alpha\beta)^{-1} a \text{ if and only if, for some } b, c\beta^{-1}b \text{ and } b\alpha^{-1}a.$$

Referring again to the definition of composite, we note that

$$(\alpha\beta)^{-1} = \beta^{-1}\alpha^{-1}; \tag{1.2.5}$$

that is, we have proved the following rule.

The reversal rule.[4] The inverse of the composite $\alpha\beta$ of α and β, is the composite of the inverses of α and β *but in reversed order.*

The reversal rule has a simple expression in terms of array representation. Let M_α, M_β, and $M_{\alpha\beta}$ be the arrays of α, β and $\alpha\beta$, respectively. Then, from the preceding discussion on inverse relations, we know that $M_{\alpha^{-1}} = M'_\alpha$, $M_{\beta^{-1}} = M'_\beta$, and $M_{(\alpha\beta)^{-1}} = M'_{\alpha\beta}$; from this it is immediately clear that we may express the reversal law, $M_{(\alpha\beta)^{-1}} = M_{\beta^{-1}}M_{\alpha^{-1}}$, in terms of the transpose arrays

$$M'_{\alpha\beta} = M'_\beta M'_\alpha. \tag{1.2.6}$$

The inverses of relations are so common in everyday experience that they are denoted by specific names in natural languages. For example, the relations "child of" and "parent of" are inverses of each other.

We have noted the great generality of the concept of binary relation. What we have actually accomplished is the establishment of links between elements of two sets, without specifying additional properties of this set of links. When properties are added, a more complex structure usually results. Several possibilities exist for the addition of structure to binary relations. Of all the specializations of binary relations we shall examine two in further detail because of their great importance. The first places a restriction on the type of relation alone and yields the class of *functions from A to B*. The second specialization places a restriction on the sets alone by requiring that $A = B$; we obtain thereby the class of *relations on a set A*. These two classes will be the subject of the next two sections.

---

4. Note that the "reversal rule" embodies a theorem.

EXERCISES 1.2

1. Give the graphical representation of the relation α from $A \equiv \{1,2,3\}$ to $B \equiv \{a,b,c\}$.
$$R_\alpha \equiv \{(1,a), (1,b), (2,a), (2,c), (3,a), (3,b), (3,c)\}$$

2. Let M_α and M_β be the arrays given below. What is $M_\alpha M_\beta$?

$$M_\alpha = \begin{bmatrix} 1 & 0 & 1 \\ 1 & 1 & 0 \\ 1 & 0 & 1 \end{bmatrix} \qquad M_\beta = \begin{bmatrix} 1 & 1 & 0 & 0 & 1 \\ 1 & 0 & 1 & 0 & 0 \\ 1 & 1 & 1 & 1 & 1 \end{bmatrix}$$

3. What is the inverse relation of the relation α from A to B such that $R_\alpha = \varnothing$ ($R_\alpha \subseteq A \times B$)?

4. Let α and β be two relations from the set of integers I to itself, defined by $R_\alpha \equiv \{(i,2i)|i \in I\}$ and $R_\beta \equiv \{(j,3j)|j \in I\}$. What is the composite relation $\alpha\beta$?

5. Let A, B, C, D be sets. Show that the following identities hold.
   a) $(A \cup B) \times C = (A \times C) \cup (B \times C)$
   b) $(A \cap B) \times (C \cap D) = (A \times C) \cap (B \times D)$

*6. Determine whether or not the cartesian product of two countable sets is countable. (See exercise 1.1.11e.)

7. Let α and β be two given relations from A to B such that $R_\alpha \subseteq R_\beta$. (a) Show that for any other relation γ from B to C we must have $R_{\alpha\gamma} \subseteq R_{\beta\gamma}$. (b) Similarly, if γ is a relation from C to A, then we have $R_{\gamma\alpha} \subseteq R_{\gamma\beta}$.

8. Show that for any binary relations α and β from A to B, we have
   a) $(\alpha^{-1})^{-1} = \alpha$
   b) $\alpha^{-1} = \beta^{-1}$ if and only if $\alpha = \beta$
   c) $\alpha^{-1} \subseteq \beta^{-1}$ if and only if $\alpha \subseteq \beta$

9. Define the complement of a relation α, denoted by $\bar{\alpha}$, as the relation such that $a\,\bar{\alpha}\,b$ if and only if $(a,b) \notin R_\alpha$. Show that $(\bar{\alpha})^{-1} = \overline{(\alpha)^{-1}}$.

10. Let α and β be binary relations from A to B, and γ a binary relation from C to A. Denote by $\alpha \cup \beta$ and $\alpha \cap \beta$ the relations described by $R_\alpha \cup R_\beta$ and $R_\alpha \cap R_\beta$. Prove the following identities.
   a) $(\alpha \cap \beta)^{-1} = \alpha^{-1} \cap \beta^{-1}$
   b) $(\alpha \cup \beta)^{-1} = \alpha^{-1} \cup \beta^{-1}$
   c) $\gamma(\alpha \cup \beta) = \gamma\alpha \cup \gamma\beta$
   d) $R_{\gamma(\alpha \cap \beta)} \subseteq R_{\gamma\alpha} \cap R_{\gamma\beta}$

1.3 FUNCTIONS [MAPPINGS]

A very important special case of binary relations from a set A to a set B occurs when each element $a_i \in A$ appears exactly once as the left element of the ordered pairs $(a_i,b_j) \in R_\alpha$. In the graphical representation of this relation, each vertex of A has one and only one outgoing arc; in the tabular representation each row contains one and only one 1 (note that relation α of examples 1.2.1 and 1.2.2 does not meet these requirements). This type of relation, f, is called a transformation, or map or function, and although the notations $a_i f b_j$ or $R_f \subseteq A \times B$ are legitimate for this type of relation, the uniqueness of the right element of each pair is better evidenced by the following equivalent notations.

$R_f \subseteq A \times B$ is replaced by $f: A \to B$

$a_i f b_j$ is replaced by $f: a_i \mapsto b_j$ or $b_j = f(a_i)$

In these notations b_j is called "the value of f at a_i" or "the image of a_i under f."

Definition 1.3.1. A *function* $f: A \to B$ (or map or transformation) from a set A to a set B is a binary relation from A to B such that for each $a \in A$ there is a unique $b \in B$ for which $(a, b) \in R_f$. Alternatively, $f: A \to B$ is a rule which assigns to each element of A some element of B. Sets A and B are called, respectively, the *domain* and the *codomain* of f. The set $\{f(a) | a \in A\}$ of the images of the elements of A under f is called the *image* of A under f or the *range* of f and is denoted as $f(A)$.

We explicitly remark that the definition of function specializes the concept of binary relation in two respects: (1) by requiring that a generic element of the domain A be related to a *unique* element of the codomain B; (2) by requiring that *each* element of A be related to some element of B. When the second requirement is relaxed, a subset A' of A is the domain of f. The resulting relation is a function from a subset A' of A to B and is called a *partial function* from A to B. In some cases, given a function $f: A \to B$, we may want to construct a new function f', having as domain a subset A' of A, so that $R_{f'} = R_f \cap (A' \times B)$. Thus f', denoted by $f | A': A' \to B$, could be considered as a partial function from A to B, called the *restriction of f to A'*. Very simple examples of functions are presented in figure 1.3.1 by means of their graphical representations. Clearly each diagram meets the requirement that each vertex of the domain A has one and only one outgoing arc.

A function can be specified by means of the familiar representations (tabular and graphical) introduced in the preceding section, since a function is nothing but a special type of binary relation. However, its special character is such that the generality of the given representation, though not harmful, is excessive. Indeed,

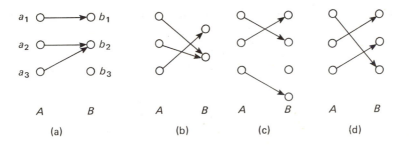

Figure 1.3.1. Diagrams of functions

because of "single-valuedness," the function is completely specified by a list of as many ordered pairs as there are elements in the domain A. This is tantamount to saying that the tabular representation of a generic function reduces to a new type of $n \times 1$ array (i.e., an array consisting of only one column, as shown in figure 1.3.2) containing the symbol $f(a_j)$ in the position corresponding to $a_j \in A$. This representation of a generic function can be shown[5] to be less complex than that of a generic relation involving sets of the same cardinality. However, some functions may be represented by a rule which produces $f(a_j)$ from a_j. For example, the function $f: x \to x^2$ for x integer is specified by the rule of multiplying x by itself to obtain $f(x)$. This parallels the analogous instance encountered in connection with those sets which can be described either as a list or by a characteristic property.

$$
\begin{array}{c|c}
a_1 & b_1 \\
a_2 & b_2 \\
a_3 & b_2
\end{array}
$$

Figure 1.3.2. Tabular description of the function of figure 1.3.1(a)

As implied by the synonyms of "function," namely, "mapping" and "transformation," a function can be effectively viewed as performing a transformation from a set to a set. We use the graphical representation as an aid to our discussion. The definition of a function f requires that each vertex of the domain A of f have exactly one outgoing arc but places no condition on the number of arcs which reach vertices of the codomain B of f. When appropriate conditions are imposed, we obtain important classes of mappings, as offered by the following definitions.

Definition 1.3.2. A function $f: A \to B$ is said to be *onto* if for any $b \in B$ there is at least one $a \in A$ such that $b = f(a)$. (Each vertex of B is reached by *at least* one arc.)

Definition 1.3.3. A function $f: A \to B$ is said to be *one-to-one* if $a_1 \neq a_2$ in A implies that $f(a_1) \neq f(a_2)$ in B; that is, f maps distinct elements of A into distinct elements of B. (Each vertex of B is reached by *at most* one arc.)

Referring to the function diagrams of figure 1.3.1, we observe that (a) is neither onto nor one-to-one; (b) is onto but not one-to-one; (c) is one-to-one but not onto; (d) is both onto and one-to-one.

Since a function is a binary relation, we can consider its inverse. If the latter happens to be a function, then we have the following definition.

---

5. See Note 1.4 at the end of the chapter for an illustrative argument.

Definition 1.3.4. A function $f: A \to B$ is said to be *invertible*[6] if its inverse f^{-1} is a function from B to A.

Invertible functions constitute a very restrictive case and are characterized by the following theorem.

Theorem 1.3.1. A function $f: A \to B$ is invertible if and only if it is one-to-one and onto.

Proof. Assume that f is one-to-one and onto. Then, in the function diagram each vertex of B has at least one incident arc (since f is onto) and at most one incident arc (since f is one-to-one), i.e., exactly one incident arc. If the arrows on the arcs are reversed, B satisfies the requirement of being the domain of a function; that is, the diagram of the inverse relation is that of a function, and f is invertible.

Conversely, assume that f is invertible. This means that the inverse relation of f is a function, that is, that each vertex of B, under f, has exactly one incident arc. But this means that each vertex of B has at least one incident arc (f is onto) and also that each vertex of B has at most one incident arc (f is one-to-one). The proof is complete.‖

The diagram of an invertible function is given in figure 1.3.1(d). Note that the existence of an invertible function between two sets is equivalent to the existence of a one-to-one correspondence between the elements of the two sets. This gives precise meaning to the notion of one-to-one correspondence introduced in Section 1.1.

If the set A is finite, an invertible function from A to A is called a *permutation*. Permutations constitute a very important set of mappings and will be discussed in more detail in the following chapters. We simply note here that the array representation of a permutation mapping is $n \times n$ (square) and contains exactly one 1 entry per row or column. We now illustrate the previously introduced concepts with various examples.

Example 1.3.1. The function p (a mnemonic for parity) yields the parity (oddness or evenness) of a nonnegative integer, that is, maps the set N of the nonnegative integers 0, 1, 2, ... (the function domain) into the set $B \equiv \{$even, odd$\}$ (the function codomain). Note that the domain is countably infinite, while the codomain is finite; furthermore, since N has an infinite number of elements, $p: N \to B$ cannot be described by a table. However, it can be described by a rule, which obtains $p(j)$ for $j \in N$. The rule is "Divide j by 2; if the remainder is 0, $p(j) =$ even; if the remainder is 1, $p(j) =$ odd." The function p is onto, but obviously not one-to-one.

---

6. An alternative terminology uses the words *surjection* for "onto function," *injection* for "one-to-one function," and *bijection* for "invertible function." Although more modern, this terminology does not seem to us to possess the mnemonic appeal of that presented above, and it will not be used in this book.

Example 1.3.2. Consider the function f from the set $A \equiv \{1,2,3,4,5,6\}$ to the set $B \equiv \{8,9,10,11,12,13\}$ specified by the list $f(1) = 10, f(2) = 13, f(3) = 9, f(4) = 12,$ $f(5) = 8, f(6) = 11.$ The functions $f: A \rightarrow B$ is onto and one-to-one, hence invertible. An alternative description of f is the following: $f(j), j \in A$, is obtained by adding 7 to the residue of $3j$ modulo 7. (The residue of $3j$ modulo 7 is the remainder of the division of $3j$ by 7.)

Example 1.3.3. The familiar functions of a real variable, encountered in analysis, are of the type $f: R \rightarrow R$, where R is the noncountable set of the real numbers. For example, $f(x) = x$ is onto and one-to-one, hence invertible; $f(x) = e^x$ is one-to-one but not onto. Because the range of f excludes 0 and the negative real numbers, its inverse, $\log_e x$ is a partial function on the set R, but it is an invertible function on the set of positive real numbers.

EXERCISES 1.3

1. Determine which one of the following sets defines a function.
   a) $\{(i, i^2) | \ i$ is integer $\geqslant 1\}$.
   b) $\{(a, (b,c)), (a, (c,b)), (b, (c,d))\}$ for $A \equiv \{a,b,c,d\}$.

2. Let A and B be two given sets such that $\#(A) = m$, $\#(B) = n$, and $m \neq n$.
   a) If $f: A \rightarrow B$, under what condition can f be a one-to-one function? Can f be an onto function?
   b) What is the maximum number of elements that a subset C of $A \times B$ may have in order to define a function?

3. Show that the elements of the set $A \times B$ are in a one-to-one correspondence with the elements of the set $B \times A$.

4. Determine which one of the following functions satisfies the properties of being one-to-one or onto.
   a) Let f be a function from P, the set of positive integers to itself such that, for each $n \in P, f(n) = n + 1.$ (f is called the *successor function*.)
   b) Let f be a function from the set $K \equiv \{1,2,\ldots,k\}$ of positive integers less than or equal to k to itself such that

$$f(i) = \begin{cases} i + 1, & \text{for } 1 \leqslant i < k \\ 1, & i = k. \end{cases}$$

   c) Let $f: P \rightarrow \{0,1,2,3\}$ such that for each $i \in P$,

$$f(i) = \begin{cases} 0, \text{ if } i \text{ is divisible by 3} \\ 1, \text{ if } i \text{ is divisible by 7} \\ 2, \text{ if } i \text{ is divisible by 21} \\ 3, \text{ otherwise.} \end{cases}$$

   d) Let f be the function from the set of rational numbers R to itself such that for each $i \in R, f(i) = i^3.$

5. Let $f: A \rightarrow B$ and $g: C \rightarrow D$ be two given functions. Show that $R_f = R_g$ if and only if $A = C$ and $f(a) = g(a)$ for each $a \in A$.

6. Show that $f: A \rightarrow B$ is onto if and only if f^{-1} satisfies the condition: for each $b \in B$, there is at least one $a \in A$ such that $bf^{-1}a$.

7. a) How many functions from a set of three elements to a set of two elements are onto?
   b) How many functions from a set of three elements to a set of four elements are one-to-one?

8. Show that if $f: A \rightarrow B$ is a given function and $A_1 \subseteq A_2 \subseteq A$, then $f(A_1) \subseteq f(A_2)$, where $f(A_i) \equiv \{f(a) | a \in A_i\}$, $i = 1, 2$. (Recall that $\{f(a) | a \in A_i\}$ is read: "the set of $f(a)$ such that a belongs to A_i".)

9. a) Show that a function $f: A \rightarrow B$ is one-to-one if and only if it maps proper subsets into proper subsets; that is, $A_1 \subset A_2 \subseteq A$ implies that $f(A_1) \subset f(A_2)$.
   b) Show that a function $f: A \rightarrow B (A \neq \varnothing)$ is one-to-one if and only if there exists a partial function $g: B \rightarrow A$ such that $g(f(a)) = a$ for all $a \in A$.

10. Let $f: A \rightarrow B$ be a given function and $A' \subseteq A$, $B' \subseteq B$. Show that
    a) $A' \subseteq f^{-1}(f(A'))$
    b) $f(f^{-1}(B')) = B'$
    where $f^{-1}(B') = \{a | a \in A$ and $f(a) \in B'\}$. (Recall that f^{-1} is the inverse binary relation of f, considered as a binary relation.)

11. Let $f: A \rightarrow B$ and $g: B \rightarrow C$ be two given functions. Show that the composite fg of f and g is again a function from A to C. (Note that, because of our choice of notation, $fg(a) = g(f(a))$.)

12. Which of the following properties are preserved under composition of functions?
    a) one-to-one   b) onto   c) invertibility

13. Show that any function $f: A \rightarrow B$ can be represented as the composition of two functions g and h, $f = gh$, where g is onto and h is one-to-one. (*Hint:* Consider a collection of subsets $F \equiv \{A_1, A_2, \ldots\}$ of A such that each A_i has the property that for any two arbitrary elements a, b in A_i, $f(a) = f(b)$. Let $g: A \rightarrow F$ and $h: F \rightarrow B$.)

1.4 RELATIONS ON A SET

When in the specification of a relation α, the two terms of the cartesian product are the same set A, that is, $R_\alpha \subseteq A \times A$, we say that we have a *relation on a set A*. The importance of this specialization of binary relations will be amply demonstrated by the various examples that follow.

An immediate consequence of the choice $B = A$ is that the array of the tabular representation is square; i.e., it has identical numbers of rows and columns. For this reason, we can also identify two diagonals in the array. Of interest is the so-called *main diagonal* of the array, which runs from the upper left to the lower right corner.

The graphical description of a relation on A is a collection of arcs between vertices of A. Because of their outstanding importance as models of various real-world systems, the class of such graphical structures has received the name of (*directed*) *graphs*. Graphs will be analyzed in detail in the next chapter in relation to their remarkable geometric and topological properties; here they are viewed primarily as a descriptive tool. An arc from vertex a_i to vertex a_j of A is represented by the concise notation

$$(\overrightarrow{a_i, a_j}).$$

Example 1.4.1 The relation α on $A \equiv \{a_1, a_2, a_3, a_4, a_5\}$, described in set notation as $R_\alpha \equiv \{(a_1, a_1),\ (a_1, a_2),\ (a_2, a_3),\ (a_2, a_5),\ (a_3, a_2),\ (a_3, a_4),\ (a_5, a_1),\ (a_5, a_4)\}$, has tabular and graphical representations as shown in figure 1.4.1.

The structure of a relation on a set is expressed in terms of properties that the relation exhibits. We now analyze some of these properties in considerable detail.

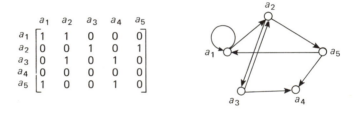

Figure 1.4.1. Representations of a relation on a set

First, we need some nomenclature as a premise for this analysis. Appealing temporarily to intuition and without dwelling on its algebraic significance, we say that a relation α on A *covers* a relation β on A if $R_\beta \subseteq R_\alpha$. The intuitive significance of the word "cover" should be apparent since, if we superimpose the array of α on that of β, all 1's of β are indeed covered by 1's of α; similarly, if we draw the diagrams of α and β so that corresponding vertices are superimposed, the arcs of α cover the arcs of β.

Definition 1.4.1 (Reflexivity). A relation α on a set A is said to have the *reflexive property* (to be reflexive) if, for every $a \in A$, $a \, \alpha \, a$.

The tabular representation of a reflexive relation contains 1's in each position of the main diagonal. The graphical representation shows, for each vertex, an outgoing arc which is incident on the vertex itself, technically denoted as a *loop* (see figure 1.4.2a). The loop is the characteristic graphical component of reflexivity.

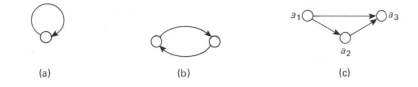

(a)   (b)   (c)

Figure 1.4.2. Illustration of the graphical characteristic components of the reflexive, symmetric, and transitive properties

Definition 1.4.2 (Symmetry). A relation α on a set A is said to have the *symmetric property* (to be symmetric) if, for a and b in A, $a \, \alpha \, b$ implies $b \, \alpha \, a$.

The array of a symmetric relation α clearly coincides with its transpose since there is a 1 in the ith position of the jth row if and only if there is a 1 in the ith position of the jth column; formally, for a symmetric relation α,

$$M_\alpha = M_\alpha'. \tag{1.4.1}$$

The graph of a symmetric relation α is such that all its arcs occur in pairs, as shown in figure 1.4.2(b). The *symmetric pair* is the characteristic graphical component of symmetry. It is readily realized that, since the array $M_{\alpha^{-1}}$ of α^{-1}, the inverse of the relation α, coincides with M_α', the property of symmetry is formally expressed by $M_\alpha = M_\alpha' = M_{\alpha^{-1}}$; α and α^{-1} coincide.

When not all arcs of the graph of α form symmetric pairs, α is said to be *asymmetric*. However, when no symmetric pair exists, we have antisymmetry.

Definition 1.4.3 (Antisymmetry). A relation α on a set A is said to have the *antisymmetric property* (to be antisymmetric) if, for a and b in A, $a \, \alpha \, b$ and $b \, \alpha \, a$ imply $a = b$.

A simple example of an antisymmetric relation is the relation "less than or equal to" defined on the set of integers.

Definition 1.4.4 (Transitivity). A relation α on a set S is said to have the *transitive property* (to be transitive) if, for a, b, and c in A, $a \, \alpha \, b$ and $b \, \alpha \, c$ imply $a \, \alpha \, c$.

The characteristic graphical component of transitivity is the three-arc structure in figure 1.4.2(c). It means that, if there exists a path consisting of *two* arcs traversable in the direction of the arrows from vertex a to vertex c, there must also be an arc from a to c. It is also simple to give a characterization of transitivity on the array of the relation. All paths consisting of two arcs are found by applying the relation α twice, that is, by the composite $\alpha\alpha$. We realize that α is transitive if and only if each arc of the graph of $\alpha\alpha$ is an arc of the graph of α, that is, if and only if α covers $\alpha\alpha$. We notice that, for a transitive α, there may be arcs of the graph of α which are not arcs of the graph of $\alpha\alpha$. This happens, for example, when a vertex of the graph of α has no outgoing arc. The graph of figure 1.4.2(c) describes a transitive relation α on $\{a_1, a_2, a_3\}$. The graph of $\alpha\alpha$ contains only the arc $\overrightarrow{(a_1, a_3)}$.

The properties of reflexivity, symmetry, and transitivity constitute an interesting set of attributes for the categorization of relations on a set. In order to gain more insight into their significance and to relate them to real-world situations, we shall now present and discuss various examples of relations on a set. In the interests of better organization, each relation is accompanied by a triplet (R^*, S^*, T^*), in which the symbol R^* stands for either R or $\bar{R}$; similarly, S^* and T^* stand for S or $\bar{S}$ and T or $\bar{T}$, respectively. The symbol R represents the statement, or proposition, "(the relation is) reflexive," and $\bar{R}$ represents the statement "(the relation is) not reflexive." (The latter statement, however, does not necessarily mean that there is no $a \in A$ for which $a \alpha a$.) Analogous meanings pertain to S and T for symmetry and transitivity, respectively. There are eight possible cases, which we now consider, one at a time.

There exist relations which are $(\bar{R}, \bar{S}, \bar{T})$; that is, they possess none of the three properties, reflexivity, symmetry, and transitivity. This class of relations, because of the lack of structure, is very large. An important sub-class of this class consists of the "predecessor-successor" relations, a typical graph of which (see figure 1.4.3(a)) is an example of an important class of graphs, called *trees*, to be studied in Chapter 2. A real-world relation falling in this category is that between father and sons.

A relation which is $(\bar{R}, \bar{S}, T)$, that is, transitive but not reflexive or symmetric, is the relation "greater than" over the set P of the positive integers. The graph of this relation over the set $A \equiv \{1, 2, 3, 4\}$ is shown in figure 1.4.3(b).

Consider the set I of the integers. Two consecutive integers are related in such a fashion that the absolute value of their difference is 1. Defining "adjacency" as the relation α on I such that, for $i, j \in I$, $i \alpha j$ means $|i - j| = 1$, we recognize that α is not reflexive or transitive, but it is certainly symmetric; i.e., it is an example of an $(\bar{R}, S, \bar{T})$ relation. Its graph is shown in figure 1.4.3(c) for the set $A \equiv \{1, 2, 3, 4\}$, a subset of I. Another relation in the same class is defined by the set $\{(n, -n) | n$ is an integer$\}$.

The graph of an $(\bar{R}, S, T)$ relation is shown in figure 1.4.3(d). One real-world relation in this category is brotherhood.

An example of a relation on a set which is reflexive but not symmetric or transitive $(R, \bar{S}, \bar{T})$ is supplied by the notion of legal representation. An individual is obviously a legal representative of himself, but he may also entrust authority of representation to another individual. It is clear, however, that delegation is not reciprocal or transferable. A graph of such a relation on a set of three elements is shown in figure 1.4.3(e).

It is not hard to envision an $(R, \bar{S}, T)$ relation. An example of this class is the relation of set inclusion (definition 1.1.3), a typical graph of which appears in figure 1.4.3(f). It is worth stressing that set inclusion is more than simply nonsymmetric; since no symmetric pair of elements exists, inclusion is antisymmetric. This is an instance of the important class of relations that are reflexive, transitive, and antisymmetric, technically called *partial orderings*. These relations, which induce a very special structure on sets, will be studied in detail in Chapter 4.

In the set I of integers, consider as related to a given integer i all the integers j for which $|i - j| \leqslant m$, for some fixed integer m. The ensuing relation α, named "bounded difference," is clearly reflexive and symmetric, but not always transitive. Suppose that $m = 2$; then for $1 \alpha 2$ and $2 \alpha 3$ we also have $1 \alpha 3$, but for $1 \alpha 2$ and $2 \alpha 4$ we do *not* have $1 \alpha 4$. "Bounded difference" belongs to the class of $(R, S, \overline{T})$ relations, which are formally classified as *compatibility* relations and which will be analyzed extensively in the following section. A typical graph of

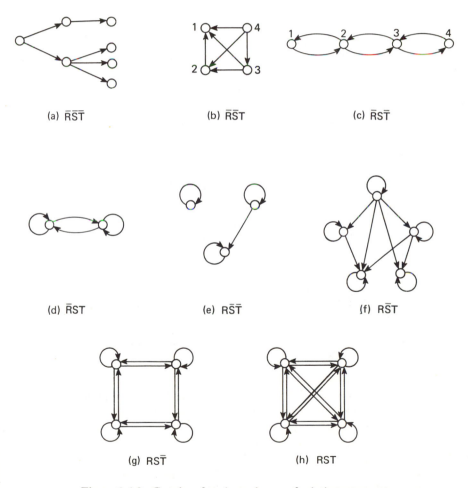

(a) $\overline{R}\overline{S}\overline{T}$ (b) $\overline{R}\overline{S}\overline{T}$ (c) $\overline{R}\overline{S}\overline{T}$

(d) $\overline{R}ST$ (e) $R\overline{S}\overline{T}$ (f) $R\overline{S}T$

(g) $RS\overline{T}$ (h) RST

Figure 1.4.3. Graphs of various classes of relations on a set

one such relation on a four-element set is shown in figure 1.4.3(g). A real-world example of such a relation is friendship (if we can accept as a convention the idea that one is a friend of oneself!).

Finally, we consider relations that are (R, S, T), that is, reflexive, symmetric, and transitive (figure 1.4.3(h)). A real-world example is interchangeability (for example, of spare parts for an engine). Another example is the concept of identity on a set A: each element of A is identical to itself and only to itself. The relation of identity on a set A is symbolically denoted by e_A.

Besides providing familiarity with the formal concepts presented and connecting them to common experience, the foregoing discussion shows the independence of the three properties of reflexivity, symmetry, and transitivity. In fact, since the eight cases—$(\bar{R}, \bar{S}, \bar{T})$, $(\bar{R}, \bar{S}, T)$, $(\bar{R}, S, \bar{T})$, $(\bar{R}, S, T)$, $(R, \bar{S}, \bar{T})$, $(R, \bar{S}, T)$, $(R, S, \bar{T})$, (R, S, T)—are all the possible cases for (R^*, S^*, T^*), we conclude that none of the three properties is implied by the other two.

The concept of the inverse of a relation on a set A follows, with no additional specifications, from the general concept of the inverse of relations between two sets. We note that the inverse of a reflexive relation is also reflexive; indeed, by reversing the arrow on the directed edge of a loop, we still have a loop. Similarly, we recognize that if the arrows on all the arcs of the graph of a relation are reversed, symmetric pairs transform into symmetric pairs, and so do the configurations of figure 1.4.2(c), which are characteristic of transitivity. This very simple observation is worth expressing as a far-reaching theorem.

Theorem 1.4.1. The reflexive, symmetric, antisymmetric, and transitive properties of relations on a set are preserved through inversion: If the relation α on A has one such property, so does α^{-1}.

Before closing this section, we note one interesting special case of relations on a set A, namely, the relations on A which also obey the requirements of a function. Alternatively, this class of relations, called *functions on a set*, may be viewed as a special case, functions from A to A. We realize that the graphs of functions on a set have the characteristic property that each node has exactly one outgoing arc.

EXERCISES 1.4

1. Determine whether each of the following relations, defined by ordered pairs (m, n) of positive integers, has each of the properties: reflexivity, symmetry, antisymmetry, and transitivity.

   a) m is divisible by n

   b) $m + n \geqslant 50$

   c) $m + n$ is even

   d) $m + n$ is odd

   e) mn is even

   f) m is a power of n

   g) $m + n$ is a multiple of 3

   h) m is greater than n

2. Let A be a nonempty set. Show that the empty set $\varnothing$ considered as a relation on A is not reflexive but is symmetric and transitive.

3. Let α be a relation on a set A. Is the relation defined by $R_\alpha \cap R_{\alpha^{-1}}$ symmetric?

4. Determine which of the reflexive, symmetric, antisymmetric, and transitive properties is preserved under composition of relations on a set.

5. Recall that e_A denotes the identity relation on A. Show that a relation α on A is reflexive if and only if $R_{e_A} \subseteq R_\alpha$.

6. Denoting with e_A the identity relation on A, show that a relation α on A is antisymmetric if and only if $R_\alpha \cap R_{\alpha^{-1}} \subseteq R_{e_A}$.

7. A relation α on A is said to be *irreflexive* if for each $a \in A$, $(a,a) \notin R_\alpha$. Show that α is irreflexive if and only if $R_\alpha \cap R_{e_A} = \varnothing$.

8. Show that a nonempty symmetric and transitive relation cannot be irreflexive. (See exercise 1.4.7 for definition of irreflexivity.)

1.5 THE RELATIONS OF COMPATIBILITY AND EQUIVALENCE

Let us imagine some kind of international convention where people of different countries meet. All the conferees naturally speak their own mother tongue, but it lies in the character of the meeting that most of them are also fluent in some other language. We would note groups of two or more persons engaged in conversation. We would further note that every dialogue is in a single language; that is, nobody is acting as interpreter. In order to sustain a dialogue, two persons of a group must use a language of which both have knowledge. Let us now examine a possible formal model for this (only slightly artificial) situation. Assume that six persons—whom we may denote as set $A \equiv \{a_1, a_2, a_3, a_4, a_5, a_6\}$—collectively have knowledge of five languages, which we may denote as set $B \equiv \{b_1, b_2, b_3, b_4, b_5\}$. The knowledge of languages is viewed as a relation α from A to B, which has the tabular representation shown in figure 1.5.1(a). In this scheme, the possibility for two members a_i and a_j of A to communicate is expressed by the fact that a_i knows some language b_k (that is, $a_i \alpha b_k$), which in turn "is known" by a_j (that is, $b_k \alpha^{-1} a_j$, since "to be known by" is the inverse relation of "to know"). Therefore, communicability between a_i and a_j may be formally expressed by the composite of α and α^{-1}, that is, by $a_i(\alpha\alpha^{-1})a_j$. It is worth pointing out that (α, α^{-1}) is always a composable pair for any α. The array $M_{\alpha\alpha^{-1}}$ is shown in figure 1.5.1(b). From (1.2.2) and (1.2.4) we have

$$M_{\alpha\alpha^{-1}} = M_\alpha M_{\alpha^{-1}} \quad \text{(the array of the composite relation),}$$
$$M_{\alpha^{-1}} = M'_\alpha \quad \text{(the array of the inverse relation).}$$

Thus, we obtain

$$M_{\alpha\alpha^{-1}} = M_\alpha M'_\alpha$$

and leave as an exercise the verification of the correctness of the array in figure

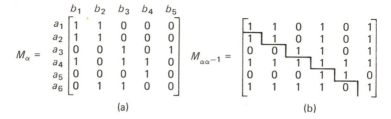

Figure 1.5.1. Arrays of the relations α and $\alpha\alpha^{-1}$

1.5.1(b). Note that $\alpha\alpha^{-1}$ is a relation on the set A. We shall now analyze the properties of $\alpha\alpha^{-1}$ for arbitrary α. First, $\alpha\alpha^{-1}$ is reflexive, since all the symbols of the main diagonal of $M_{\alpha\alpha^{-1}}$ are 1's. Second, $\alpha\alpha^{-1}$ is symmetric. Indeed, from the reversal law (1.2.5) we have $(\alpha\alpha^{-1})^{-1} = (\alpha^{-1})^{-1}\alpha^{-1}$; but obviously $(\alpha^{-1})^{-1} = \alpha$, since transposing the transpose of an array yields the original array (or equivalently, twice reversing the direction of the arrow of an arc restores the original arc). We therefore can write

$$(\alpha\alpha^{-1})^{-1} = \alpha\alpha^{-1}.$$

That is, $\alpha\alpha^{-1}$ and its inverse coincide, and that coincidence is precisely the characterization of a symmetric relation (see Section 1.4). Finally, inspection of figure 1.5.1(b) quickly reveals that $\alpha\alpha^{-1}$ is not transitive. In the figure we find several counterexamples to the hypothesis of transitivity—for example, $a_1(\alpha\alpha^{-1})a_4$ and $a_4(\alpha\alpha^{-1})a_3$—but a_1 and a_3 are not related through $\alpha\alpha^{-1}$. The relation just examined is an example of a very important class specified by the following definition.

Definition 1.5.1. A *compatibility relation* on a set A is a relation which has the reflexive and symmetric properties (the usual symbol of a compatibility relation is γ).

The nature of the preceding example motivates the term "compatibility." Indeed, the possibility of communicating is consistent with the usual connotation of being compatible.

The preceding discussion shows that for any relation α from A to B which relates *each* element of A to at least one element of B (a rather large class of relations α), there exists a compatibility relation $\alpha\alpha^{-1}$. We note that, for an arbitrary α, $\alpha\alpha^{-1}$ is symmetric; in order that $\alpha\alpha^{-1}$ be reflexive, every a in A must be related to some b in B. Therefore we have a sufficiency condition for the construction of a compatibility relation. To prove the converse, i.e., that each compatibility relation can be expressed as the composite $\alpha\alpha^{-1}$ for a suitable α, we require the additional notions presented below.

Hereafter a general compatibility relation is denoted by γ, although we shall continue to refer to our running example illustrated in figure 1.5.1. With reference

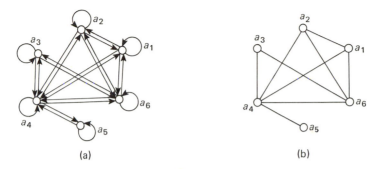

$$(a) \qquad\qquad\qquad (b)$$

Figure 1.5.2. Directed and undirected graphs for compatibility relation

to that example, the directed graph for $\gamma = \alpha\alpha^{-1}$ is shown in figure 1.5.2(a). Note that in the directed graph of γ each vertex has a loop (reflexivity), and all other arcs occur in symmetric pairs (symmetry). Therefore we may simplify the graphical representation by omitting the loops and by replacing a symmetric pair with a single *undirected* edge (an edge having no arrow). The resulting graphical structure —here viewed simply as the description of a compatibility relation—is known as an *undirected graph*, and it will be studied for its geometric properties in the next chapter. An undirected edge, or simply *edge*, between vertices a_i and a_j of A is represented by the concise notation $\overline{(a_i, a_j)}$. Figure 1.5.2(b) shows the undirected graph of the compatibility relation $\gamma = \alpha\alpha^{-1}$. We see the effectiveness of the un-directed edges in evidencing the symmetric relation. We also observe that the array representation of figure 1.5.1(b) is redundant. In fact, complete information about $\alpha\alpha^{-1}$ is provided by the symbols in the $(n-1)n/2$ positions below the main diagonal, since those in the main diagonal are 1's and those above the main diagonal are obtained by symmetry. For this reason, the "staircase array" shown in the lower left portion of figure 1.5.1(b) is frequently used to describe a compatibility relation.

We shall now analyze sets of persons such that mutual dialogue among pairs of persons of the set is possible. For obvious reasons we term such sets *compatibility classes*, or simply *compatibles*.

Definition 1.5.2. Given a compatibility relation γ on A, a *compatibility class*, or *compatible*, induced by γ, is a subset C of A, such that, for any members a_1 and a_2 of C, $a_1 \gamma a_2$. (Note that a_1 and a_2 may coincide; that is, a compatible may consist of just one element.)

By inspection of figure 1.5.2(b), we see that $\{a_4, a_6\}$ is a compatible because $a_4 \gamma a_6$. We can also adjoin to $\{a_4, a_6\}$ any one of the three elements a_1, a_2, a_3, because each of them is related, under γ, to both a_4 and a_6. Suppose that we now form the compatible $\{a_1, a_4, a_6\}$. We note further that a_2 can be adjoined to form the compatible $\{a_1, a_2, a_4, a_6\}$, because $a_2 \gamma a_1$, $a_2 \gamma a_4$, $a_2 \gamma a_6$. However, we recognize that no larger compatible exists which contains $\{a_1, a_2, a_4, a_6\}$. The same holds for the two-element compatible $\{a_4, a_5\}$ or the three-element compatible $\{a_3, a_4, a_6\}$. These examples point to the fact that, for each C_1, there is some compatible C_2 (possibly identical to C_1 itself) which contains it, such that there is no compatible properly containing C_2. A compatible like C_2 falls in the following category.

Definition 1.5.3. Given a compatibility relation γ on A, a *maximal* compatibility class, or *maximal* compatible, is a compatibility class which is not *properly* contained in any other compatibility class.

We may determine all the maximal compatibles induced by γ on the set A. In our running example, by inspection of figure 1.5.2(b), we obtain $\{a_4, a_5\}$, $\{a_3, a_4, a_6\}$, $\{a_1, a_2, a_4, a_6\}$. The *family of sets* just given is characterized by the following definition.

Definition 1.5.4. The *family of sets* $C_\gamma(A)$ is the collection of all and only the maximal compatibles induced by γ on A and is termed a *complete cover* of A with respect to γ.

The intuitive justification of the phrase "complete cover" will be evident from the following lemma.

Lemma 1.5.1. If γ is a compatibility relation on a finite set A and C is a compatibility class, then there is some maximal compatible C' such that $C \subseteq C'$.

Proof. Since A is finite, the elements of A can be numbered; that is, $A \equiv \{a_1, a_2, \ldots a_n\}$. We define a sequence $C_0 \subset C_1 \subset C_2 \subset \cdots$ of compatibles by the following rule: $C_0 = C$.

Given C_i, then C_{i+1} is defined as $C_i \cup \{a_j\}$, where j is the smallest integer for which $a_j \notin C_i$ and C_{i+1} is a compatible. If there is no such j, then C_i is the last compatible of the sequence and $C_i = C'$. Thus C' is compatible and maximal. Since $C \subseteq C'$, the lemma is proved.‖

Thus each element of A belongs to at least one member of $C_\gamma(A)$—that is, it is "covered"—and all the maximal compatibles belong to $C_\gamma(A)$. This explains the choice of the term "complete cover." The complete cover $C_\gamma(A)$ is an important concept, for there is a one-to-one correspondence between a relation γ on A and $C_\gamma(A)$. In other words, we can uniquely determine one from the other. The demonstration of this fact will refer to the undirected graph of the relation γ on the set A. First, we observe that, from the definition of complete cover, each undirected graph of a relation γ on A determines a unique $C_\gamma(A)$.

Thus the relation between the set of undirected graphs and the set of complete covers is a function. Second, we show that there are no two distinct graphs, pertaining to relations γ and γ' on the same A, such that $C_\gamma(A) = C_{\gamma'}(A)$. The hypothesis of distinct γ and γ' means that there are at least two elements $a_i, a_j \in A$, such that, although $a_i \, \gamma \, a_j$ *is* true, $a_i \, \gamma' \, a_j$ is *not* true. From $a_i \, \gamma \, a_j$ it follows, from lemma 1.5.1, that there is a pair $\{a_i, a_j\}$ in some member C of $C_\gamma(A)$. Since $C_\gamma(A) = C_{\gamma'}(A)$, there is a member C' of $C_{\gamma'}(A)$ which is identical to C; hence it contains the same pair $\{a_i, a_j\}$. But this indicates that $a_i \, \gamma' \, a_j$ *is* true, a contradiction. Therefore we have proved the next theorem.

Theorem 1.5.2. There is a one-to-one correspondence between compatibility relations γ on A and complete covers $C_\gamma(A)$.

The previous demonstration implicitly offers a method for obtaining γ from $C_\gamma(A)$. Indeed, the graph of γ can be constructed as follows: For every member C_k of $C_\gamma(A)$, draw an edge between a_i and a_j for every pair $a_i, a_j \in C_k$. (In carrying out this procedure, obviously, we may try to draw some edges more than once, but we certainly draw them all.)

Example 1.5.1. For the set $A \equiv \{a_1, a_2, a_3, a_4, a_5\}$ let the complete cover $C_\gamma(A) \equiv \{\{a_1, a_5\}, \{a_4, a_5\}, \{a_1, a_2, a_3\}, \{a_2, a_3, a_4\}\}$ be given. From $C_\gamma(A)$ we readily obtain the graph of figure 1.5.3, in which the elements that are members of the maximum compatibles are also indicated.

One can carry out the method outlined simply by scanning the list of the members of $C_\gamma(A)$. Unfortunately, the converse operation, the derivation of $C_\gamma(A)$ from γ, is not so simple. In fact, we might obtain maximum compatibles by inspection of the graph, and we might even obtain their complete set. For large sets A, however, we cannot rely on the efficacy of graphical methods to obtain $C_\gamma(A)$. A systematic method for accomplishing this purpose requires more complicated procedures whose presentation is beyond the scope of this book although the concepts on which these procedures are based will be presented in Chapter 5.

At this point, it is appropriate to observe that an arbitrary collection of subsets of A, such that each element $a \in A$ belongs to some member C of the collection, does not automatically qualify as the complete cover of some compatibility relation. Indeed, an obvious necessary condition for a complete cover $C_\gamma(A)$ is that no member of it be contained in any other member since a complete

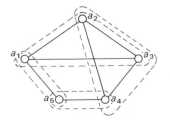

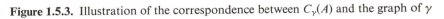

Figure 1.5.3. Illustration of the correspondence between $C_\gamma(A)$ and the graph of γ

cover is a collection of *maximal* compatibles. But this condition is not sufficient. Consider the collection $\{a_1, a_5\}$, $\{a_4, a_5\}$, $\{a_1, a_2\}$, $\{a_1, a_3\}$, $\{a_2, a_3, a_4\}$ of subsets of $A \equiv \{a_1, a_2, a_3, a_4, a_5\}$. This collection satisfies the aforementioned condition and defines the same graph as that defined in example 1.5.1. However, we easily realize that the given collection of subsets does not consist entirely of maximal compatibles; indeed, $\{a_1, a_2\}$ and $\{a_1, a_3\}$ belong to the maximal compatible $\{a_1, a_2, a_3\}$.

We are now in a position to demonstrate the assertion that for any compatibility relation γ on A, we can construct a relation α from A to some set B such that $\gamma = \alpha\alpha^{-1}$. Let β be the complete cover $C_\gamma(A)$; for $a \in A$ and $C \in C_\gamma(A)$, the relation is defined as follows:

$$a \; \alpha \; C \text{ if and only if } a \in C.$$

We note that this relation α is specified for each $a \in A$ (since, by lemma 1.5.1, each a belongs to at least one maximal compatible). We now claim that $\gamma = \alpha\alpha^{-1}$. Assume that $a_i \; \gamma \; a_j$. Then there is a maximal compatible C, such that $a_i \in C$ and $a_j \in C$, or, in terms of α, $a_i \; \alpha \; C$ and $a_j \; \alpha \; C$ (see figure 1.5.4). The notation $a_j \; \alpha \; C$ may be rewritten as $C\alpha^{-1}a_j$; from $a_i \; \alpha \; C$ and $C\alpha^{-1}a_j$ we obtain $a_i(\alpha\alpha^{-1})a_j$; namely, each pair of γ is also a pair of $\alpha\alpha^{-1}$. Conversely, assume that $a_i(\alpha\alpha^{-1})a_j$. This means that there is some C in $C_\gamma(A)$, such that $a_i \; \alpha \; C$ and $C\alpha^{-1}a_j$; thus a_i and a_j belong to C, that is, $a_i \; \gamma \; a_j$. This shows that each pair of $\alpha\alpha^{-1}$ is also a pair of γ, whence $\alpha\alpha^{-1} = \gamma$. We state this result as a theorem (characterization theorem of compatibility relations).

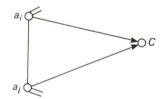

Figure 1.5.4. Illustration of $\gamma = \alpha\alpha^{-1}$

Theorem 1.5.3. A relation γ on a set A is a compatibility relation if and only if there is at least one relation α from A to some set B, such that $\gamma = \alpha\alpha^{-1}$, with the additional condition that for each $a \in A$ there is some $b \in B$ such that $a \; \alpha \; b$.

The reader must have noticed that elements of A may belong to more than one maximal compatible (in the previous example, a_3 and a_2 belong to both $\{a_1, a_2, a_3\}$ and $\{a_2, a_3, a_4\}$). This is an important consequence of the lack of transitivity. Assume that, for $a_1, a_2, a_3 \in A$, $a_1 \; \gamma \; a_2$ and $a_2 \; \gamma \; a_3$ *are* true but that $a_1 \; \gamma \; a_3$ is *not* true; then a_1 and a_2 will belong to some maximal compatible C_1, and a_3 and a_2 will belong to some maximal compatible C_2, and $C_1 \neq C_2$; otherwise a_1 and a_3 would be in the same compatible. Hence C_1 and C_2 are distinct, but they share at least the element a_2.

A compatibility relation γ, in addition to having reflexivity and symmetry, has the transitive property within a compatibility class. In fact, for $a_1, a_2, a_3 \in C$, we have, from the definition of compatible, that $a_1 \gamma a_2$, $a_2 \gamma a_3$, and $a_1 \gamma a_3$, which proves transitivity. What is now the consequence of requiring that the transitive property hold not only within compatibles but within the entire set A? This requirement produces a very important specialization of compatibility relations, as expressed by the following definition (recall the case RST in Section 1.4).

Definition 1.5.5. An *equivalence relation* on a set A is a relation which has the reflexive, symmetric, and transitive properties. (There are two commonly used symbols of an equivalence relation: $\equiv$ and ε.)

Since equivalence is a special case of compatibility, all the previously illustrated concepts and results apply to it. However, transitivity brings about some very special properties, which we now analyze.

Given an equivalence relation ε on A, we may form its maximal compatible set $C_\varepsilon(A)$, which we call the set of "equivalence classes."

Definition 1.5.6. A maximal compatible of an equivalence relation ε on A is termed an *equivalence class*. (Equivalence classes are usually denoted by the capital letter E.)

We shall now prove a very important result concerning the collection $C_\varepsilon(A)$.

Lemma 1.5.4. For an equivalence relation ε, any two distinct members E_1 and E_2 of $C_\varepsilon(A)$ have no common element of A; that is, they are disjoint.

Proof. We shall prove the lemma by showing that if two classes E_1 and E_2 have some common element, they coincide. Assume that $a \in E_1 \cap E_2$. If E_1 contains no other element but a, then $E_1 \subseteq E_2$; otherwise, for every other element $a_1 \in E_1$, we have $a \, \varepsilon \, a_1$, and since E_2 is maximal, we also have $a_1 \in E_2$. We conclude that an arbitrary member of E_1 is also a member of E_2; that is, $E_1 \subseteq E_2$. Similarly, we can show that $E_2 \subseteq E_1$, thus obtaining $E_1 = E_2$.$\|$

We remember that, by lemma 1.5.1, for a compatibility relation γ, each element a of A belongs to at least one maximal compatible induced by γ on A. In the case of equivalence, we have just shown that two distinct equivalence classes have no common elements; thus we have the following immediate result.

Theorem 1.5.5. Each element $a \in A$ belongs to one and only one equivalence class induced by an equivalence relation ε on A.

Such a subdivision of A conforms to the following definition.

Definition 1.5.7. A *partition* of a set A is a collection of subsets of A, such that each $a \in A$ belongs to exactly one of these subsets. A partition of A into two classes is called a *dichotomy*.[7]

It then follows that the complete cover of an equivalence relation is a partition. But we can also make a stronger statement: Every partition of a set A qualifies as a complete cover of an equivalence relation. To prove this we must show that a partition uniquely specifies an equivalence relation ε whose complete cover $C_\varepsilon(A)$ coincides with the partition itself. Given a partition of A, define a relation ε as follows:

$a_i \; \varepsilon \; a_j$ if and only if a_i and a_j belong to the same subset C in the partition.

Clearly ε is a compatibility relation because, for any pair a_i, a_j in C, we have $a_i \; \varepsilon \; a_j$ and $a_j \; \varepsilon \; a_i$. Furthermore, for a_i, a_j, a_k in A, if $a_i \; \varepsilon \; a_j$ and $a_j \; \varepsilon \; a_k$, then a_i, a_j and a_k belong to the same subset C; hence $a_i \; \varepsilon \; a_k$ (transitivity). Therefore ε is an equivalence relation. Finally, we claim that each class C of the original partition is a maximal compatible. First, C is by hypothesis a compatible. Second, assume that C is not maximal; namely, there is an a in A but not in C, such that for every $c \in C$, we have $a \; \varepsilon \; c$. But $a \; \varepsilon \; c$ implies $a \in C$, a contradiction. Therefore we have proved the following corollary.

Corollary 1.5.6. There is a one-to-one correspondence between equivalence relations on A and partitions of A.

Example 1.5.2. Consider the set of integers I. For $i \in I$, compute r_i as the remainder of the division of i by 5. Then we have $i = 5q + r_i$, with unique q and $r_i < 5$ (that is, $r_i = 0, 1, 2, 3, 4$). We now define an equivalence relation $\equiv$ on I, such that $i \equiv j$ if and only if $r_i = r_j$. We remark that $r_i = r_i$; hence $i \equiv i$ (reflexivity). If $i \equiv j$, then $r_i = r_j$ and $r_j = r_i$, whence $j \equiv i$ (symmetry). Finally, if $i \equiv j$ and $j \equiv k$, then $r_i = r_j = r_k$, whence $i \equiv k$ (transitivity). We have an equivalence relation $\equiv$, which induces a partition of I into five (one for each value of r_i) equivalence classes $E_0, E_1, \ldots, E_4$, namely, $E_i \equiv \{j \mid j = 5k + i, k = 0, 1, \ldots\}$.

To close the discussion, let us consider the specialization of theorem 1.5.3 to an equivalence relation ε on A. As we noted, the complete cover $C_\varepsilon(A)$ is a partition; hence each $a \in A$ belongs to exactly one member of $C_\varepsilon(A)$, which is equivalent to saying that α is a function from A to $C_\varepsilon(A)$. Conversely, assume that α is a function from A to B for some B. The composite $\alpha\alpha^{-1}$, besides being reflexive and symmetric, is transitive (figure 1.5.5). For a_i, a_j and $a_k \in A$, if $a_i(\alpha\alpha^{-1})a_j$, there must be a $b_r \in B$ for which $a_i \; \alpha \; b_r$ and $a_j \; \alpha \; b_r$. Similarly, if $a_j(\alpha\alpha^{-1})a_k$, there must be a $b_s \in B$ for which $a_j \; \alpha \; b_s$ and $a_k \; \alpha \; b_s$ (figure 1.5.5(a)). But α is a function;

---

7. This word comes from the Greek words δίχα (dicha), "in two," and τομή (tomé), "cut, division."

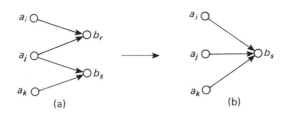

Figure 1.5.5. Illustration of the proof of theorem 1.5.7

hence $a_j \, \alpha \, b_r$ and $a_j \, \alpha \, b_s$ imply $b_r = b_s$, a unique image (figure 1.5.5(b)). Therefore, from $a_i \, \alpha \, b_r$ and $a_k \, \alpha \, b_r$ (or $b_r \alpha^{-1} a_k$), we obtain $a_i(\alpha \alpha^{-1}) a_k$; that is, $\alpha \alpha^{-1}$ is an equivalence relation. This is summarized as a characterization theorem of equivalence relations.

Theorem 1.5.7. A relation ε on a set A is an equivalence relation if and only if there is at least one function α from A onto some set B for which $\varepsilon = \alpha \alpha^{-1}$.

A natural choice for the previous function α is the "membership function" of an element of A in its equivalence class, that is, a function $E: A \to C_\varepsilon(A)$, such that $E(a)$ is the unique equivalence class in $C_\varepsilon(A)$ containing the element a in A.

_____

EXERCISES 1.5

1. Determine which of the following sets define an equivalence relation and which define a compatibility relation.

   a) $A \equiv \{(x, y) | x$ and y are people of the same age$\}$
   b) $B \equiv \{(x, y) | x$ and y are cousins$\}$

2. Let α be a relation defined on $A \equiv \{1, 2, 3, 4, 5, 6\}$ such that $R_\alpha = R_{e_A} \cup R_\beta \cup R_{\beta^{-1}}$, where $R_\beta \equiv \{(1, 2), (1, 3), (2, 3), (2, 4), (3, 4), (2, 5), (4, 5), (3, 6), (4, 6)\}$. Show that α is a compatibility relation and determine the complete cover of α.

3. A family of subsets $A_1, \ldots, A_n$ of a set A is called a *cover* of A if

$$A_1 \cup A_2 \cup \cdots \cup A_n = A.$$

   a) Show that every cover C of A induces a compatibility relation α on A such that $a \, \alpha \, b$ if and only if $\{a, b\} \subseteq A_i$, for some i, $1 \leq i \leq n$.
   b) Let α be a relation given in Exercise 1.5.2. Show that α can be induced by at least two different covers on A.

4. A compatibility relation γ on a set A may have more maximal compatibles than there are elements in A. Which is the smallest cardinality for A for which this may happen? Justify your answer.

5. Let α be a relation which is not an equivalence. Give a systematic procedure to extend α so that it becomes an equivalence relation.

6. Let $A \equiv \{a_1, \ldots, a_n\}$ be a set of n elements. Let P denote the family of all subsets A_i of A, and let α be a relation on P such that $A_i \, \alpha \, A_j$ means that the elements of A_i and A_j can be placed in one-to-one correspondence. Show that α is an equivalence relation on P.

7. If α and β are compatibility relations on a set A, is the composite $\alpha\beta$ a compatibility relation?

8. Show that the identity relation, e_A, is the only equivalence relation on A which also satisfies the antisymmetric property.

9. Let α and β be two given equivalence relations on a set A.

   a) Is the composite $\alpha\beta$ an equivalence relation?
   b) If we define a relation $\gamma = \alpha \cap \beta$ on A such that $a \, \gamma \, b$ if and only if $a \, \alpha \, b$ and $a \, \beta \, b$, is γ an equivalence relation?
   c) Describe γ in part (b) in terms of partitions.

*10. Let α and β be two given equivalence relations on a set A. Show that the composite $\alpha\beta$ is an equivalence relation if and only if $\alpha\beta = \beta\alpha$. Give an example of $\alpha\beta = \beta\alpha$ and of $\alpha\beta \neq \beta\alpha$.

11. Let m be a positive integer. We define a relation θ_m such that

$$R_{\theta_m} = \{(i,j) \mid i = j + km, k \text{ an integer}\}.$$

   a) Show that θ_m is an equivalence relation.
   b) Show that if $i \, \theta_m \, i'$ and $j \, \theta_m \, j'$, then $(i+j) \, \theta_m (i'+j')$ and $ij \, \theta_m \, i' j'$.
   c) What is the equivalence relation defined by the set $R_{\theta_m} \cap R_{\theta_n}$?
   *d) Let α be the relation such that

$$a \, \alpha \, b \text{ if and only if } a^2 \, \theta_7 \, b^2.$$

   Show that α is an equivalence relation. Into how many equivalence classes does α partition the set of nonnegative integers?

12. Show that lemma 1.5.1 holds for countable set also.

1.6 BIBLIOGRAPHY AND HISTORICAL NOTES

The theory of sets originated in mathematical physics during the past century. The original "concrete" instance, the set of points, was then generalized, largely as a result of the work of Georg Cantor, to the notion of abstract sets, with no reference to the nature of their elements. Since then, the concept of sets has become one of the foundations of mathematics and logical thinking in general.

Abstract sets—as we consider them here—are discussed in a great variety of publications in connection with algebra, logic, probability, etc., but we shall limit our references to two very valuable works. The first is a very clear treatment of sets and relations, given in Chapter 1 of

H. Paley and P. M. Weichsel, *A First Course in Abstract Algebra*, Holt, Rinehart and Winston, New York, 1966.

The second is a classical, quite formal reference for the mathematically mature reader:

P. R. Halmos, *Naive Set Theory*, Van Nostrand, Princeton, 1960.

1.7 NOTES AND APPLICATIONS

Note 1.1. Finite State Machines

The reader has surely heard the phrase "digital system" many times, and he has probably already dealt with such systems, perhaps even without being fully aware of it. The phrase "digital system" is correctly interpreted to mean "finite discrete system." Disregarding momentarily the most obvious and illustrious example of all, the digital computer, we can cite a teletype or the control unit of an elevator as good examples of such a system, in that its state can at any instant be described as an element of a finite set of states. For example, the teletype system receives a stream of symbols belonging to a very simple set of four symbols {dot, dash, letter space, word space}, recognizes dot-dash strings as the representations of letters or numbers (a finite number of characters), and prints the recognized characters.

But finite discrete systems are by no means necessarily to be thought of in terms of hardware, specifically of "digital" hardware. To convince ourselves of this fact, let us consider the very familiar experience of a tennis game. A spectator of this game will do a variety of other things—aesthetically enjoy the game, emotionally participate in it, talk to other spectators, etc. But the aspect of his behavior that consists in keeping the score and ultimately declaring the winner of the game can be adequately described as a finite discrete system. Suppose that there are two players *A* and *B*, that the "game" consists of three "sets," and—to simplify the example somewhat—that no detailed account of the points made during each set is kept by our spectator (we assume that some other spectator performs this task for him). The relevant information our observer receives from the environment is either "player *A* wins the set" or "player *B* wins the set." At the completion of each set he must update the score to reflect both the outcome of the just completed set and the history of the game (i.e., the previous score). Clearly, the "situations" to be remembered are "1 *vs* 0 for player *A*," "1 *vs* 0 for player *B*," "1 even," as well as "beginning of game" and "end of game." The last two situations may be identified as the single situation "new game" if one game is followed by another. In the meantime, our scorekeeper will communicate with the environment either by saying that the game is in progress or by giving the result on completion. Therefore, the behavior of our spectator can be described by a double-entry table (see figure 1.7.1) whose rows identify the present situation and whose columns identify the messages from the environment. This table has two entries per row-column position (separated by a comma), the first specifying the situation resulting from the present situation and the present message *from* the environment, the second specifying the response *to* the environment as caused by the same stimuli (present situation and present message). Note that we have a finite number of messages, two ("*A* wins set," "*B* wins set"), a finite number of situations, four

Situation \ Message	A wins set	B wins set
New game	1 vs 0 for A, game in progress	1 vs 0 for B, game in progress
1 vs 0 for A	new game, A wins game	1 even, game in progress
1 vs 0 for B	1 even, game in progress	new game, B wins game
1 even	new game, A wins game	new game, B wins game

Figure 1.7.1

("new game," "1 *vs* 0 for A," "1 *vs* 0 for B," "1 even"), and a finite number of responses, three ("game in progress", "A wins game," "B wins game"). The process can be viewed as the reception (input) of information from the environment and the transmittal (output) of information to the environment; therefore it can be categorized as some sort of information processing.

This rather unusual description of keeping the score in a tennis game is an example of a very important class of information processing systems called *finite state machines* — "machines" because of their deterministic behavior (no uncertainty exists as to the action to be taken as a consequence of given conditions), "state" as a synonym of "situation," and "finite" because the sets of the input symbols, output symbols, and internal states are finite. Finite state machines are also called *sequential machines* in the literature.

At this point, if we consider the set of states $S \equiv \{s_1,\ldots,s_n\}$, the set of inputs $I \equiv \{i_1,\ldots,i_m\}$, and the set of outputs $Z \equiv \{z_1,\ldots,z_r\}$, we recognize that the table of figure 1.7.1 describes two functions. The first is a function of the type $S \times I \to S$, that is, a mapping of a pair (a state and an input) to a state. This function, usually denoted $\delta(s_j, i_k)$, is called the *transition function*. The second is a function of the type $S \times I \to Z$, that is, a mapping of a state and an input to an output. This function, usually denoted $\lambda(s_j, i_k)$, is called the *output function*. A natural way to describe the two functions $\delta(s_j, i_k)$ and $\lambda(s_j, i_k)$ is therefore a generalization of the specific table of figure 1.7.1. Such a table, known as a *flow table,* is shown in

State \ Input	$i_1\ i_2\ldots$	i_k	$\ldots i_m$
s_1 s_2 $\vdots$			
		$\delta(s_j, i_k),\ \lambda(s_j, i_k)$	
$\vdots$ s_n			

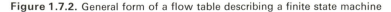

Figure 1.7.2. General form of a flow table describing a finite state machine

figure 1.7.2. Obviously, $\delta(s_j, i_k) \in S$ and $\lambda(s_j, i_k) \in Z$. It is convenient to formalize the definition of a finite state machine for future reference.

Definition 1.7.1. A *finite state machine* $\mathcal{M}$ is a system $[S, I, Z, \delta, \lambda]$, where S, I, and Z are finite nonempty sets of states, input symbols, and output symbols, respectively, $\delta : S \times I \rightarrow S$ is the transition function, and $\lambda : S \times I \rightarrow Z$ is the output function.

Each of the two functions $\delta : S \times I \rightarrow S$ and $\lambda : S \times I \rightarrow Z$ may be viewed as a collection of m functions. Specifically, we have $\delta \equiv \{\delta_1, \delta_2, ..., \delta_m\}$ and $\lambda \equiv \{\lambda_1, \lambda_2, ..., \lambda_m\}$, where $\delta_k : S \rightarrow S$ and $\lambda_k : S \rightarrow Z$ are respectively the transition and output functions associated with the input symbol i_k. In other words, we have that $\delta_k(s_j) = \delta(s_j, i_k)$ and $\lambda_k(s_j) = \lambda(s_j, i_k)$; that is, δ_k and λ_k are described by the kth column of the flow table illustrated in figure 1.7.2.

The functioning of a finite state machine $\mathcal{M}$ may be viewed as follows. Time is considered as subdivided into time units, and at each time unit, $\mathcal{M}$ is in some state $s_j \in S$, receives an input symbol $i_k \in I$, generates an output symbol $\lambda(s_j, i_k)$, and makes a transition to state $\delta(s_j, i_k)$. By $i^{(t)}$ and $z^{(t)}$ we denote the input and output symbol at time unit t, respectively. If $\mathcal{M}$ receives a sequence of input symbols $i^{(1)} i^{(2)} ... i^{(t)}$ (one symbol per time unit), it produces a sequence of output symbols $z^{(1)} z^{(2)} ... z^{(t)}$, which, clearly, depends on the state $\mathcal{M}$ is in when $i^{(1)}$ is received (called the *initial* state of $\mathcal{M}$). Let I^* denote the set of the input sequences and Z^* the set of the output sequences. Then $\mathcal{M}$, starting in a given initial state, may be considered as realizing a function f from I^* to Z^*. Note, however, that there are several restrictions on the nature of f:

1. For two sequences $x \in I^*$ and $w \in Z^*$, if $w = f(x)$, then x and w contain the same number of symbols (we say that f is *length-preserving*).

2. If the sequence x' contains t' symbols, and x contains t symbols (with $t' < t$), and the first t' symbols of x coincide with x' (we say that x' is a *prefix* of x), then the first t' symbols of $f(x)$ coincide with $f(x')$ (we say that f is *prefix-inclusive*).

In terms of input and output sequences, behaviorally a finite state machine $\mathcal{M}$ acts like a sequence *transducer*. Therefore, a function $f : I^* \rightarrow Z^*$ which is length-preserving and prefix-inclusive is called a *transduction*. This also justifies calling $\mathcal{M}$ a *sequential* machine.

Suppose now that $\mathcal{M}$ has two states s_i and s_j, such that an input sequence is transduced into the same output sequence, regardless of whether s_i or s_j is the initial state of $\mathcal{M}$, and this occurs for every input sequence. It follows that behaviorally (i.e., by observing the input and the output of $\mathcal{M}$), we are not able to tell whether $\mathcal{M}$ started in s_i or in s_j. The two states are said to be "equivalent," denoted by $s_i \, \varepsilon \, s_j$. Note that the relation ε is an equivalence relation (definition 1.5.5), since it is reflexive, symmetric, and transitive, as can be easily verified. Then, by corollary 1.5.6, the set S of the states of $\mathcal{M}$ is partitioned by the relation ε into equivalence classes, and by theorem 1.5.7, all the states in a class are mapped to a "representative" state. It follows that the given machine $\mathcal{M}$ may be replaced by a *reduced* machine $\mathcal{M}_R$, which has as many states as there are classes of equivalent states in $\mathcal{M}$.

When some of the entries of the flow table of a machine $\mathcal{M}$ (either the outputs or the transitions) are unspecified,[8] i.e., left to the designer's choice, the relation between states of $\mathcal{M}$ is a compatibility relation γ (definition 1.5.1).

The actual determination of either ε or γ from the flow table of $\mathcal{M}$ is beyond the scope of this book. Our objective was to show the reader an interesting application of the notions developed in this chapter. If his curiosity has been stimulated, he should consult one of the several good references on the subject. We shall return to the topic of finite state machines in the following chapters.

Reference

Z. Kohavi, *Switching and Finite Automata Theory,* McGraw-Hill, New York, 1970, Chapters 9 and 10.

Note 1.2. Document and Information Retrieval

Modern developed societies are characterized by a close interdependence of their constituents, whether they are individuals, businesses, corporations, governmental offices, etc. This interdependence requires an active exchange of information and the handling of extremely large volumes of data. Situations requiring such handling are innumerable. Let us consider, for example, the inventory and shipping schedules of a large company, the updating of checking accounts of a bank, or the timetable of flights of airlines members of the International Air Transport Association.

But one of the most conspicuous examples of the need for exchange of information occurs in scientific literature, whose extremely rapid growth makes it very hard for scientific investigators to keep up with developments even in their own fields. This phenomenon is often referred to as the "information explosion." The necessity of handling large volumes of data and the commercial availability of digital computers suggested very early (in the 1950's) the possibility of automating the processes of storing and retrieving information. Since then, information storage and retrieval has become an extremely important area of information processing. This brief note only aims at pointing out the relevance to the retrieval of documents of some concepts presented in this chapter.

The ideal information retrieval system would be one that provides an answer to a question, or *query*, formulated either in natural language or in a moderately formalized version thereof. For example, one such question could be "Is the theorem (theorem statement) known?" The reader easily realizes the extreme difficulty of this approach; indeed, important discoveries in manipulation of natural languages and representation of meaning (semantics) must be made before the feasibility of this approach can be established. Therefore, we must content ourselves with less ambitious goals, and leave the information in the same "capsules" into which it was placed by its originators, that is, in *documents* (books, articles, reports, etc.). With this organization, the requester will retrieve a certain set of documents, which presumably contain the answer to his query; it is still up to him to find the answer, but the system has greatly narrowed the amount of data to be searched.

How can documents be accessed and retrieved? Several methods have been proposed, of which two—and their variants—are clearly superior to the others: *coordinate indexing* and *citation indexing*.

---

8. In this case, both δ and λ are partial functions.

We have a collection of documents $D \equiv \{d_1, d_2, ..., d_N\}$ (N may be many thousands). In coordinate indexing, each document d_i is labeled by a set of identifiers (also called descriptors, index terms, keywords, etc.). For example, this note may have the following index terms: information retrieval, coordinate indexing, citation indexing, direct file, inverse file. Therefore we have a set of index terms $K \equiv \{k_1, k_2, ..., k_n\}$ (n may be a few thousands), and indexing is a relation α from D to K. We may store this relation in a computer memory as a list (d_1, K_1), (d_2, K_2), ..., (d_N, K_n), where $K_i \subseteq K$ is the set of the index terms assigned to d_i (a good figure for the average cardinality of K_i is about ten). Such organization is usually referred to as a *direct file*. Documents so indexed may be retrieved in the following way. The query itself is converted into a set of keywords from K. Let such set be $Q \subseteq K$. (Usually the cardinality of Q is comparable to that of the K_i's.) Next, we examine in sequence the pairs (d_i, K_i) of the direct file and count the number n_i of keywords common to K_i and Q; that is, we determine the cardinality of $K_i \cap Q$. Finally, we rank the n_i's in decreasing order and retrieve, for example, all the documents d_i whose n_i are above a certain value. Numerous variants exist of this basic elementary procedure at all levels: in the formalization of the query, in the scoring of the match between query and document, and in the retrieval criterion. Their consideration largely exceeds the scope of this note, however. We shall mention only an important variant of the file organization. When the computer system allows the accessing of data in parallel (so-called *random access*, as distinct from *serial access*), we need not read the complete file to perform the retrieval task. In fact, we could store the inverse indexing relation α^{-1} as a list (k_1, D_1), (k_2, D_2), ..., (k_n, D_n), referred to as an *inverse file*, where $D_i \subseteq D$ is the set of documents indexed by the keyword k_i. Therefore, if the query is the keyword set $(k_{s_1}, k_{s_2}, ..., k_{s_r})$, we shall retrieve only the sets $D_{s_1}, D_{s_2}, ..., D_{s_r}$, and for each d_i member of $D_{s_1} \cup D_{s_2} \cup \cdot \cdot \cup D_{s_r}$, we compute in how many of the sets $D_{s_1}, ..., D_{s_r}$ it occurs. From there on, everything is as it was before.

In citation indexing, the indexes are not assigned to a document as descriptors of its content but are mechanically derived from the document itself; they are the documents cited by the document under consideration. Therefore citation indexing is a relation β on the set D. Storing of β may be accomplished essentially as in coordinate indexing. However, retrieval is in general more complex because a set D' of documents may be retrieved on the basis of the ratio of arcs to vertices in the subgraph of β restricted to D'. Citation indexing, used alone, does not seem to be very satisfactory, but when used in conjunction with coordinate indexing, either it may serve as a confirming device or it may reach documents that are very relevant to the query yet inaccessible with index terms alone.

Reference

G. Salton, *Automatic Information Organization and Retrieval*, McGraw-Hill, New York, 1968.

Note 1.3. Russell's Paradox

The notion of set, arising in such a natural way from our experience and seemingly so basic to our way of reasoning, is perfectly acceptable in instances like "the set of red apples" or "the set of positive integers less than 100." These specifications of sets, however, are still quite "concrete" since they explicitly state the characteristic property of the set. Although these descriptions are quite useful because they provide an intuitive support for more formal considerations, the mathematician tends to consider a set as a collection of "abstract" objects represented by symbols (actually, he tends to identify these objects, unspecified, with the symbols themselves, which are tangible). This important abstraction step is subtly deceptive and has led to celebrated logical paradoxes, which may shake one's confidence in formal argument.

One such paradox, discussed by Bertrand Russell, concerns the notion of "set of all sets." Consider the set S defined as follows:

$$S \equiv \{x \mid x \text{ is a set and } x \text{ does not belong to } x\}. \qquad (1.7.1)$$

That is, S is the set of all sets x, such that x is not a member of itself; for example, the set of all houses is a set but it is not a house, so it belongs to S. Since S itself is a set, we may ask the question "Does S belong to S?" Suppose that S satisfies the characteristic property of S (that is, S *belongs* to S). Then, replacing S for x in (1.7.1), we have "S is a set and S *does not belong* to S." Conversely, suppose that S *does not belong* to S; that is, S does not satisfy the characteristic property. This characteristic property consists of two statements, and since S satisfies the first, it must violate the second, or "S *belongs* to S." In both cases we reach the paradoxical situation that premise and conclusion are mutually contradictory.

Substantial speculative effort has been devoted in this century to identifying a tacit assumption that acts as the hidden cause of the paradox and to eliminating it from formal thinking. To this aim, Russell himself and Whitehead proposed the so-called *theory of logical types*, which introduces a hierarchical ordering of sets. Each set is assigned an integer, called "level," and sets with the same level are of the same logical type. Levels are assigned as follows: Objects are "sets" of level 0, and a set is of level ℓ if the highest level of its members is $(\ell - 1)$: Thus a set a belongs to another set A if an only if a is of a level lower than A. In this manner the "set of all sets" disappears, being sent away to level infinity. This is a way to avoid causing the characteristic property of a set S to be tested on S itself—a vicious circle, considered by Russell and Whitehead the source of the paradox.

Note 1.4. On the Complexity of Representation of Binary Relations and Functions

In Section 1.3 we claimed that the list representation of a function is less complex than the tabular representation of a general binary relation for the same pair of sets. Although a deep approach to such problems is well beyond the scope of this book, we offer the following approximate argument to give some flavor of the concepts involved. For brevity, let the phrase "*b*-question" denote a question which can receive only a "yes" or "no" answer. We assume that a satisfactory measure of the complexity of the representations being considered can be given by the number of *b*-questions which must be asked of a reliable informant in order to identify the relation between two known sets. By reference to the array, it becomes clear that a generic *b*-question consists of the identification of an array position, and that the symbols 1 and 0 correspond, respectively, to "yes" and "no" answers (yes for presence, no for absence of a directed edge). It then follows that a total of nm b-questions are needed to identify a generic relation. For a function, however, assume first that a one-to-one correspondence has been established between the elements of the codomain B and the integers 0, 1, 2, ..., $m - 1$. Then for each element of A we need to know only one such integer in the range $0 \leqslant j \leqslant m - 1$. How many b-questions are then required for this purpose? This can be readily established as follows: We ask if j is odd (a b-question). If j is odd, the informant forms $(j - 1)/2$ (an integer) and sets $j' = (j - 1)/2$. Otherwise he sets $j' = j/2$.

Subsequently, we subject j' to the same b-question, and so on recursively until the informant reaches 0. Since at each b-question the integer is at worst halved, we need at most $\lceil \log_2 m \rceil$ questions, where $\lceil \ \rceil$ denotes the "least integer not less than." We conclude that at most $n\lceil \log_2 m \rceil < nm$ b-questions are needed to identify a generic function, which substantiates our claim on complexities.

The investigation of a quantitative measure of representation complexities (coding) is an important area of information theory. For an in-depth exposure to these concepts, the reader should consult the following works.

N. Abramson, *Information Theory and Coding*, McGraw-Hill, New York, 1963, Chapters 1, 2, 3, and 4.

R. B. Ash, *Information Theory*, Interscience Publishers, New York, 1965, Chapters 1 and 2.

CHAPTER 2

GRAPHS

2.1 DIRECTED GRAPHS

There are many occasions in our everyday life when we are confronted with various types of diagrams, such as city maps, organizational charts, electrical networks, and chemical bonds. All these diagrams in one way or another indicate how the constituents of various structures are related or connected. Consider a sports tournament in which there are four teams, a, b, c, and d. The situation to be represented is that teams a and b, a and c, and c and d have met and that a defeated b, c defeated a and d defeated c (ties are not allowed). This can be represented by the diagram in figure 2.1.1. It immediately follows from our discussions in the preceding chapter that figure 2.1.1 is the graphical representation of the binary relation $\{(a,b), (c,a), (d,c)\}$ on the set $\{a,b,c,d\}$.

We arrive again at the familiar connection between graphs and binary relations on a set. As we said in Section 1.4, the two classes of structures, directed graphs and binary relations on a set, actually coincide. However, we also mentioned that graphs are remarkably important as models of real systems, and for this reason, their geometric properties are worth investigating. That is the objective of this chapter.

Suppose now that we want to provide a graphical representation, analogous to the one of figure 2.1.1, for a similar yet somewhat more complex situation.

Figure 2.1.1. Diagrammatical illustration of a sports tournament

Specifically, the four teams, a, b, c, and d are in a tournament organized as two consecutive round robins; each two teams meet twice (the typical scheme of national league soccer championships in Europe and Latin America). We shall still assume that ties are not allowed. Therefore, there will be a total of 12 matches, 6 for each round robin. Each match is going to be represented as a segment in the diagram, with an arrow pointing to the vertex of the losing team. Diagrams for the possible outcomes of the first round robin, the second round robin, and the complete tournament are respectively given in the three parts of figure 2.1.2.

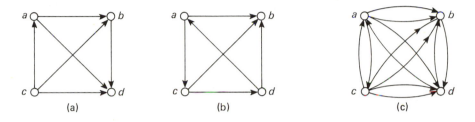

Figure 2.1.2. Diagrammatical illustrations of a double round-robin sports tournament

Note that parts (a) and (b) are descriptions of the binary relations "defeat." If a pair, say (a,b), is in the relation, then a defeated b. But in order to summarize the tournament, we merge (or superimpose) these two diagrams, obtaining figure 2.1.3(c). This new diagram no longer describes a binary relation on a set (by means of the binary relation "to defeat" there is no way to express "a defeated b twice"). Rather, it results from the superposition of two graphs, each of which describes a binary relation.

Another interesting example is a road map of a region where there are several urban centers (towns) connected by highways. A model, whose structure is extremely close to that of the system it represents, is a diagram whose vertices, the towns, are connected by arcs; and an arc from a to b indicates that there is exactly one lane of highway from a to b. A fragment of one such diagram is shown in figure 2.1.3. Here again, to represent a multiple-lane highway we draw

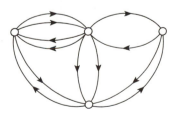

Figure 2.1.3. Diagram of a road map

more than one concurrent arc between two vertices. Thus we see that in various situations the need arises for more general structures than the graphs introduced in the previous chapter. This need motivates the following definition.

Definition 2.1.1. A *multiple graph* (or *multigraph*) G consists of a set V of vertices, a set A of arcs, and a function φ from the set A to the set $V \times V$.

The symbolic representation for G is $[V, A, \varphi]$. Arcs of G will be usually denoted by Greek letters with an arrow on top. If $\varphi(\vec{\alpha}) = (u, v)$, then the two notations $\vec{\alpha}$ and $\overrightarrow{(u, v)}$ will be used interchangeably when no ambiguity arises.

Definition 2.1.1. deserves some comment. First of all, graphs are defined in terms of geometric constituents (arcs and vertices); vertices are usually denoted by the symbols $v_1, v_2, \ldots$, and arcs by the symbols $\vec{\alpha}_1, \vec{\alpha}_2, \ldots$ (this new representation of arcs, different from the one adopted in the previous chapter, is expedient in the present framework).

Second, an arc $\vec{\alpha}$ connects a unique vertex v_1 *to* a unique vertex v_2 (not vice versa!); that is, it corresponds to a unique *ordered* pair (v_1, v_2) or, equivalently, $\varphi : A \to V \times V$ is a function. In general, this function is arbitrary. If for a given ordered pair (v_1, v_2) there are k distinct $\vec{\alpha}$ in A for which $\varphi(\vec{\alpha}) = (v_1, v_2)$, then there are k arcs from v_1 to v_2; k is said to be the *multiplicity* of the pair (v_1, v_2), or the multiplicity of connections from v_1 to v_2. However, when φ is a one-to-one function (definition 1.3.3) and $\vec{\alpha}_1$ in A corresponds to (v_1, v_2), there is no other $\vec{\alpha}_2$ in A corresponding to the same (v_1, v_2); namely, an arbitrary ordered pair (v_1, v_2) has a multiplicity of at most 1. The corresponding class of graphs is characterized in the following definition.

Definition 2.1.2. A *simple graph*[1] is a multigraph $G = [V, A, \varphi]$ such that φ is one-to-one: each pair (v_1, v_2) in $V \times V$ has a multiplicity of at most 1.

The simple graph is an old acquaintance of ours. Indeed, for two arbitrary vertices v_1 and v_2, either there is an arc from v_1 to v_2 or there is not. Simple graphs coincide with the graphs used in Chapter 1 to describe arbitrary relations on a set. Therefore simple graphs can also be denoted by the symbol $[V, \rho]$ where ρ is a binary relation on V.

Example 2.1.1. An important application of the simple graph is the description of procedures, typically computer programs, which consist of simple steps to be executed in an appropriate sequence. Such a graph is universally referred to as a *flowchart*, since it pictorially describes the flow of action (in jargon, "flow of control") in the execution of the program. Vertices represent operations, and customarily they are drawn as boxes of various shapes containing highly compact descriptions of the operations to be performed. The most common shapes of the boxes are rectangular (to denote an operation, such as calculating $a + b$) and

---

1. Sometimes also called a *linear graph*.

diamond (to denote a branching—that is, a test of a condition, such as $a \geqslant 0$ or $a > 0$, and the labeling of the outgoing arcs according to the result of the test). The relation ρ on the vertices V is the relation "to be executed next"; rectangular boxes have one outgoing arc, whereas diamond-shaped boxes have two or three outgoing arcs. Figure 2.1.4 shows a simple program for the solution of the equation $x^3 = 2^x$, its flowchart, and its more abstract graph representation.

1. Start
2. Set x equal to a given constant c
3. Compute $y = 2^{x/3}$
4. Compute $|x - y|$
5. Is $|x - y| \leqslant 10^{-4}$? If yes, stop; otherwise set $x = y$ and return to step 3.

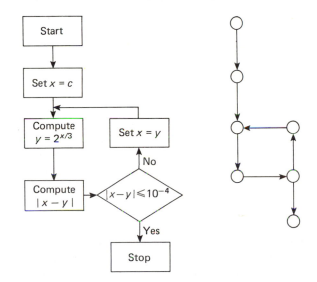

Figure 2.1.4. A program, its flowchart, and the graph structure of the latter

We recall that binary relations on a set $V \equiv \{v_1, \ldots, v_n\}$, and equivalently simple graphs $[V, \rho]$, could be represented by an $n \times n$ array M. Let $(M)_{ij}$ denote the symbol (usually called "entry") in the array M at the intersection of the row of v_i with the column of v_j. We recall that the entry $(M)_{ij}$ is a symbol 1 if there is an arc from v_i to v_j and is a symbol 0 otherwise. If we now no longer consider 1 and 0 as mere symbols but rather as integers, then $(M)_{ij}$ gives exactly the multiplicity of the pair (v_i, v_j)! This naturally suggests the introduction of the *connection array*, which is an array description of multiple graphs, constructed using $(M)_{ij}$, which now may be any nonnegative integer, to express the multiplicity of the connections from v_i to v_j. This results in the following alternative definition of multiple graphs.

Definition 2.1.3. (equivalent to definition 2.1.1) A multiple graph G consists of a set V of vertices and a function ψ from the set $V \times V$ to the set N of nonnegative integers, $\psi(v_1, v_2)$, expressing the multiplicity of (v_1, v_2).

Example 2.1.2. For $V \equiv \{v_1, v_2, v_3, v_4\}$ the multigraph in figure 2.1.5(a) is represented by the array of figure 2.1.5(b).

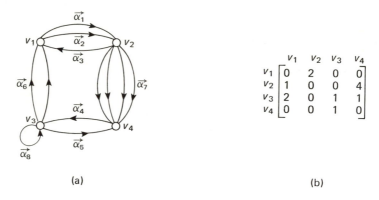

(a)   (b)

Figure 2.1.5. A multigraph and its connection array

For an analysis of the geometric properties of multigraphs, we shall need some additional notions, which we now define.

Definition 2.1.4. Let $G = [V, A, \varphi]$ be a multigraph. Given an arc $\vec{\alpha} = \overrightarrow{(u, v)}$ in A, the vertices u and v in V are called the *origin* and the *terminus* of $\vec{\alpha}$, respectively. A path P of G is a sequence $(\vec{\alpha}_1, \vec{\alpha}_2, \ldots)$ of arcs such that for every pair, $\vec{\alpha}_i, \vec{\alpha}_{i+1}$, of consecutive arcs in P the terminus of the first arc, $\vec{\alpha}_i$, and the origin of the second arc, $\vec{\alpha}_{i+1}$, coincide.

In figure 2.1.5(a), we see that the sequence $(\vec{\alpha}_4, \vec{\alpha}_8, \vec{\alpha}_6)$ is a path. By letting $\vec{\alpha}_i = \overrightarrow{(v_i, v_{i+1})}$ in the preceding definition, we can interpret a path P as a sequence $(v_1, v_2, \ldots)$ of vertices. For example, the sequence of vertices of the path $(\vec{\alpha}_4, \vec{\alpha}_8, \vec{\alpha}_6)$ is (v_4, v_3, v_3, v_1). Various types of paths are introduced in the following definition.

Definition 2.1.5. A path P is called a *simple* path if it does not traverse the same arc twice. It is called an *elementary* path if it does not traverse the same vertex twice. A *circuit* is a finite path such that the origin of its first arc coincides with the terminus of its last arc. Finally, a *simple circuit* is a circuit which is also a simple path. The number of arcs of a *finite* path P is called the *order* of P. A circuit of order 1 is called a *loop*.

In figure 2.1.5(a) the sequence $(\vec{\alpha}_7, \vec{\alpha}_4, \vec{\alpha}_8, \vec{\alpha}_6)$ is a simple path of order 4. However, it is not an elementary path, since vertex v_3 has been traversed more than once. On the other hand, the path $(\vec{\alpha}_4, \vec{\alpha}_6)$ is both simple and elementary. The sequence $(\vec{\alpha}_1, \vec{\alpha}_7, \vec{\alpha}_4, \vec{\alpha}_6)$ forms a simple circuit. Finally, $\vec{\alpha}_8$ is a loop.

_ _

EXERCISES 2.1

1. Let G be the multigraph given in figure 2.1.6.
   a) Give the connection array of G.
   b) Give a simple path of maximum order in G.
   c) List all elementary paths of G.
   d) List all simple circuits of G.

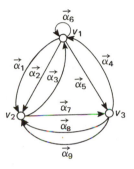

Figure 2.1.6

2. Let G be a finite multigraph with n vertices and m arcs.
   a) What is the maximum order of simple paths in G?
   b) What is the maximum order of elementary paths in G?
   c) How many elementary paths can G have?

3. Prove the equivalence of the two definitions (2.1.1 and 2.1.3) of multigraphs.

4. A corporation has k positions available and there are k persons, each of whom qualifies for one or more of the available positions. Give a graph-theoretical formulation of this problem.

_ _

2.2   REACHABILITY AND STRONG-CONNECTEDNESS

In a grid of urban streets, assume that the segment of road between any two intersections is one-way. Then, given two arbitrary points a and b in this grid (with no restriction, we may assume that points a and b are intersections), we may ask several interesting questions: Is there a path of one-way streets from a to b? What is the order of the shortest such path? How many paths of order k—that is, k blocks long—exist from a to b? These types of questions motivate the following definition.

Definition 2.2.1. A vertex v is said to be _reachable_ (or _accessible_) from another vertex u in a multigraph $G = [V, A, \varphi]$ if $u = v$ or there is a path from u to v.

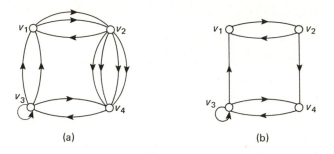

Figure 2.2.1. A multigraph and its associated simple graph

We note that the concept of reachability depends on the existence of at least *one* path. If there is more than one path from one vertex u to another vertex v, the additional paths do not affect the reachability from u to v. For any pair of vertices u and v in G for which there exist arcs $\overrightarrow{(u,v)}$, let us eliminate all but one of the arcs. The resulting simple graph G^* is equivalent to the original multigraph G as far as reachability is concerned (G^* is called the simple graph *associated* with G). This is illustrated in figure 2.2.1. The connection arrays of the graphs in figure 2.2.1 are given in figure 2.2.2. We can see in figure 2.2.2 that in the connection

$$
M = \begin{array}{c} \\ v_1 \\ v_2 \\ v_3 \\ v_4 \end{array}
\begin{array}{cccc} v_1 & v_2 & v_3 & v_4 \\
\left[\begin{array}{cccc} 0 & 2 & 0 & 0 \\ 1 & 0 & 0 & 4 \\ 2 & 0 & 1 & 1 \\ 0 & 0 & 1 & 0 \end{array}\right] \end{array}
\qquad
M^* = \begin{array}{c} \\ v_1 \\ v_2 \\ v_3 \\ v_4 \end{array}
\begin{array}{cccc} v_1 & v_2 & v_3 & v_4 \\
\left[\begin{array}{cccc} 0 & 1 & 0 & 0 \\ 1 & 0 & 0 & 1 \\ 1 & 0 & 1 & 1 \\ 0 & 0 & 1 & 0 \end{array}\right] \end{array}
$$

(a)   (b)

Figure 2.2.2. The connection arrays of G and G^* of figure 2.2.1

array of the simple graph G^* an entry $(M^*)_{ij}$ is 1 if the corresponding entry $(M)_{ij}$ in the connection array of the original graph G is a positive integer. This remark embodies a procedure for constructing the simple graph G^* from a given graph G. The procedure depends on the fact that a vertex v is reachable from u in G if and only if v is reachable from u in the associated simple graph G^*.

As pointed out previously, a simple graph describes a binary relation on a set. Therefore the connection array of G^* is just the tabular representation of some binary relation $\rho\colon (u,v) \in R_\rho$ means the existence of an arc (a path of order one!) from u to v. This remark can be generalized into a more fundamental statement using a preparatory result.

Definition 2.2.2. The $(k+1)$-th power, ρ^{k+1}, of a binary relation ρ on a set V is the composite relation $\rho\rho^k(k > 1)$. Similarly, $M_\rho^{k+1} = M_\rho M_\rho^k$.

We then prove the following lemma.

Lemma 2.2.1. Given a binary relation ρ on V, we have the identity

$$\rho\rho^k = \rho^k \rho. \tag{2.2.1}$$

Proof. We use induction on k. If $k = 1$, then $\rho\rho = \rho\rho$ and the statement is proved. Assume now that $k > 1$, and that $\rho\rho^{k-1} = \rho^{k-1}\rho$. Then we have

$$\rho\rho^k = \rho(\rho\rho^{k-1}) = \rho(\rho^{k-1}\rho) = (\rho\rho^{k-1})\rho = \rho^k\rho,$$

where the first identity follows from definition 2.2.2, the second from the inductive hypothesis, the third from the associative property of composition of relations (theorem 1.2.1), and the last from definition 2.2.2.‖

As a straightforward consequence we have a corollary.

Corollary 2.2.2. Given a binary relation ρ on V,

$$M_\rho^k = M_{\rho^k}. \tag{2.2.2}$$

We are now prepared for the following important theorem.

Theorem 2.2.3. Let $G = [V, \rho]$ be a simple graph describing a binary relation ρ on the set V. Then there is a path of order n $(n \geqslant 1)$ from a vertex u to a vertex v in G if and only if $(u, v) \in R_{\rho^n}$.

Proof. We will prove the theorem by induction. It is clear that the arc $\overrightarrow{(u,v)}$ exists in G if and only if the pair (u, v) is in R_ρ. Let us assume that the lemma holds for some $k \geqslant 1$. Let $P = \overrightarrow{(u_0, u_1)} \overrightarrow{(u_1, u_2)} \cdots \overrightarrow{(u_{k-1}, u_k)} \overrightarrow{(u_k, u_{k+1})}$ be a path of order $(k + 1)$ in G. Since $\overrightarrow{(u_0, u_1)} \overrightarrow{(u_1, u_2)} \cdots \overrightarrow{(u_{k-1}, u_k)}$ is a path of order k from u_0 to u_k in G, $u_0 \rho^k u_k$ follows from the inductive hypothesis. Since $u_k \rho u_{k+1}$, by composition of relations, $u_0(\rho^k \rho) u_{k+1}$. But, by lemma 2.2.1 and definition 2.2.2, $\rho^k \rho = \rho^{k+1}$, whence $u_0 \rho^{k+1} u_{k+1}$. Thus $(u_0, u_{k+1}) \in R_{\rho^{k+1}}$. The argument is clearly reversible; hence the theorem is proved.‖

The statement $v_i \rho^k v_j$ is equivalent to saying that in M_{ρ^k} the entry $(M_{\rho^k})_{ij}$ is 1. But by (2.2.2) the same argument applies to M_ρ^k, whence the following interesting corollary.

Corollary 2.2.4. Given a simple graph $[V, \rho]$, the array M_ρ^k (kth power of the array M_ρ) describes the relation on V: "There is at least one path of order exactly k" $((M_\rho^k)_{ij} = 1$ if and only if there is at least one path of order k from v_i to $v_j)$.

Clearly, it is meaningful to speak of reachability only when $n = \#(V) > 1$. Corollary 2.2.4 suggests that in order to compute the reachability among vertices of G, it is necessary to compute the arrays M_ρ^k for $k = 1, 2, \ldots$. Must this process continue indefinitely? Fortunately we need to compute M_ρ^k only for $k \leqslant n - 1$, because any path P of order greater than $n - 1$ starting from a vertex u must use a vertex, say w, more than once. If the sequence of arcs in P between the first occurrence and the last occurrence of w is removed, the remaining sequence of arcs is still a path from u to v. This procedure may be repeated as long as the resulting path is of order greater than $n - 1$. Therefore we can conclude that if there is a path from u to v in G, then there is a path of order less than or equal to $n - 1$ from u to v in G. This motivates the following definition.

Definition 2.2.3. Let G be a given multigraph and $G^* = [V, \rho]$ the associated simple graph. The *reachability array*, $M(G)$, of G is defined as follows: $(M(G))_{ij}$ is 1 if there is at least one $k(1 \leqslant k \leqslant n - 1)$ for which $(M_\rho^k)_{ij} = 1$; otherwise it is 0.

It follows from previous discussion that indeed $(M(G))_{ij}$ is 1 if and only if the vertex v_j is reachable from the vertex v_i.

The reachability array of the graph given in figure 2.1.5(a) is shown below. This array is rather special since all its entries are 1's. In other words, every

$$\begin{bmatrix} 1 & 1 & 1 & 1 \\ 1 & 1 & 1 & 1 \\ 1 & 1 & 1 & 1 \\ 1 & 1 & 1 & 1 \end{bmatrix}$$

vertex of the graph is reachable from every other vertex, as is clearly demonstrated in figure 2.1.5(a). Graphs of this type are models of many real-world structures, such as communication networks and streets of a city. We identify this class of graphs by defining it formally.

Definition 2.2.4. A multigraph G is said to be *strongly connected* if for each pair of vertices v_1 and v_2 in G, there is a path from v_1 to v_2.

Strong-connectedness is equivalent to mutual reachability for any two vertices in the graph. Strong-connectedness will now be characterized in terms of geometric properties of the graph.

Definition 2.2.5. Let $G = [V, A, \varphi]$ be a multigraph, and $\{V_1, V_2\}$ be a dichotomy of the vertex set V. The set of arcs of G joining vertices of V_1 and vertices of V_2 is called the *cut-set* of G relative to the dichotomy $\{V_1, V_2\}$.

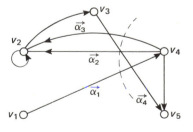

Figure 2.2.3. A cut-set of a multigraph

Example 2.2.1. In the graph of figure 2.2.3 consider the dichotomy $\{\{v_1, v_2, v_3\}, \{v_4, v_5\}\}$. The cut-set relative to this dichotomy is the set of arcs $\{\vec{\alpha}_1, \vec{\alpha}_2, \vec{\alpha}_3, \vec{\alpha}_4\}$.

Theorem 2.2.5. A multigraph $G = [V, A, \varphi]$ is strongly connected if and only if in the cut-set corresponding to every dichotomy $\{V_1, V_2\}$ of V, there exists at least one arc from V_1 to V_2 and at least one arc from V_2 to V_1.

Proof. Suppose that $\{V_1, V_2\}$ is a dichotomy of the vertex set of a strongly connected graph G. For any two vertices $u \in V_1$ and $v \in V_2$, there is a path P_1 from u to v and a path P_2 from v to u. This implies that there is an arc from V_1 to V_2 and an arc from V_2 to V_1.

Conversely, suppose that G is not strongly connected. Then there exist two vertices, v_1 and v_2, in V such that either v_1 is not reachable from v_2 or v_2 is not reachable from v_1. Without loss of generality, we assume that v_1 is not reachable from v_2 and show that it is possible to construct a dichotomy $\{V_1, V_2\}$ such that the corresponding cut-set does not contain any arc from V_2 to V_1. Let V_2 be the set of vertices which are reachable from the vertex v_2, and V_1 the set of all the other vertices in V. The sets V_1 and V_2 are both nonempty since $v_1 \in V_1$ and $v_2 \in V_2$. We claim that the cut-set relative to the dichotomy $\{V_1, V_2\}$ cannot contain any arc from V_2 to V_1, for otherwise there would exist a vertex in V_1 which is reachable from v_2, contradicting the assumption that V_2 contains all the vertices reachable from v_2.$\|$

EXERCISES 2.2

1. Draw the simple graph associated with the multigraph G of figure 2.2.4.

    b) Is the given graph G strongly connected?
    c) List all possible cut-sets of the multigraph G.

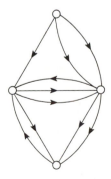

Figure 2.2.4. A multigraph G

2. a) Give a systematic procedure for obtaining the reachability array of a multigraph.
   b) Apply the procedure obtained in (a) to compute the reachability array of the multigraph given in figure 2.2.4.

3. Show that a multigraph G without isolated vertices is strongly connected if and only if there is a circuit in G which includes each arc at least once. (An isolated vertex is one which is neither the origin nor the terminus of any arc.)

4. Let G be a multigraph. Define the *distance*, $d(u, v)$, from vertex u to vertex v as the order of the shortest path from u to v. By convention, $d(u, u) = 0$ and $d(u, v) = \infty$ if there is no path from u to v. Show that $d(u, v)$ satisfies the following condition (triangle inequality):

$$d(u, v) + d(v, w) \geqslant d(u, w).$$

5. Define the *diameter* of a finite strongly connected multigraph G to be the number

$$\delta = \max_{u, v} d(u, v).$$

Find the diameters of the graphs shown in figure 2.2.5.

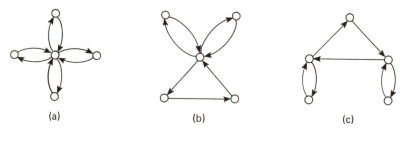

(a)   (b)   (c)

Figure 2.2.5

6. A strongly connected multigraph G is said to be *minimally connected* if the graph obtained by deleting any arc from G is no longer strongly connected.
   a) Exhibit a minimally connected graph.
   b) Show that if G is minimally connected and G' is a strongly connected subgraph of G, the graph G'' obtained from G by "collapsing" G', that is, by identifying all vertices of G' and deleting all arcs in G', is minimally connected.

7. A multigraph G is said to be *uniformly connected* if there exists an integer $p > 0$ such that for each pair of vertices u and v in G, there is a path of order p connecting u to v. Exhibit a uniformly connected multigraph.

*8. Let G be a finite multigraph. For each vertex u in G, let

$$l(u) = \max_{v \in G} d(u, v).$$

Let c denote

$$\min_{u \in G} l(u).$$

If c is a finite number, any vertex v in G such that $l(v) = c$ is called a *center* of G.

a) Find a graph G having more than one center.

b) Show that a multigraph G has a center if and only if for every pair of vertices u and v in G, there exists a vertex w (possibly $w = u$ or $w = v$) such that there is a path from w to u and a path from w to v. (In case $w = u$ or $w = v$, we allow the path of order 0, that is, $d(u,u) = 0$).

2.3 UNDIRECTED GRAPHS

Often the situation to be modeled does not require the representation capabilities of directed graphs. Let us return to the example given at the beginning of Section 2.1, that of four teams, a, b, c and d, in a sports tournament. If we are interested in recording only the matches and not their outcomes, then the relation between teams a, b, c, and d becomes symmetric and is described by the graph of figure 2.3.1(a)—that is, if a meets b, then also b meets a. As noted in Section 1.5, the

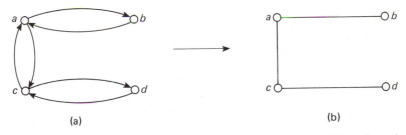

(a)   (b)

Figure 2.3.1. A symmetric (directed) graph and its equivalent undirected graph

graph of symmetric relations can be considerably simplified by replacing a symmetric pair of arcs with a single (undirected) *edge* (figure 2.3.1(b)). This step prepares for the following definition, which closely parallels definition 2.1.1.

Definition 2.3.1. A *multiple undirected graph* (or *undirected multigraph*) G consists of a set V of vertices, a set E of edges, and a function φ from the set E to the set of *unordered* pairs $\langle u,v \rangle$ for $u \in V$ and $v \in V$.

The symbolic representation of G is $[V, E, \varphi]$. Edges of G will usually be denoted by Greek letters. If $\varphi(\alpha) = \langle u,v \rangle$, then the notations α and $\overline{(u,v)}$ will be used interchangeably.

We can now give the definitions of notions relevant to the geometrical properties of undirected multigraphs.

Definition 2.3.2. Let $G = [V, E, \varphi]$ be an undirected multigraph. Given an edge $\alpha = \overline{(u,v)}$ in E, the vertices u and v in V are called the *terminals* of α. A *chain C* of G is a sequence of edges $(\overline{(v_0,v_1)}, \overline{(v_1,v_2)}, \ldots, \overline{(v_{n-2},v_{n-1})}, \overline{(v_{n-1},v_n)}, \ldots)$. A chain is said to be *simple* if none of its edges is repeated, and *elementary* if the vertices $v_0, v_1, \ldots, v_n, \ldots$ appear only once in the chain. A *cycle* is a finite chain $((\overline{v_0, v_1}), \ldots, (\overline{v_{n-1}, v_n}))$ for which v_0 coincides with v_n, and no other vertex is transversed more than once. A cycle which is a simple chain is called a *simple cycle*.

We consider again the example of a grid of urban streets. If all the streets allow two-way traffic, then we are interested only in itineraries *between* two places without concern about direction. This viewpoint motivates the following definition.

Definition 2.3.3. An undirected multigraph G is said to be *connected* if for each pair of vertices v_1 and v_2 in G, there is a chain between v_1 and v_2 (if G consists of a single vertex, it is trivially connected).

Although chains do not involve directions, directions of edges are involved whenever one traces a chain in a graph. Furthermore, since each edge may be thought of as a pair of arcs in opposite directions, it is conceivable that connectedness properties of undirected multigraphs may in fact be derivable from directed multigraphs.

Definition 2.3.4. Let G be a given undirected multigraph. Then the corresponding *adjoint* multigraph, G^A, is obtained from G by replacing each edge by a pair of arcs in opposite directions.

It follows from the definition that the adjoint of the undirected graph in figure 2.3.1(b) is the multigraph in figure 2.3.1(a). Furthermore, since every chain in an undirected multigraph G is associated with a circuit in its adjoint multigraph G^A, we have the following simple result.

Theorem 2.3.1. An undirected multigraph G is connected if and only if its adjoint multigraph G^A is strongly connected.

This theorem together with theorem 2.2.5 gives another characterization of connectedness (theorem 2.3.2), whose proof is similar to that of theorem 2.2.5 and hence is left to be done as exercise 2.3.4.

Theorem 2.3.2. An undirected multigraph G is connected if and only if its vertices cannot be partitioned into two nonempty subsets such that both terminals of every edge of the graph belong to the same subset.

Consider now a connected undirected multigraph representing a communication system. We see that vertices and edges of the graph representing such a system may be stations (or cities) and telephone lines, respectively. Although it is possible to isolate two vertices by cutting all the lines of communication between them, many vertices may be cut off from each other by the removal of a single vertex. For example, in figure 2.3.2(a) it is necessary to remove two edges to isolate any two vertices; however, the removal of vertex e will cause vertices c and d to become disconnected from a and b. In other words, removal of such a vertex will

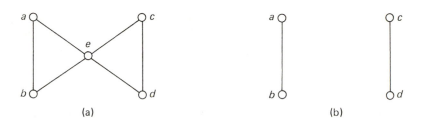

(a)   (b)

Figure 2.3.2. A graph and its components after the removal of a cut point

cause the rest of the graph to become disconnected. To understand the very special role played by vertices like e in the graph of figure 2.3.2(a), we need to delve into the details of the notion of connectedness a little more. We start with the following definition.

Definition 2.3.5. An undirected multigraph $G' = [V', E', \varphi']$ is called a *subgraph* of an undirected multigraph $G = [V, E, \varphi]$ if $V' \subseteq V$, and if E' (a subset of E) consists of all the edges in E joining vertices in V'. (This means that φ' is the restriction of φ to the domain E'.) If E' is a subset of all edges joining vertices in V', then G' is called a *partial subgraph* of G.

Note that a subgraph G' of G is entirely specified by its vertex set, since its edge set is determined by it. It follows from definition 2.3.5 that graph G_1 in figure 2.3.3(a) is a subgraph of the graph G in figure 2.3.2(a), whereas graph G_2 in figure 2.3.3(b) is not, since edge (e, d) is missing from the latter. Graph G_2 is a partial subgraph of G.

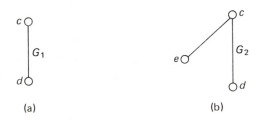

(a)   (b)

Figure 2.3.3. Graphical illustrations of the definition of a subgraph

We also note that subgraphs of a connected undirected multigraph need not be connected, as shown by the subgraph in figure 2.3.2(b). By contrast, consider the following definition.

Definition 2.3.6. A *component* $G' = [V', E', \varphi']$ of an undirected multigraph $G = [V, E, \varphi]$ is a connected subgraph of G such that no vertex in V' is connected in G to any vertex outside V'.

If a graph contains more than one component, it cannot be connected, since vertices belonging to different components are not connected. This fact is stated as a theorem.

Theorem 2.3.3. An undirected multigraph is connected if and only if it consists of only one component.

In the case of a finite simple graph, the following result gives the minimum number of edges a graph must have in order to be connected.

Theorem 2.3.4. A connected undirected simple graph having n vertices must have at least $n - 1$ edges ($n \geqslant 1$).

Proof. We use induction on the number of vertices of a given graph. As the basis of induction, we see that the trivially connected graph consisting of a single vertex has no edge. Assume that the theorem holds for graphs with k vertices. Consider a connected graph G with $k + 1$ vertices. Let v be an arbitrary vertex in G having m incident edges. Let G' be the graph obtained from G by deleting the vertex v and all the m edges incident on it. Clearly, G' must be a graph consisting of l connected components $G_1, G_2, \ldots, G_l$ for some $l \leqslant m$. Let each G_i contain n_i vertices. By the inductive hypothesis, each G_i must have at least $n_i - 1$ edges, and hence G' must have at least

$$\sum_{i=1}^{l} (n_i - 1) = \sum_{i=1}^{l} n_i - l = n - 1 - l$$

edges. But then G must have at least $n - l - 1 + m \geqslant n - 1$ edges.$\|$

Definition 2.3.7. If $G = [V, E, \varphi]$ is a connected graph, then a vertex v is called a *cut point* of G if the subgraph obtained by deleting v is not connected.

Clearly, the vertex e in the graph of figure 2.3.2(a) is a cut point of the graph. The following theorem gives the characterization properties of cut points.

Theorem 2.3.5. A vertex v is a cut point of a connected undirected multigraph if and only if there exist two vertices u and w such that every chain joining u and w passes through v.

Proof. If the vertex v is a cut point of a connected graph $G = [V, E, \varphi]$, then the subgraph G' having as vertex set all the vertices of G except v contains at least two components. Let V_1 and V_2 be two distinct components of G', and let u and w be two vertices such that $u \in V_1$ and $w \in V_2$. Since G is connected, there is a chain μ in G joining u and w. However, u and w are not connected in G', since they belong to different components. Therefore, μ must pass through the vertex v. Since μ is arbitrary, this fact holds for all chains joining u and w.

Conversely, if every chain joining two vertices of a graph passes through a vertex v, then these two vertices are certainly disconnected in any subgraph whose vertex set does not contain v. Therefore, the vertex v is a cut point of the graph.$\|$

The previous results have obvious practical applications. For example, if you were entrusted with the design of a national defense communication network you would know that the network should not contain any cut point since it would be too vulnerable. Of equal importance is the problem of designing a backup network so that communication will not be stopped after a surprise attack by the enemy. It turns out that this problem can be examined in graph-theoretic terms, as discussed in Note 2.5 on dominating and independent sets at the end of this chapter.

EXERCISES 2.3

1. An undirected multigraph G is given in figure 2.3.4.
    a) Find a simple chain which is not an elementary chain in G.
    b) Find a simple cycle in G.
    c) Is G connected?
    d) How many components does G have?

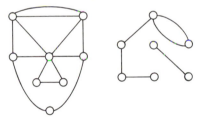

Figure 2.3.4. An undirected multigraph

2. An undirected multigraph G is given in figure 2.3.5.

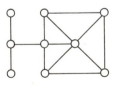

Figure 2.3.5. An undirected multigraph

    a) List all the cut points (if any) of G.
    b) Give a subgraph of G which is *maximal* in the sense that its vertex set is not properly contained in the vertex set of another subgraph different from G itself.
    c) Give a partial subgraph of G.
    d) Draw the adjoint multigraph of G.

3. Prove theorem 2.3.1.

4. Prove theorem 2.3.2.

5. Show that a finite graph with n vertices is connected if and only if every pair of vertices is connected by a chain of order less than or equal to $n - 1$.

6. Show that a connected undirected simple graph remains connected after removal of an edge if and only if the edge is contained in some cycle.

7. Show that if α_1 and α_2 are two distinct edges of a connected graph without loops, one can construct an elementary chain which begins with α_1 and terminates with α_2.

8. A simple graph is said to be *bipartite* if its vertices can be partitioned into two disjoint sets V_1 and V_2 so that every edge has one terminal vertex in V_1 and the other in V_2.

   a) Show a nontrivial application of bipartite graphs.

   b) Show that a graph is bipartite if and only if the totality of its edges forms a cut-set.

2.4 PATH PROBLEMS

We recall that an interesting problem for children is to draw a figure without lifting pencil from paper and without retracing lines. After several attempts, we admit that it is not possible to draw the diagram of figure 2.4.1 according to the given rules. On the other hand, it is possible to draw the diagram of figure 2.4.2.

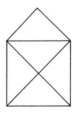

Figure 2.4.1   **Figure 2.4.2**

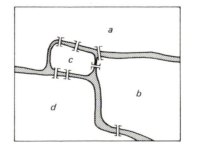

Figure 2.4.3. The Königsberg bridge problem

It is therefore natural to inquire whether there is a criterion by which one can determine which graphs can and which cannot be drawn by one continuous movement of a pencil. In fact, the theory of graphs began when the celebrated mathematician Euler encountered a problem of this type (the Königsberg bridge problem) in 1736. In Königsberg there were seven bridges over the river Pregel, as shown in figure 2.4.3. Euler wondered whether it was possible to begin walking at any of the four land areas, cross each bridge exactly once, and return to the starting point. This problem is equivalent to the problem of drawing figure 2.4.4 with a continuous movement of the pencil without retracing (and obviously, the route must terminate at its initial point). In figure 2.4.4(b) we present the drawing of figure 2.4.4(a) as an undirected graph.

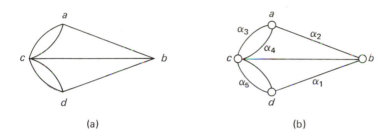

(a)   (b)

Figure 2.4.4. The undirected multigraph of the Königsberg bridge problem and an equivalent line drawing

Returning now to the Königsberg bridge problem we discover that we cannot draw figure 2.4.4(a) as a continuous line. We cannot proceed beyond a vertex if all edges incident on that vertex have been traced previously, and consequently we cannot traverse the remaining edges of the graph. An example of this fact is the graph given in figure 2.4.5, which can be traced by traversing edges α_2, α_3, and α_4 in this sequence. Note that after such an itinerary, there is no edge for exiting from vertex a. It is natural to attribute this impasse to the odd number of edges incident on this vertex. In fact, when the number of incident edges is even, one may form pairs of them and use one member of the pair for entering and the other member for leaving the vertex. Thus the number of edges incident on a vertex

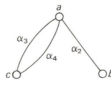

Figure 2.4.5. A partial subgraph of the multigraph in figure 2.4.4

may be relevant to the solution of the Königsberg bridge problem. This motivates us to give the following definition.

Definition 2.4.1. Let G be an undirected multigraph. The *degree* of a vertex v in G is the number of edges which are incident on it. If G' is a (directed) multigraph, then the *incoming degree* of a vertex v' is the number of arcs that terminate on it, and the *outgoing degree* of v' is the number of arcs that issue from it. A vertex of degree 0 is called an *isolated vertex*.

To illustrate the concepts just introduced, we see that the degrees of vertices v_1, v_2, and v_3 in figure 2.4.6(a) are respectively 4, 1, and 0. Hence vertex v_3 is an

(a)   (b)

Figure 2.4.6. Graphical illustrations of degrees of vertices

isolated vertex. The incoming degrees of vertices v_1 and v_2 in figure 2.4.6(b) are 2 and 0, respectively, whereas their respective outgoing degrees are 1 and 2.

We are now ready to solve the problem posed at the beginning of this section. First we formalize the notion of drawing a figure without lifting a pencil.

Definition 2.4.2. An *eulerian chain* (*cycle*) in an undirected multigraph is a chain (cycle) that uses every edge once and only once.

We are now ready to establish a condition that provides the answer to our original problem.

Theorem 2.4.1. A undirected multigraph $G = [V, E, \varphi]$ without isolated vertices has an eulerian cycle if and only if it is connected and contains no vertices of odd degree.

Proof. If G contains an eulerian cycle, G is connected, because it does not contain isolated vertices by hypothesis, and an eulerian cycle contains all the edges of the graph. Since all the vertices of G are traversed by the eulerian cycle, every time we traverse a vertex we use a *pair* of edges, one for entering it, the other for exiting from it. By definition 2.4.2, every edge incident on a vertex is used once and only once by the eulerian cycle; hence each vertex has an even degree.

Conversely, suppose that G is connected and has no vertex of odd degree. We can easily see that such a graph with fewer than three edges must contain an eulerian cycle. We shall now assume that this holds for such graphs with less than m edges ($m > 2$) and prove that it also holds for graphs with m edges. Let G have m edges and satisfy the hypothesis of the theorem. We construct a chain by starting at an arbitrary vertex u of G and by adding edges to it so that no edge is used twice (that is, no edge is "traversed" twice). Any time we add a new edge to the chain under construction, we reach some new vertex w. Note that an odd number of edges incident on w belong to the chain if and only if $w \neq u$; indeed, reaching a vertex requires one edge and passing through a vertex requires two edges, and for u an extra edge was needed to initiate the chain. Since the degree of w is even, a new edge can be added to the chain being constructed. When this is no longer possible, we must have $w = u$; the chain is a cycle C. However, it is possible that C does not contain all the edges of G (for example, the cycle $\alpha_1 \, \alpha_2 \, \alpha_3 \, \alpha_5$ in figure 2.4.4(b)). Consider now the graph G' obtained from G by removing all edges in C. Let $G_1, \ldots, G_n$ be components of G' containing edges. We note that each G_j ($j = 1, \ldots, n$) contains fewer than m edges and has no vertex of odd degree; in fact, all the vertices of G have even degree by hypothesis, and removal of the edges of the chain C decreases by *two* the degree of some vertices. Thus, by inductive hypothesis, they contain eulerian cycles $C_1, \ldots, C_n$ because G_j contains less than m edges ($j = 1, 2, \ldots, n$). Since G is assumed to be connected, the cycle C must meet with each $C_i (1 \leqslant i \leqslant n)$ at one vertex, say u_i. If we use the notation $C[a,b]$ to denote that portion of the cycle C starting with some vertex a and ending with the vertex b, then the chain

$$C_1, \, C[u_1, u_2], \, C_2, \ldots, C_n, \, C[u_n, u_1]$$

(graphically illustrated in figure 2.4.7) is clearly an eulerian cycle containing all the vertices in G. This proves the sufficiency condition of the theorem.$\|$

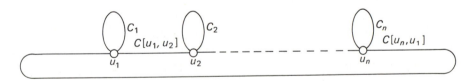

Figure 2.4.7. Graphical illustration of the eulerian cycle

This theorem, as expected, provides a negative answer to the Königsberg bridge problem. Indeed, the undirected multigraph in figure 2.4.4 has vertices of odd degree. In general, theorem 2.4.1 offers a condition for tracing a connected figure with a continuous "closed" trace, whose starting and ending points coincide. If we remove the requirement that the trace be closed, the notion of eulerian cycle is replaced by that of eulerian chain. The resulting condition is expressed by the following corollary, whose proof mirrors the one of theorem 2.4.1 and is therefore omitted.

Corollary 2.4.2. An undirected multigraph $G = [V, E, \varphi]$ without isolated vertices has an eulerian chain which is not a cycle if and only if it is connected and contains exactly two vertices of odd degree.

Analogous notions can be defined for directed multigraphs with no additional complications. When drawing a connected figure without lifting the pencil, we are really tracing a path, and in so doing, we implicitly assign directions to edges. However, this subject is not pursued further.

A closely related problem consists of finding a path in a graph which passes through each vertex of the graph exactly once. This problem was proposed by another celebrated mathematician, W. R. Hamilton, when he invented a game using a regular solid dodecahedron whose vertices are labeled with the names of various cities. Players are supposed to travel "around the world" by going through each vertex once and returning to the starting place. We shall formalize this concept with the following definition.

Definition 2.4.3. A *hamiltonian path* (*circuit*) in a multigraph G is a path (circuit) which passes through each of the vertices in G exactly once.

Although the concepts of eulerian chain and hamiltonian path are quite similar, there is no simple criterion to determine the existence of a hamiltonian path except for some special classes of graphs, one of which is given in the following definition.

Definition 2.4.4. A multigraph G is said to be *complete* if every pair of vertices in G is joined by at least one arc.

Theorem 2.4.3. Every complete multigraph G contains a hamiltonian path.

Proof. Since vertices are the essential objects under consideration in a hamiltonian path, we may, without loss of generality, assume that G is a simple graph $[V, \rho]$. Starting with an arbitrary arc in G, we will construct a path which will traverse all the vertices. Suppose that a path P is constructed such that the vertices it traverses form the sequence $(u_1, u_2, \ldots, u_k)$. If a vertex v is not traversed by this path, we can extend P to include v, because for every $j = 1, 2, \ldots, k$, by the hypothesis of completeness, either $\overrightarrow{(v, u_j)}$ or $\overrightarrow{(u_j, v)}$ exists—possibly both arcs exist. [We recall that the existence of $\overrightarrow{(v, u_j)}$ is denoted by $(v, u_j) \in R_\rho$.] If, for some $j \leqslant k - 1$, $(u_j, v) \in R_\rho$ and $(v, u_{j+1}) \in R_\rho$, then the path $\overrightarrow{(u_1, u_2)} \cdots \overrightarrow{(u_j, v)} \overrightarrow{(v, u_{j+1})} \cdots \overrightarrow{(u_{k-1}, u_k)}$ can be formed and contains v. If the previous condition is not met for any j, then either $(u_j, v) \in R_\rho$ holds for every j or $(v, u_j) \in R_\rho$ holds for every j; in the first case, we form the path $P\overrightarrow{(u_k, v)}$, in the second case the path $(v, \overrightarrow{u_1})P$, both of which contain v. All vertices of G can therefore be traversed by systematically extending the path.‖

The problem of finding a hamiltonian circuit can be viewed as a special case of the so-called "traveling salesman problem." A salesman must visit a number of towns, and given the distance between various pairs of towns, he would like to find the shortest route so that he reaches each town once and returns to the point of departure. In graph-theoretic terminology, the problem is to find the shortest hamiltonian circuit of a graph when each edge of the graph is assigned a length. Since no efficient general algorithm is known for the solution to this problem, we will not pursue it further here. A note on the difficulty of this problem appears at the end of Chapter 6. We will now consider a simpler related problem, namely, that of finding the "shortest path" from one place to another, given the distances between pairs of places. Let G be a given finite multigraph such that a nonnegative number $w(\vec{\alpha})$ is assigned to each of its arcs $\vec{\alpha}$ indicating the *length* of $\vec{\alpha}$. The following algorithm[2] finds a path P from a vertex u_0 to another vertex v_0 such that the sum of the distances of its arcs is minimal. The execution of the algorithm consists of a sequence of iterations. Associated with each iteration are a function from V to the set of the nonnegative real numbers and a special subset of V called the *candidate set*; specifically, λ_k and V_k are respectively the function and candidate set associated with the kth iteration. We begin by setting $\lambda_1(u_0) = 0$ and $\lambda_1(v) = \infty$, for $v \neq u_0$ and $V_1 \equiv \{u_0\}$. At the kth iteration we first determine a vertex $u^* \in V_k$ such that $\lambda_k(u^*)$ is the minimum of the values $\lambda_k(v)$, for $v \in V_k$. If $u^* = v_0$, then the algorithm terminates and $\lambda_k(v_0)$ is the length of a shortest path from u_0 to v_0. If $u^* \neq v_0$, then for every $v \in V$ such that an arc $\vec{\alpha} = (\overrightarrow{u^*, v})$ exists, we select the lesser of the values $\lambda_k(v)$ and $\lambda_k(u^*) + w(\vec{\alpha})$ as $\lambda_{k+1}(v)$; for every v such that an arc $(\overrightarrow{u^*, v})$ does not exist, we set $\lambda_{k+1}(v) = \lambda_k(v)$. In addition, the candidate set V_{k+1} consists of the vertices v for which $(\overrightarrow{u^*, v})$ exists and of the vertices in V_k, after removal of u^*. We are now ready to begin the $(k + 1)$-st iteration. Note that at each iteration we discard one vertex (the vertex u^*). This shows that the algorithm terminates, either when $u = v_0$ or when $V_k \equiv \{v_0\}$. So, at the completion, the value $\lambda_k(v_0)$ is the length of a shortest path from u_0 to v_0. The actual path can be obtained by backtracking through the vertices of the path. Starting from v_0, we look for a vertex v_1 such that $\lambda_k(v_1) = \lambda_k(v_0) - w(\vec{\alpha})$ for some $\vec{\alpha}$ which goes from v_1 to v_0. Likewise, there must exist a vertex v_2 and an arc $\vec{\alpha}_2$ such that $\lambda_k(v_2) = \lambda_k(v_1) - w(\vec{\alpha}_2)$. Since the sequence $\lambda_k(v_0), \lambda_k(v_1), \lambda_k(v_2), \ldots$ is strictly decreasing, we shall have $v_n = u_0$ at some stage, and the path $P = (u_0, v_{n-1}, \ldots, v_1, v_0)$ has length $\lambda_k(v_0)$.

That the procedure given above is indeed a valid algorithm is quite intuitive since at each step we construct a shortest path from u_0 to $u^* \in V_k$. The formal proof of validity is left as an exercise.

Example 2.4.1. The procedure above is demonstrated by figure 2.4.8; part (b) is the processed graph of part (a), and it indicates that the shortest path between u_0 and v_0 has length 16. We see that there are two shortest paths represented by the

---

2. The word algorithm, which means "procedure," comes from the name of the celebrated Uzbek mathematician Mohamed Ibn-Musa Al-Khowarizmi (ninth century).

sequences of vertices $(u_0, u_1, u_2, u_3, u_4, u_5, v_0)$ and $(u_0, u_7, u_8, u_9, u_{10}, u_{11}, v_0)$, respectively.

Four successive iterations are illustrated in detail.

1. $\lambda_1(u_0) = 0$, $\lambda_1(v) = \infty$ for $v \neq u_0$; $V_1 \equiv \{u_0\}$.

2. $\lambda_2(u_1) = \min(w(\overrightarrow{u_0, u_1}), \lambda_1(u_1)) = \min(1, \infty) = 1$, $\lambda_2(u_7) = \min(3, \infty) = 3$,
   $\lambda_2(u_0) = 0$, $\lambda_2(v) = \infty$ for $v \neq u_0, u_1, u_7$.
   $V_2 \equiv \{u_1, u_7\}$, $u^* = u_1$.

3. $\lambda_3(u_2) = \min(1 + 1, \infty) = 2$, $\lambda_3(u_7) = 3$, $\lambda_3(u_0) = 0$,
   $\lambda_3(u_1) = 1$, $\lambda_3(v) = \infty$ for $v \neq u_0, u_1, u_2, u_7$.
   $V_3 \equiv \{u_2, u_7\}$, $u^* = u_2$.

4. $\lambda_4(u_3) = \min(2 + 5, 2 + 6, \infty) = 7$, $\lambda_4(u_7) = 3$,
   $\lambda_4(u_0) = 0$, $\lambda_4(u_1) = 1$, $\lambda_4(u_2) = 2$,
   $\lambda_4(v) = \infty$ for $v \neq u_0, u_1, u_2, u_3, u_7$. $V_4 \equiv \{u_3, u_7\}$, $u^* = u_7$.

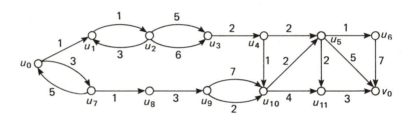

(a)

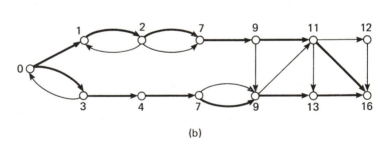

(b)

Figure 2.4.8. Graphical illustration of the shortest path algorithm

A special case of the shortest path problem is that in which the distance assigned to every arc is 1. In other words, the problem reduces to finding a path from one vertex to another such that the order of the path is minimal. (A further simplification of the problem would be to find any path from one vertex to another.) We would now like to illustrate that many puzzles can be treated as path problems for graphs.

Example 2.4.2. Consider the well-known puzzle of the wolf, the cabbage, and the goat. A wolf, a goat, and a cabbage are on one bank of a river; a ferryman can take only one of them at a time across the river. Since, for obvious reasons, neither the wolf and the goat, nor the goat and the cabbage can be left alone without a guard (the ferryman), how can the ferryman get all of them across the river? The problem can be solved by drawing a graph, as shown in figure 2.4.9, where the

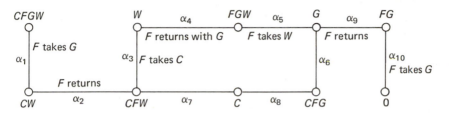

Figure 2.4.9. Graphical representation of a puzzle

letters C, F, G, and W stand for cabbage, ferryman, goat, and wolf, respectively, and each vertex of the graph represents a possible situation on the starting bank of the river (for example, FGW means that ferryman, goat, and wolf are on the given river bank). There are 16 possible combinations (the incredulous reader may enumerate them). However, since the combinations CG, GW, and CGW are not allowed, FW, FC, and F are not allowed either. For example, ferryman and wolf (FW) on the boat would mean that cabbage and goat (CG) were left alone. Hence there are only 10 vertices in figure 2.4.9. The problem then, in terms of graphs, is to find a path from $CFGW$ to 0 in figure 2.4.9, where 0 represents the situation that the ferryman has taken all across. Clearly, $(\alpha_1, \alpha_2, \alpha_3, \alpha_4, \alpha_5, \alpha_9, \alpha_{10})$ is such a path.

Example 2.4.3. A problem closely related to the shortest path problem is the so-called *scheduling problem*. Given a set of tasks and the resources to perform them, we wish to find the most efficient use of the resources. As an illustration, consider that two workers (resources) A and B are assigned to assemble a car (target task). For simplicity we shall assume that there are seven component tasks that need to be performed.

1. Assemble the body.
2. Assemble the engine.
3. Install the seats in the body.
4. Install the engine.
5. Install the battery.
6. Mount the tires.
7. Test the car on the road.

In order to have a complete understanding of how the whole task is to be completed, we need to assign different tasks to A and B and describe certain relationships among these tasks. We assign to A the tasks 1, 3, 5, and 7 and to B the remaining tasks. The relationships are as follows:

a) Tasks 1 and 2, 3 and 4, and 5 and 6 can be done concurrently.

b) Tasks 1 and 2 must be completed before task 4 can be started.

c) Task 1 must precede tasks 3, 4, 5, and 6.

d) Task 7 cannot be started until all other tasks have been completed.

e) Task 5 cannot be started until task 4 is completed.

Based on the relationship between tasks and the assignment of tasks to A and B, we can express the order of tasks to be performed by A and B in the linear graphs shown in figure 2.4.10. Taking these graphs and incorporating in them the

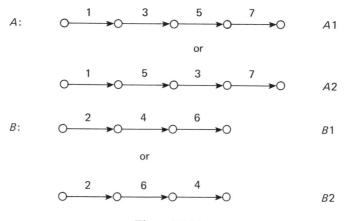

Figure 2.4.10

relationships between tasks, we have the possible "schedules" described graphically in figure 2.4.11, where broken lines indicate "dummy" tasks. [A dummy task is one that requires no time to complete. It is usually assigned in such a way as to ensure that the schedule is followed within the given constraints. For example, in figure 2.4.11(a), the broken line from vertex a to vertex b represents a dummy task that guarantees that task 4 will not be started before task 1 is completed, as required by the constraint given in relationships (b).]

Now we want to measure the efficiency of the various schedules in terms of time, according to the estimates of time (in some unit) for completing each task, as shown in figure 2.4.12.

The time required for completion of the task according to schedules (a), (b), (c), and (d) would be 270.5, 275.5, 280.5 and 285.5, respectively. Schedule (a)

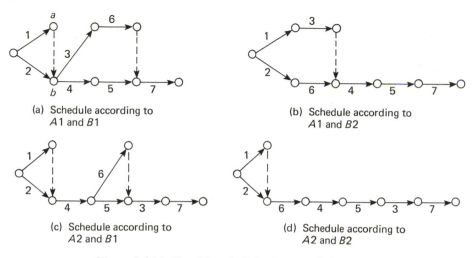

(a) Schedule according to
    A1 and B1

(b) Schedule according to
    A1 and B2

(c) Schedule according to
    A2 and B1

(d) Schedule according to
    A2 and B2

Figure 2.4.11. Possible schedules for completing the job

takes the least time, which coincides with that of path 2457 in figure 2.4.11(a):
such path is called a *critical path*. (See Note 2.2 at the end of this chapter for a
more general analysis of the critical path method.)

task \ time	A	B
1	50	
2		200
3	10	
4		50
5	.5	
6		5
7	20	

Figure 2.4.12

EXERCISES 2.4

1. For each vertex of the graph of figure 2.4.13, find the incoming and outgoing degrees.

Figure 2.4.13

2. a) Construct an undirected multigraph with four vertices, each of degree 3.
   b) Is it possible to construct an undirected multigraph of four vertices with respec-
      tive degrees 1, 2, 3, and 4?

3. Does the graph in figure 2.4.14 have an eulerian cycle? Why?

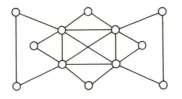

Figure 2.4.14

4. Let G be the directed graph given in figure 2.4.15.

   a) Is G complete?
   b) Does it have a hamiltonian path?

Figure 2.4.15

5. Show that the graph of figure 2.4.16 has no hamiltonian path.

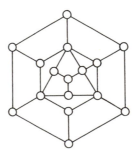

Figure 2.4.16

(*Hint:* pick any vertex and label it *A*. Label the adjacent vertices *B*. Repeat this
procedure until all vertices are labeled. A hamiltonian path must traverse the
vertices so that their labels alternate.)

6. Apply the algorithm for finding the shortest path from v to u in figure 2.4.17.

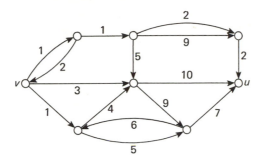

Figure 2.4.17

7. Let $G = [V, A, \varphi]$ be a given directed multigraph. Let $\delta^+(v)$ and $\delta^-(v)$ denote, respectively, the incoming and outgoing degrees of a vertex $v \in V$, and show that

$$\sum_{v \in V} \delta^+(v) = \sum_{v \in V} \delta^-(v) = \#(A).$$

8. Show that in any undirected multigraph, there is an even number of vertices which have odd degrees.

9. Prove corollary 2.4.2.

*10. Define a covering C of an undirected multigraph $G = [V, E, \varphi]$ as a partition P of the set E such that each member of P is either a chain or a cycle. A covering is said to be *minimal* if the corresponding partition has the minimal number of classes. Show that if a connected undirected multigraph G has $2n$ vertices of odd degree (where $n \geqslant 1$), then every minimal covering of G consists of n chains, each of which joins two vertices of odd degree.

*11. Show that if a connected undirected multigraph contains at least four vertices of odd degree, then it must have at least two distinct minimal coverings (see exercise 2.4.10).

12. Show that every cut-set of an undirected multigraph with an eulerian cycle must contain an even number of edges (cut-set of undirected graphs is defined as for directed graphs, see def. 2.2.5).

2.5 TREES

Superficially similar to the traveling salesman problem is the so-called *minimal connector problem*, a form of which is as follows: We wish to connect a certain number of cities by communication lines so that the total length of the latter is minimized. (This implies that no superfluous connections are introduced.) Before pursuing this problem further, we observe that if we think of the cities as vertices and the communication lines as edges of a graph, then the graph must be connected, since each city must be able to communicate with every other city. Furthermore, the graph cannot contain any cycles because it has no superfluous connections. This type of graph, called a *tree*, constitutes an important class of graphs, encompassing such familiar representations as family trees and some chemical structures.

Definition 2.5.1. A *tree* is a connected undirected graph with no cycles. A tree consisting of an isolated vertex is said to be a *degenerate* tree.

Some nondegenerate trees with less than 6 vertices are given in figure 2.5.1.

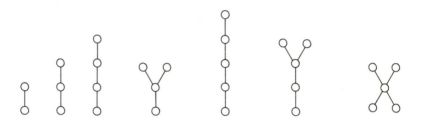

Figure 2.5.1. Some nondegenerate trees with less than 6 vertices

There are many different ways of defining a tree; some are given in the following theorem.

Theorem 2.5.1. Let G be an undirected graph. Then the following statements are equivalent.

1. G is a nondegenerate tree.
2. Every pair of vertices is connected by one and only one chain.
3. G is connected, but if any edge is deleted, the resulting graph is no longer connected.
4. G contains no cycles, and if an edge is added which joins two vertices, one and only one cycle is formed.

Proof. We need to show that any two statements of the theorem mutually imply each other. This can be achieved more economically by demonstrating that statement (i) implies statement $(i+1)$ for $i = 1, 2, 3$, and that statement (4) implies statement (1).

Statement (1) implies (2). Let G be a nondegenerate tree and u, v two distinct vertices of G. Since G is connected, there is a chain C between u and v. However, there cannot exist another chain C' different from C between u and v, since otherwise the chain CC' would form a cycle, contradicting the fact that G contains no cycles.

Statement (2) implies (3). Suppose that an edge $\alpha = \overline{(u,v)}$ is deleted from G. Since α is the only chain between u and v by assumption, its deletion separates u and v; that is, G becomes disconnected.

Statement (3) implies (4). Let u and v be two vertices of G. Let G' be a graph obtained by adjoining an edge $\alpha = \overline{(u,v)}$ to G. Since G is assumed to be connected, there is a chain C between u and v, and hence $C\alpha$ is a cycle in G'. Suppose that, after α is adjoined to G, there is another cycle S distinct from $C\alpha$ in G'. Since G does not contain any cycle, α must belong to S; that is, $S = \alpha C'$, where C' is a chain. But then CC' is a cycle (figure 2.5.2) in G, contradicting the assumption. Hence, $C\alpha$ must be the only cycle in G'. Hence G' contains no cycles.

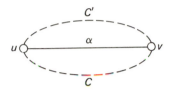

Figure 2.5.2. Graphical illustration of the proof that statement (3) implies statement (4) in theorem 2.5.1

Statement (4) implies (1). Since the addition of any edge joining two vertices in G will produce a cycle, every pair of vertices in G must necessarily be connected by a chain, and hence G is connected; G is nondegenerate since at least two vertices exist.‖

We observe that none of the graphs in figure 2.5.1 contains a cycle and further that each has at least two vertices of degree 1. We now formally prove this characteristic property of nondegenerate trees.

Lemma 2.5.2. A nondegenerate tree contains at least two vertices of degree 1.

Proof. We shall prove the lemma by induction. As the basis of induction, we see that the tree with exactly two vertices contains exactly one edge and that each vertex is of degree 1. Assume that the lemma holds for trees with k or fewer vertices for some $k \geqslant 2$. Let T be a tree with $k + 1$ vertices, and suppose that T contains less than two vertices of degree 1. Let $\alpha = \overline{(u,v)}$ be an arbitrary edge of T. By property (3) of theorem 2.5.1, we see that the graph T', obtained from T by deleting the edge α, must be disconnected. Since T does not contain any cycles, the graph T' must contain exactly two components, say G_1 and G_2. There are two possibilities: (1) G_1 is a degenerate tree and G_2 is a nondegenerate tree; (2) G_1 and G_2 are both nondegenerate trees. These two cases are illustrated in figure 2.5.3. In case (1), since G_1 consists of an isolated vertex v_1, we see that G_2 must contain two or more vertices of degree 1 by the inductive hypothesis. But then T must contain at least two vertices of degree 1, since the addition of α_1 to T'_1 can eliminate

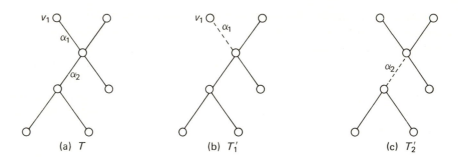

Figure 2.5.3. A tree T and trees T_1' and T_2' resulting from T by deleting edges α_1 and α_2, respectively

at most one vertex of degree 1 from T_1', and v_1 is a vertex of degree 1 in T. In case (2), we observe that G_1 and G_2 each must contain at least two vertices of degree 1 by the inductive hypothesis, and since the addition of α to T_2' can eliminate at most one vertex of degree 1 from each of G_1 and G_2, T_2' must have at least two vertices of degree 1. Therefore, the lemma holds for all nondegenerate trees.‖

If we restrict ourselves to a discussion of finite trees only (i.e., trees having a finite number of vertices) we can obtain several equivalent definitions of a tree.

Theorem 2.5.3. Let G be a finite graph containing n vertices. Then the following statements are equivalent.

1. G is a tree.
2. G contains no cycles and has $n - 1$ edges.
3. G is connected and has $n - 1$ edges.

Proof. We again adopt the method employed in the proof of theorem 2.5.1; that is, we show that the three statements cyclically imply each other.

Statement (1) implies (2). We will prove this by induction on the number of vertices. Clearly, a tree with one vertex contains no edge. Suppose that (2) holds for all trees with $k \geqslant 1$ vertices. Let G' be a tree with $k + 1$ vertices. By lemma 2.5.2, G' contains a vertex v of degree 1. Let G'' be the tree obtained from G' by deleting the vertex v and the edge incident on v. Since G'' is a tree with k vertices, it has $k - 1$ edges by the inductive hypothesis. But this implies that G' contains k edges, and hence (2) holds for all trees.

Statement (2) implies (3). Suppose that G is not connected. Since G has no cycles and has only $n - 1$ edges, it contains a finite number, say k, of components, each of which is a tree. Let the k trees be $T_1, \ldots, T_k$ such that T_i has n_i vertices. By (2), each T_i has $n_i - 1$ edges. Since $\sum_{i=1}^{k} n_i = n$, and since $\sum_{i=1}^{k} (n_i - 1) = n - k$, we see that $(n - k) = (n - 1)$ and k must equal 1. That is, G contains only one component, and hence by theorem 2.3.3 it must be connected.

Statement (3) implies (1). Suppose that G is connected with $n - 1$ edges but is not a tree; that is, G contains cycles. Since the deletion of an edge from a cycle will not affect the connectedness of the graph, we can delete at least one edge from each cycle of the graph. The resulting tree G' will have n vertices with no more than $n - 2$ edges. But we know (theorem 2.3.4) that a connected undirected graph with n vertices must have at least $n - 1$ edges; hence G' is not connected, and that fact contradicts the hypothesis that G' is a tree. Hence G cannot contain any cycle.‖

The "minimal connector problem" can now be rephrased in graph-theoretic terms as that of finding, for a given undirected graph G, a tree T whose vertices are all the vertices of G and whose edges are a subset of the edges of G of minimal total length [where the length of an edge α is, as usual, a nonnegative number $w(\alpha)$]. Such a tree T is a member of the following important class.

Definition 2.5.2. A *spanning tree* of a connected undirected multigraph $G = [V, E, \varphi]$ is a tree $T = [V, E', \varphi']$, where $E' \subseteq E$ and where φ' is the restriction of φ to the domain E'.

Note that the vertex sets of G and T coincide, and that T is *not* a subgraph of G (recall definition 2.3.5; in fact, a subgraph of G having the same vertex set as G is G itself). Furthermore, there are, in general, several choices for the set E'; they are all consistent with the definition of a tree. For the undirected multigraph G given in figure 2.5.4(a), the graphs in parts (b) and (c) are spanning trees of G, whereas the graph in part (d) is not.

Clearly, given a graph G and the length of each edge of G, the solution of the minimal connector problem is a spanning tree of G of minimal total length. Such a tree is referred to as a *minimal spanning tree*. We now present and discuss an algorithm for the construction of a minimal spanning tree T of a given undirected multigraph G. We begin the construction by selecting an edge α_1 such that $w(\alpha_1)$ is minimum: let G_1 be the graph consisting of α_1 and its two vertices.

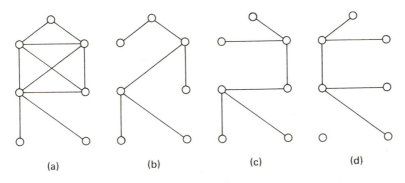

(a)   (b)   (c)   (d)

Figure 2.5.4. Illustration of the spanning tree concept

In each subsequent step we construct a graph G_i by adding to G_{i-1} an edge α_i (and its two terminal vertices if they are not already in G_{i-1}) such that $w(\alpha_i)$ is minimum over the edges not belonging to G_{i-1}, and such that G_i does not contain any cycles. (Note that G_i may consist of more than one component, because α_i need not use a vertex of G_{i-1}. See the following example.) The algorithm stops with the construction of G_{n-1}. Since, by the mechanics of the algorithm, G_{n-1} contains n vertices, $(n-1)$ edges, and no cycle, by theorem 2.5.3, statement (2), it is a tree. Moreover, the vertices of G_{n-1} are by construction a subset of the vertices of G, but since both graphs have the same number of vertices, n, their vertex sets coincide. It follows from definition 2.5.2 that G_{n-1} is a spanning tree of G.

To show that G_{n-1} is a *minimal* spanning tree, let us suppose that a spanning tree T of G exists such that $W(T) = \sum_{\alpha \in T} w(\alpha)$ is minimum. Notationally, for a given graph G, $G + \alpha$ and $G - \beta$ denote the graphs obtained by adding an edge α to G and removing an edge β from G, respectively. Let α_i be the first edge in the sequence $(\alpha_1, \alpha_2, \ldots, \alpha_{n-1})$ of the edges of G_{n-1} such that α_i is not in T. Then the graph $T + \alpha_i$ contains exactly one cycle C, by theorem 2.5.1. Since G_{n-1} has no cycle, there is an edge β in C such that β is not in G_{n-1}. The graph T_1 obtained from $T + \alpha_i$ by eliminating the edge β is again a spanning tree of G, as shown in figure 2.5.5(b), and since T has minimum total length by assumption, we conclude that

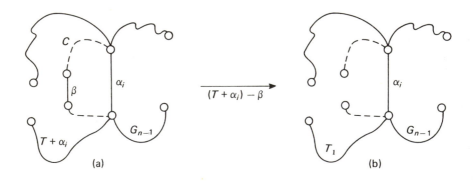

Figure 2.5.5. Graphical illustration of the transformation from the graph $T + \alpha_i$ to the tree T_1

$w(\alpha_i) \geqslant w(\beta)$. But the *strict* inequality violates the algorithm rules in the construction of G_i; thus we must have $w(\alpha_i) = w(\beta)$ and $W(T_1) = W(T)$. If all edges of T_1 are also edges of G_{n-1}, then by the fact that each spanning tree of G must contain $n-1$ edges, we must conclude that $T_1 = G_{n-1}$, and hence $W(G_{n-1}) = W(T_1) = W(T)$. Otherwise we repeat the same procedure on T_1 to obtain a spanning tree T_2 such that T_2 has one more edge in common with G_{n-1} and

$W(T_1) = W(T_2)$. Since G is finite, we see that there exists a $k \geqslant 1$ such that $T_k = G_{n-1}$, and

$$W(T) = W(T_1) = W(T_2) = \cdots = W(T_k) = W(G_{n-1}).$$

This proves the fact that G_{n-1} is indeed a minimal spanning tree of G.‖

Example 2.5.1. The algorithm above is illustrated by the graph G in figure 2.5.6. A step-by-step illustration of the algorithm for the selection of edges of G to obtain a minimal spanning tree of G is given in figure 2.5.7. The total length of the

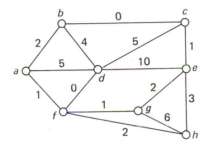

Figure 2.5.6. The graph G

resulting minimal spanning tree T is 7. Needless to say, T is not the only minimal spanning tree of G_1. As the readers may have noticed, the choices are not unique in steps (c), (d), (f), and (g).

Thus far we have been considering undirected trees, since our main emphasis has been on connectedness. However, many very familiar structures—families, organizations, classification schemes in natural history—are such that the relationships among their elements are not symmetric. Thus it is desirable to introduce directions on the edges of a tree, that is, to replace edges with arcs. An important class of directed trees, that describing predecessor-successor relations (Section 1.4), is expressed by the following definition.

Definition 2.5.3. A *rooted tree* R is obtained from a tree T by specifying as the root a special vertex v of T such that each chain in T *between* v and some other vertex u is replaced by a path in R *from* v to u. The order of the path from the root v to u is called the *level* of u. For every arc (u, u'), u and u' are a *predecessor-successor* pair.

Since a tree T contains a unique chain between any two vertices, directions are assigned to edges in a unique way once the root is specified. The directions of the arcs need not be graphically indicated if one chooses a representation which implicitly specifies them as follows: An arc $\overrightarrow{(u,v)}$ is always drawn so that u is closer to the top of the page and v is closer to the bottom. For example, the rooted

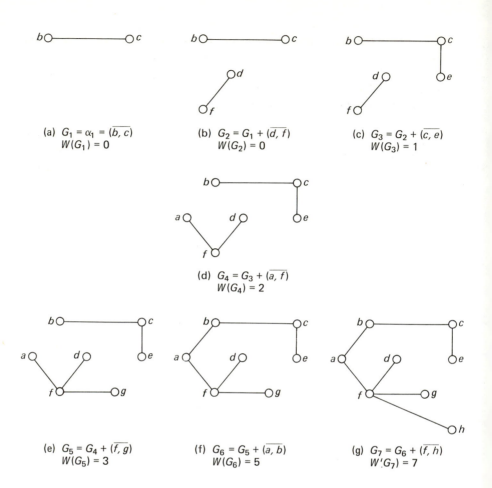

Figure 2.5.7. Graphical illustration of the minimal spanning tree algorithm

tree of figure 2.5.8(a) is unambiguously drawn as in figure 2.5.8(b). Note that in this representation, which is the most common, the root v is the topmost vertex, in contradiction of the usual meaning of the word "root." It is also worth mentioning that removal of the arcs issuing from the root of R results in components $R_1, \ldots, R_k$ ($k \geqslant 1$), such that any R_j with at least two vertices is in turn a rooted tree; an R_j consisting of exactly one vertex (that is, a degenerate tree) is called a *terminal* (or a *leaf*). This remark not only permits the *analysis* of a rooted tree R in terms of smaller components, rooted trees and terminals, but also permits the reverse procedure, that is, the synthesis of R from smaller components.

In general, structures whose formal models are rooted trees may appear, at first sight, completely unrelated. For example, all three diagrams in figure 2.5.10 have the structure of the rooted tree in figure 2.5.9. Figure 2.5.10(a) may be regarded as an outline of an algebraic formula involving "nested parentheses." Figure 2.5.10(b) is a special case of the concept of "nested sets," i.e., a collection of sets in which either any two are disjoint or one is a subset of the other. Figure 2.5.10(c) is the representation of "indentations" (we may think, for example, of the articles of a law and of their subdivisions).

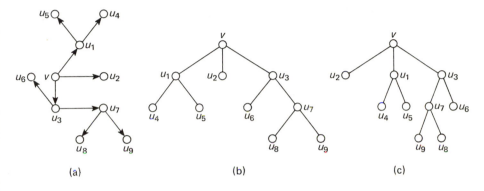

(a)   (b)   (c)

Figure 2.5.8. Equivalent representations of a rooted tree

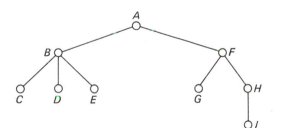

Figure 2.5.9. A rooted tree

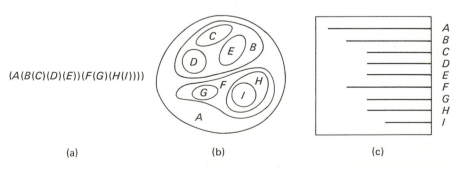

(a)   (b)   (c)

Figure 2.5.10. Different structures represented by the rooted tree in figure 2.5.9

It is clear that these different concrete instances can all be represented by the same abstract structure, a rooted tree.

Example 2.5.2. An algebraic expression involving only addition and multiplication can be represented by a rooted tree, which describes the sequence of operational steps required for evaluating the given expression for specific values of the variables. For example, in figure 2.5.11 a very simple tree is exhibited, whose

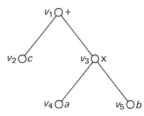

Figure 2.5.11. Rooted tree representation of the algebraic expression $E = c + ab$

vertices v_1, v_2, v_3, v_4, and v_5 are labeled with the symbols $+$, c, $\times$, a, b, respectively. $E(v)$ is interpreted as the result of the algebraic expression represented by the tree whose root is v. Furthermore, terminals are labeled with *variables* (a, b, c) and intermediate vertices with *operation symbols* $(+, \times)$ specifying the rule of combination. Therefore $E(v_1) = E(v_2) + E(v_3)$; but $E(v_2) = c$ and $E(v_3) = E(v_4) \times E(v_5) = ab$, whence, by substitution, $E(v_1) = c + ab$.

From definition 2.5.3, it appears that there is more than one way to draw a rooted tree according to the given convention. In fact, if the segment uv drawn on the page is to represent the arc $\overrightarrow{(u,v)}$, this convention prescribes a *vertical* (i.e., up-down) ordering of the points u and v but says nothing as to their *horizontal* (i.e., left-right) ordering. Therefore parts (b) and (c) of figure 2.5.8 are both representations of the rooted tree of part (a). However, there are structures which are not correctly described unless additional restrictions are applied to rooted trees.

Definition 2.5.4. An *oriented rooted tree* R is a rooted tree such that the set of arcs issuing from any vertex of R is a *sequence* (i.e., an ordered set).

When representing graphically an oriented rooted tree R, we draw the arcs issuing from a vertex v of R so that their counterclockwise succession around v is consistent with their ordering. Since the structure of an oriented rooted tree R is very rigid, we can also represent R by a set of sequences of integers as follows: A vertex v is assigned the sequence $S(v)$, which is the catenation of the sequence $S(u)$ and the integer k, where u is the predecessor of v and $\overrightarrow{(u,v)}$ is the kth element in the ordered set of arcs issuing from u. Thus the rooted tree of figure 2.5.8(b), considered now as an oriented rooted tree, can have its vertices labeled as in figure 2.5.12. The set of all the integer sequences thus obtained is called the *representation set* of the oriented rooted tree.

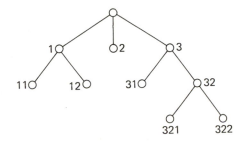

Figure 2.5.12. An oriented rooted tree and its representation set {1, 2, 3, 11, 12, 31, 32, 321, 322}

Example 2.5.3. The representation just given of an oriented rooted tree may also be seen, reciprocally, as an important application of oriented rooted trees for describing codes. A *code* is a function from a set $A \equiv \{a_1, \ldots, a_n\}$ of messages onto a set of finite sequences of symbols drawn from a set $B \equiv \{b_1, \ldots, b_J\}$ called the *alphabet*. These sequences, called code words, are used to represent the messages. The code is described by an oriented rooted tree as follows: If the longest code word contains l symbols, we construct a tree T such that its terminal vertices are of level l and the arcs issuing from each nonterminal vertex are labeled by the ordered sequence $(b_1, b_2, \ldots, b_J)$. There is a unique path from the root to any given vertex, and this path identifies a unique sequence of symbols of B. A code is defined when a subset of the vertices of T are labeled by the symbols $a_1, \ldots, a_n$. For $A \equiv \{a_1, a_2, a_3, a_4\}$ and $B \equiv \{0, 1\}$, the tree of figure 2.5.13 describes the code $a_1 \rightarrow 010$, $a_2 \rightarrow 10$, $a_3 \rightarrow 11$, $a_4 \rightarrow 001$.

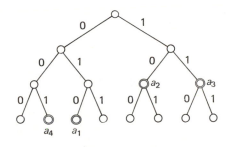

Figure 2.5.13

Example 2.5.4. A very interesting and important instance of oriented rooted trees is offered by the parsing of sentences of natural languages. For illustrative purposes, the parsing of the English sentence "The man eats the apple" is given in figure 2.5.14. It needs no additional comment because of its self-explanatory character.

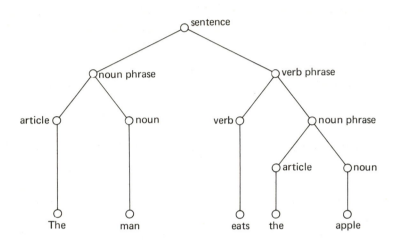

Figure 2.5.14. The parsing tree of the sentence "The man eats the apple"

Example 2.5.5. Another important case is the representation of an algebraic expression involving not only addition and multiplication but also subtraction and division. If an intermediate vertex is marked with either $(-)$ or $(/)$, the order of the arcs issuing from it cannot be altered since $(a - b)$ is different from $(b - a)$, and similarly, a/b is different from b/a. Therefore an algebraic expression is, in general, adequately represented by an oriented rooted tree. In figure 2.5.15 we exhibit the oriented rooted tree representation of the algebraic expression $c(a - b) + (c + d)/(a - b)$.

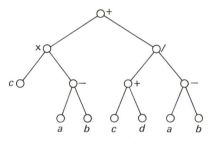

Figure 2.5.15. Oriented rooted tree representation of the algebraic expression $c(a - b) + (c + d)/(a - b)$

Example 2.5.6. Oriented rooted trees can be used to represent the process of making decisions in activities which involve a large number of conditions. We will illustrate with an extremely simple example. Consider the situation in which a person wants to cash a check at a bank. The bank usually has a list of questions to ask before any action is taken. A possible sequence of questions follows.

Q1: Do you have an account here?

Q2: Does your account have sufficient money to cover the check at your bank?

Q3: Does the balance of your account exceed $200?

Based on the answers to the three questions above, one of the following actions may be taken.

A1: Cash the check.

A2: Cash the check but charge a fee for handling.

A3: Refuse the check.

The decision to pick a particular action can be made easily by tracing the path of the *sequential decision tree* given in figure 2.5.16. Each vertex represents a question and each arc issuing from the vertex represents a yes (Y) or no (N) answer to the question.

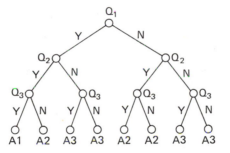

Figure 2.5.16. A sequential decision tree

_____

EXERCISES 2.5

1. Draw all the distinct trees having six vertices.

2. Draw all the spanning trees of the graph in figure 2.5.17.

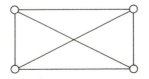

Figure 2.5.17

3. Obtain all minimal spanning trees of the graph in figure 2.5.6.

4. Draw an oriented rooted tree representation of the arithmetic expression

$$(a + b)(cd + f)/(a + c)e.$$

(See example 2.5.5.)

*5. Suppose that you are given eight balls and told that all but one of them are of the same weight. You are also supplied with an equal-arm scale and told you can use it at most three times. Devise a procedure for identifying, with the aid of the scale, the odd ball and for determining whether it is heavier or lighter than the others. Draw the sequential decision tree. (See example 2.5.6.)

6. Show that an undirected multigraph G contains a nondegenerate spanning tree if and only if G is connected and has at least two vertices.

7. Let $G = [V, E, \varphi]$ be a connected undirected multigraph. Define the concept of cut-set for G similar to that in definition 2.2.5. Show that every cut-set in G has an edge in common with every spanning tree in G.

8. What is the rooted tree structure corresponding to the representation set {1, 2, 3, 4, 11, 12, 21, 22, 23, 31, 41, 42, 111, 112, 113, 114, 211, 221, 222, 223, 231, 232, 311, 312, 313, 314, 411, 412, 421, 422, 423, 424, 3121, 3122, 3123, 3124, 31211, 31212}? (See figure 2.5.12.)

- -

2.6 PLANARITY AND COLORATION PROBLEMS

In cartography, it is desirable to color different regions of a map in such a way that no two adjacent regions are given the same color. One of the famous problems in coloring a map asks whether a map of a country, divided into counties, can

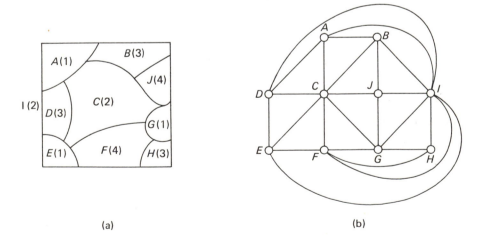

(a)   (b)

Figure 2.6.1. A map colored with four colors {1, 2, 3, 4} and its corresponding graph

be colored with only four distinct colors so that no two adjacent counties have the same color.

If each county on a map (figure 2.6.1(a)) is replaced with a vertex and two vertices are joined by an edge whenever their corresponding counties have a common boundary, the graph obtained is that shown in figure 2.6.1(b). Thus the coloration problem of a map is certainly graph-theoretic. We note that the graph corresponding to a map has the special property that it can be drawn so that edges cross only at vertices, as illustrated by the graph in figure 2.6.1(b).

Definition 2.6.1. A finite undirected multigraph is *planar* if it can be drawn on a plane in such a way that no two of its edges intersect except, possibly, at vertices.

The graph given in figure 2.6.2(a) is planar since it can be redrawn as in figure 2.6.2(b). However, figure 2.6.2(c) is not a planar graph; the proof of this fact will be given later. We now formalize the notion of coloring the vertices of a graph.

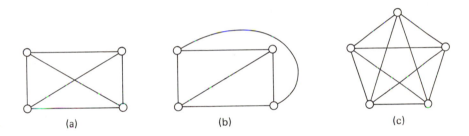

(a)   (b)   (c)

Figure 2.6.2. Planar and nonplanar graphs

Definition 2.6.2. An undirected multigraph $G = [V, E, \varphi]$ is said to be *n-colorable* if there is a function f mapping V onto the set $\{1, 2, \ldots, n\}$ such that if two vertices u and v are joined by an edge, then $f(u) \neq f(v)$.

In other words, the definition above says that if each vertex of G can be assigned one of the n colors $\{1, 2, \ldots, n\}$ such that no two vertices of the same color are joined by an edge, then G is called *n-colorable*. For example, the graph in figure 2.6.1(b) is 4-colorable with the function f defined as shown in figure 2.6.3. The problem of the possibility of coloring a map with only four colors is now stated as a conjecture.

The Four-Color Conjecture: Every planar graph is 4-colorable.

Although there appears to be a consensus among mathematicians that this conjecture is true, no one has yet come up with a proof (but fallacious proofs are proposed quite frequently). However, it is possible to show that every planar graph is 5-colorable. This fact is given as a note at the end of the chapter. A few properties of graph coloration are given in the following.

V	f = color
A	1
B	3
C	2
D	3
E	1
F	4
G	1
H	3
I	2
J	4

Figure 2.6.3

Theorem 2.6.1. If the maximum degree of the vertices in an undirected multigraph G is n, then the smallest number of distinct colors needed to properly color G is less than or equal to $n + 1$.

Proof. We may, without loss of generality, assume that $G = [V, E, \varphi]$ is connected; otherwise one can color each connected component separately. We define a function $f: V \rightarrow \{1, 2, \ldots, n + 1\}$ as follows:

   i) Select an arbitrary vertex $v \in V$ and define $f(v) = 1$.
   ii) For each vertex $u \neq v$, we define $f(u)$ to be the smallest positive integer such that $f(u) \neq f(u')$ for each u' which is adjacent to u and for which f has been defined. Since each vertex has at most n vertices adjacent to it, we see that f can be defined for all vertices of G.$\|$

Theorem 2.6.2. An undirected multigraph is 2-colorable if and only if it contains no cycles of odd length.

Proof. Suppose that G is a undirected multigraph which contains no cycles of odd length. We may assume again that G is connected, as we did in the proof of theorem 2.6.1. Then the vertices of G can be colored by two colors as follows:

   i) An arbitrary vertex v is colored 1; that is, $f(v) = 1$.
   ii) If for some vertex u, $f(u) = 1$, then for a vertex w adjacent to u, we define $f(w) = 2$; if $f(u) = 2$, then we define $f(w) = 1$.

Since G is connected, the function f is defined for all vertices eventually. A vertex w cannot be assigned both colors because this would imply that w is in a cycle of odd length. The graph is therefore 2-colorable.

Conversely, if a graph is 2-colorable, then the colors will alternate when any cycle of the graph is traversed. But this implies that the cycle cannot be of odd length, and hence all cycles in G must be of even length.$\|$

Corollary 2.6.3. A tree is 2-colorable.

We will now return to planar graphs and investigate their characterizing proper-ties. A planar graph such as the one in figure 2.6.4 partitions the plane into "regions," one of which, region I, is not bounded. In fact, since no two edges of a planar graph can intersect, every planar graph determines a set of regions, which are now defined precisely.

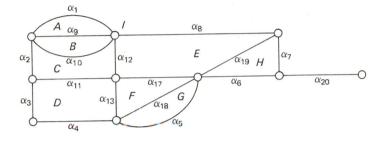

Figure 2.6.4. A planar graph

Definition 2.6.3. Let G be a connected planar graph. A *region* of G is a domain of the plane surrounded by edges of the graph such that any two points in it can be joined by a line not crossing any edge. The edges "touching" a region contain a simple cycle called the *contour* of the region. Two regions are said to be *adjacent* if the contours of the two regions have at least one edge in common.

It follows from the definition above that the contour of the region C is the cycle $(\alpha_2, \alpha_{11}, \alpha_{12}, \alpha_{10})$. We see also that regions C and D are adjacent, whereas regions C and H are not. Furthermore, since in a planar graph, by definition, there is one and only one region which is *infinite* in the sense that the area deter-mined by its contour is infinite, all other regions are finite. Thus region I is the infinite region of the graph in figure 2.6.4.

The following formula, derived by Euler, provides a necessary condition for the planarity of graphs.

Lemma 2.6.4. If a connected planar graph has n vertices, m edges, and r regions, then

$$n - m + r = 2. \text{ (Euler formula)} \qquad (2.6.1)$$

Proof. We construct a spanning tree T of G. Since T has n vertices, it must have $n - 1$ edges, by theorem 2.5.3. Since G is planar, the addition to T of any edge not already in T increases the number of regions by 1. Hence the number of finite regions, $r - 1$, must be equal to the number of edges not already included in T, which is $m - (n - 1)$. Therefore, $r - 1 = m - n + 1$, and we have $n - m + r = 2.\|$

Corollary 2.6.5. If G is a connected simple planar graph without loops and has n vertices, $m \geqslant 2$ edges, and r regions, then the following inequality holds:

$$\tfrac{3}{2}r \leqslant m \leqslant 3n - 6. \tag{2.6.2}$$

Proof. The inequalities clearly hold for $r = 1$ since $m \geqslant 2$. For $r > 1$, let N be the count of edges in the contours of finite regions. Since G is simple, each finite region is bounded by at least 3 edges, whence $N \geqslant 3(r-1)$. But, in a planar graph, an edge belongs to the contours of at most two regions, and at least 3 edges touch the infinite region: whence $N \leqslant 2m - 3$. Thus we have $3r \leqslant 2m$, or $\tfrac{3}{2}r \leqslant m$. But by lemma 2.6.4, we have $n - m + \tfrac{2}{3}m \geqslant 2$ or $m \leqslant 3n - 6$. Hence we have $\tfrac{3}{2}r \leqslant m \leqslant 3n - 6.\|$

The corollary above is a useful test of whether some simple graphs are candidates for being planar, as demonstrated by the following examples.

Example 2.6.1. Consider the graph modeling the following situation. We have three houses, a, b, and c, for which we wish to make connections for water, gas, and electricity. Can we make the connections so that no two pipelines cross each other except at their initial and terminal points? The corresponding graph, called the *utility graph*, is given in figure 2.6.5. We wish to show that the utility graph is not planar. If we assumed it to be planar, then by Euler's formula, we would have $r = 2 - n + m = 2 - 6 + 9 = 5$. We note that a region R in this graph must be bounded by four or more edges, because, if it is bounded by three edges, the contour of the region must contain three vertices, of which two are of the same kind: houses or supply stations. However, two vertices of the same kind are not joined by an edge, and hence there is no region bounded by three edges. Using the same argument as in the proof of corollary 2.6.5, we must have $2m \geqslant 4r$. However, this leads to a contradiction, since $2m = 18$ and $4r = 20$.

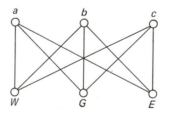

Figure 2.6.5. The utility graph

Example 2.6.2. We wish to show that the complete graph with five vertices, shown in figure 2.6.6 and called the *star graph*, is not planar.

If we assumed the graph to be planar, then since $n = 5$ and $m = 10$, we would have to have $10 = m \leqslant 3n - 6 = 9$, by (2.6.1). But this is clearly a contradiction; thus the star graph cannot be planar.

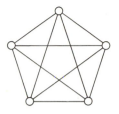

Figure 2.6.6. The star graph

Since the planarity of a graph is not affected if an edge is replaced by a path of arbitrary order, the utility graph and the star graph allow us to define a whole family of nonplanar graphs by simply inserting as many vertices as we wish on each edge to define other nonplanar graphs, as shown in figure 2.6.7. The two types of graphs are nonplanar, as previously shown. But the converse, proved by Kuratowski, also holds and is stated as a theorem below. The proof of the theorem is beautiful but very long and hence is omitted here. Interested readers will find helpful references in Section 2.7.

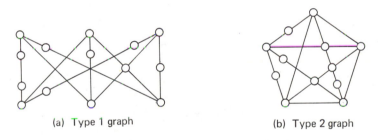

(a) Type 1 graph   (b) Type 2 graph

Figure 2.6.7. Basic nonplanar graphs

Theorem 2.6.6. (Kuratowski) A necessary and sufficient condition for a graph to be planar is that it contains no partial subgraphs of either type 1 or type 2.

Example 2.6.3. An application of the concepts of planarity and coloring arises in the design of printed circuits, which are made by etching or depositing metallic connections on the surface of nonconducting material. Since the printed connections are not insulated, they cannot cross each other except at the junctions. Since not all circuits correspond to planar graphs, it is sometimes necessary to decompose the physical circuit into several parts, each of which corresponds to a planar graph, realize each part as a separate layer, and connect all layers at the junctions. For example, if we redraw the utility graph, as shown in figure 2.6.8(a), we may realize it as a circuit by assigning edge 5 to one layer and all other edges to another layer, as shown in figure 2.6.8(b). A systematic method for decomposing a graph into layers of planar components is given in Note 2.6.

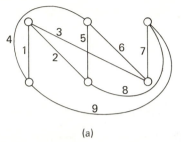

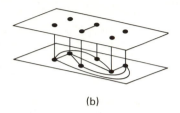

(a)

(b)

Figure 2.6.8. Decomposition of the utility graph into planar components

EXERCISES 2.6

1. Determine which of the graphs in figure 2.6.9 are planar.

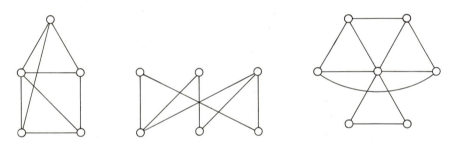

Figure 2.6.9

2. Redraw figure 2.6.4 so that region B becomes the infinite region.
3. Which is the minimum number of colors required for coloring the graph in figure 2.6.10?

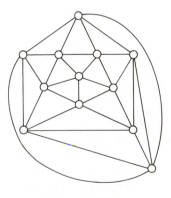

Figure 2.6.10

4. Show that in a simple planar graph without loops, there is a vertex with degree equal to or less than 5. (*Hint:* use corollary 2.6.5.)

*5. Show that simple planar graphs without loops can be drawn on the plane, using straight line segments as edges.

————————————————————————————

2.7 BIBLIOGRAPHY AND HISTORICAL NOTES

Although the term "graph" became popular in the mathematical sciences only after publication of König's work in 1936, many of the well-known problems in graph theory had been observed by Euler and others in the eighteenth century. The development of graph theory is intimately connected with that of combinatorial analysis, and in recent times, graph theory has become an undisputed mathematical tool in the area of the behavioral sciences.

The literature in graph theory is very extensive. However, we shall list only a few books here, in order not to overwhelm the reader. A clearly written and lucid book is

C. Berge, *The Theory of Graphs and Its Applications* (translated by Alison Doig), Wiley, London, 1964.

The reader may also find very helpful the chapters on graphs in

C. L. Liu, *Introduction to Combinatorial Mathematics*, McGraw-Hill, New York, 1969.

A variety of interesting applications of graph theory can be found in Chapter 6 of

R. G. Busacker and T. L. Saaty, *Finite Graphs and Networks: An Introduction with Applications*, McGraw-Hill, New York, 1965.

2.8 NOTES AND APPLICATIONS

Note 2.1. Directed Labeled Graphs and State diagrams of Finite State Machines

Consider the three relations "father of," "uncle of," and "brother of" defined on a set of four persons $\{a, b, c, d\}$, described by the simple graphs in figure 2.8.1. In order to represent all three relations by a single graph, we can superimpose the three graphs of figure 2.8.1 and label the arcs according to which relation the arc is representing. Hence, if we let α, β, and γ denote the relations of "father of," "uncle of," and "brother of," respectively, we can replace the three graphs of figure 2.8.1 by the graph in figure 2.8.2, whose arcs are labeled. This example gives motivation for the following definition.

Definition 2.8.1. An *arc-labeled directed graph* (or simply a *labeled graph*) is a system $[V, \alpha_1, \alpha_2, \ldots, \alpha_n]$, where V is a set of vertices and each α_i is a binary relation on V.

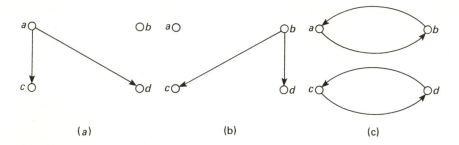

Figure 2.8.1. Graphical representations of three relations: (a) father of; (b) uncle of; (c) brother of

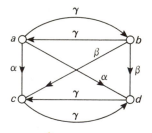

Figure 2.8.2. An arc-labeled graph

It is easy to see that if $G = [V, \alpha_1, \ldots, \alpha_n]$ is a labeled graph, then it may also be described by a $k \times k$ array, where $k = \#(V)$, such that each entry is a subset of the set $\{\alpha_1, \alpha_2, \ldots, \alpha_n\}$. In other words, if the (i, j)th entry is a set $\{\alpha_{l_1}, \ldots, \alpha_{l_p}\}$, then $v_i \alpha_{l_m} v_j$, for each m such that $1 \leqslant m \leqslant p$. For example, the array representation of the labeled graph in figure 2.8.3(a) is given in figure 2.8.3(b).

We shall now illustrate an important application of labeled graphs for the description of finite state machines.

We recall from Note 1.1 that a finite state machine $\mathcal{M}$ is a quintuple, $\mathcal{M} = [S, I, Z, \delta, \lambda]$. (The reader is referred to definition 1.7.1 for the meaning of the symbols S, I, Z, δ, and λ.) Each input $i \in I$ will cause the machine to change its

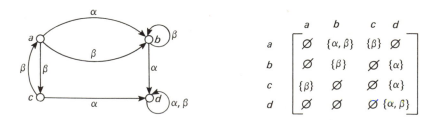

Figure 2.8.3. (a) A labeled graph; (b) its array representation

state, say, from state s_k to state s_h, and produce an output symbol $z \in Z$. This fact can be represented by a labeled graph, for which the vertex set is the set of states, and an arc $(\overrightarrow{s_k, s_h})$ is labeled by the symbols i/z, if $\delta(s_k, i) = s_h$ and $\lambda(s_k, i) = z$. Indeed, if we define a relation $\alpha(i/z)$ for each input-output pair (i, z) such that

$$R_{\alpha(i/z)} \equiv \{(s_k, s_h) | \delta(s_k, i) = s_h \text{ and } \lambda(s_k, i) = z\},$$

then every finite state machine $\mathcal{M} = [S, I, Z, \delta, \lambda]$, with $I \equiv \{i_1, ..., i_m\}$ and $Z \equiv \{z_1, ..., z_r\}$, can be represented by a labeled graph $G(\mathcal{M}) = [S, \alpha(i_1/z_1), \alpha(i_1/z_2), ..., \alpha(i_m/z_r)\}$. The graph $G(\mathcal{M})$ is called the *state graph* or the *state diagram* of $\mathcal{M}$.

The correspondence between the flow table (see Note 1.1) and the state diagram of a finite state machine is illustrated by the following example.

Let $\mathcal{M} = [S, I, Z, \delta, \lambda]$, with $S \equiv \{s_1, s_2, s_3\}$, $I \equiv \{a, b\}$, and $Z \equiv \{0, 1\}$. The functions δ and λ are described by the following flow table:

Input State	a	b
s_1	$s_2, 0$	$s_3, 0$
s_2	$s_2, 0$	$s_3, 1$
s_3	$s_1, 0$	$s_1, 1$

Formally, we can construct four relations $\alpha(a/0)$, $\alpha(a/1)$, $\alpha(b/0)$, $\alpha(b/1)$ expressed by

$$R_{\alpha(a/0)} \equiv \{(s_1, s_2), (s_2, s_2), (s_3, s_1)\}$$
$$R_{\alpha(a/1)} \equiv \varnothing$$
$$R_{\alpha(b/0)} \equiv \{(s_1, s_3)\}$$
$$R_{\alpha(b/1)} \equiv \{(s_2, s_3), (s_3, s_1)\}$$

and from these construct $G(\mathcal{M})$. Less formally, (1) we represent each state by a vertex; (2) we draw an arc from s_i to s_j if there is a transition from s_i to s_j in the flow table; (3) we label the arc with the input-output pair associated with the transition. The state graph of $\mathcal{M}$ is given in figure 2.8.4.

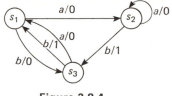

Figure 2.8.4

Reference
Z. Kohavi, *Switching and Finite Automata Theory*, McGraw-Hill, N.Y., 1970.

Note 2.2. Scheduling by PERT-CPM

Example 2.4.3 gives motivation to develop analytic techniques for scheduling different tasks of a project. In this note we discuss briefly the essentials of two very similar techniques: the *critical path method* (CPM) and the *Project Evaluation and Review Technique* (PERT).[3]

The PERT or CPM is applicable to any "scheduling network." *A scheduling network* is a simple, weighted, directed graph G without circuits which has exactly one vertex v_i, whose incoming degree is zero, and exactly one vertex v_t, whose outgoing degree is zero. The vertices of G are usually numbered. A scheduling network is the graphical representation of some project consisting of a collection of interdependent activities that must be executed in a certain order. An arc in G is used to represent an *activity*, with the head of the arrow indicating the direction of progress in the project, and the weight of each arc denoting the duration of the activity. Each vertex in G denotes an *event*, which represents the completion of some activities and the beginning of new ones. Once a scheduling network G is formed, based on the precedence relation of various activities, the application of PERT-CPM should eventually yield a schedule specifying the start and completion dates of each activity, beginning with activities initiated from v_i and ending with activities terminating at v_t in G. In the following, we will consider the concept of the *critical path* in a scheduling network.

Let G be a given scheduling network. We assume that vertices of G are labeled from 1 through n, with v_i labeled 1 and v_t labeled n. For each event i, we define the *earliest start time*, $E(i)$, of all activities emanating from event i, as follows:

i) $E(1) = 0$
ii) $E(k) = \max_i \{E(i) + w(i,k)\}$, ($k \neq 1$, all i such that $\overrightarrow{(i,k)}$ exists).

Here $w(i,k)$ is the weight of the arc $\overrightarrow{(i,k)}$. Thus $E(i)$ is the earliest time the event i can be realized. As an illustration, these numbers for the scheduling network in figure 2.8.5 are given inside the cells of type ▢ .

In G, we also define the *latest completion time*, $L(j)$, for each event j, as follows:

i) $L(n) = E(n)$
ii) $L(j) = \min_k \{L(k) - w(j,k)\}$, ($j \neq n$, all k such that $\overrightarrow{(j,k)}$ exists).

Thus the number $L(j)$ is the latest time we can reach event j without upsetting the schedule. The numbers $L(j)$ for each event j in the scheduling network in figure 2.8.5 are given inside the cells of type ▢ .

---

3. CPM was developed by E. I. du Pont de Nemours Company as an application to construction projects. PERT was developed by the U.S. Navy for scheduling the research and development activities for the Polaris missile program. It is believed that the Polaris missile program, which was completed in five years, would have taken seven years without the aid of PERT.

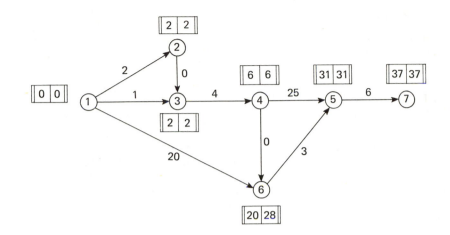

Figure 2.8.5. A Scheduling network

With each activity $\overrightarrow{(i,j)}$ we associate a number $fl(i,j)$, called the *float time* of the activity, defined by

$$fl(i,j) = L(j) - E(i) - w(i,j).$$

The float time is the maximum tolerable delay for completing an activity.

An activity is said to be *critical* if it has zero float time. A *critical path* in a scheduling network G is a path from 1 to n consisting entirely of arcs representing critical activities. Thus, a critical activity is such that a delay in its start will cause a delay of the entire project. A noncritical activity, on the other hand, is such that the time between its earliest start and its latest completion is longer than its actual duration, and the difference is the float time. In the scheduling network given in figure 2.8.5, $\overrightarrow{(1,2)}$ $\overrightarrow{(2,3)}$ $\overrightarrow{(3,4)}$ $\overrightarrow{(4,5)}$ $\overrightarrow{(5,7)}$ is the only critical path since, for event 6, $E(6) \neq L(6)$.

Once critical paths are determined, the actual schedule will depend on a number of constraints, such as resources available, cost, etc., for the scheduling of the noncritical activities. The interested reader is referred to the following book for further study:

K. G. Lockyer, *An Introduction to Critical Path Analysis*, Pitman, London, 1964.

Note 2.3. Linear Network Analysis

An important application of graphs is the analysis of linear electrical networks, typically resistive networks. We now briefly review the topic and establish its connection with the ideas presented in this chapter.

The elements of electrical circuits that we shall consider here are objects with two terminals (two pins or two wires). They are called resistors, capacitors, inductors, voltage sources (typically, batteries), current sources, etc. Each is graphically represented by a specific symbol (see figure 2.8.6) and is characterized by a

| resistor | capacitor | inductor | constant voltage source |

Figure 2.8.6. Graphical representations of the most common elements of electrical networks

relationship between the positive voltage V applied to the element's terminals and the positive current I flowing through the element. (Positive voltage and current are defined as follows: The element terminals are respectively labeled "+" and "−"; then V is positive if the voltage of + is higher than the voltage of −, and I is positive if it flows from + to −.) For example, for a resistor, this relationship is the well-known Ohm's law, $V = RI$, where R is the value of the "resistance" of the resistor. This relationship states that V is proportional to I; that is, if we plot V versus I in a plane diagram, the plot is a straight line, whence the adjective "linear" is applied to the relationship. Analogous relationships can be written for inductors and capacitors. However, we shall not consider these elements in this note, since we can illustrate an interesting application of graphs by restricting ourselves to resistors and constant voltage sources (resistive networks). A constant voltage source is characterized by the property that V is constant and I is arbitrary.

A resistive network is an assembly of resistors and voltage sources interconnected at the terminals. The idea of modeling it as a graph is natural, since the elements can be considered as arcs and the network as a function from the set A of arcs to the cartesian product $V \times V$ of the set V of vertices. Thus the network in figure 2.8.7(a) is modeled as the *directed* graph of figure 2.8.7(b); note that the *direction* in each arc establishes a reference for voltages and currents if one observes the convention of associating the symbol + with the origin and the symbol − with the terminus of the arc.

The purpose of network analysis is the computation of the voltage of each vertex with respect to the voltage of a fixed vertex (reference) and of the current flowing in each arc. The solution method is based on two laws (Kirchhoff's laws): (1) for each vertex, the sum of positive currents *entering* the vertex is zero; (2) for

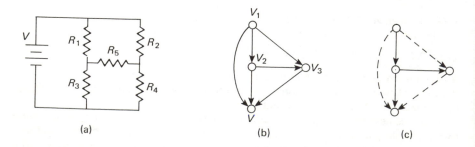

(a) (b) (c)

Figure 2.8.7. A resistive network, its graph, and one of its spanning trees

each cycle (in the graph-theoretic sense, definition 2.3.2), the sum of *positive* voltages measured across the arcs forming the cycle is zero.

We now consider an arbitrary spanning tree of the network graph (definition 2.5.2). When obtaining the tree, we consider the graph as undirected by ignoring the arrows on the arcs. We agree to call *chords* the arcs not belonging to the spanning tree (shown with broken lines in figure 2.8.7(c).) If the graph has n vertices and m edges, since the spanning tree has $(n-1)$ edges (theorem 2.5.3), there are $(m-n+1)$ chords.

We now show that all the network's unknowns—$(n-1)$ voltages and m currents—can be expressed in terms of the $(m-n+1)$ chord currents; in fact, by Kirchhoff's first law, the $(n-1)$ currents in the arcs of the spanning tree can be expressed in terms of the $(m-n+1)$ chord currents. Assuming conventionally at 0 the voltage of one vertex, we can express the voltages of the $(n-1)$ remaining vertices, by Ohm's law, in terms of the currents in the arcs of the tree, hence of the chord currents, by virtue of the previous argument.

Thus, if we can write $(n-m+1)$ independent linear equations in the $(n-m+1)$ chord currents, the analysis is complete (we assume only a modest familiarity with the theory of systems of linear equations). This is possible. Indeed, by theorem 2.5.1, statement 4, if we add an edge (a chord) to the spanning tree, a unique circuit is formed which contains the chord. For this circuit we can write an equation expressing Kirchhoff's second law, with the aid of Ohm's law; this is an equation in the network currents. We can write as many such equations as there are chords, that is, $n-m+1$. It is also easily shown that these equations are independent, thereby offering a complete solution of the network analysis problem.

Linear network analysis, which we have barely touched on, is a fundamental topic in electrical engineering and in system theory in general.

Reference

M. E. Van Valkenburg, *Network Analysis*, Prentice-Hall, Englewood Cliffs, N.J., 1955.

Note. 2.4. Representation of Data Structures in Computers

In this note we discuss briefly a way of representing the organization of data in a computer. For example, a tree may be represented as a parenthesized string (see Section 2.5) and hence can be represented as a linear array of symbols in computer memory. However, it is rather inefficient to operate on this kind of tree representation, because any deletion or addition of vertices must necessarily involve reshuffling a part of the string representation. It is more efficient to represent the tree as a list, which we now describe. Abstractly, a *list* is a section of storage subdivided into parts, called nodes. A *node* is in turn subdivided into two parts, called the *data cell* and the *pointer cell*. The function of the data cell is to store coded data; to illustrate the function of the pointer cell, we must refer to the physical implementation of a list. Physically, each node consists of one or more *consecutive* locations of a computer memory. (We assume reader awareness that the locations of a computer core memory are consecutively numbered, and that the number assigned to a location is called the *address* of that location.) The address of a node is the address of its first location; the pointer cell of a node contains the address of some other node of the list. Thus a node u can be visualized pictorially as a box with two

Figure 2.8.8. Pictorial representation of a node *u*

compartments, as shown in figure 2.8.8, where the left and right compartments denote the data and pointer cells, respectively. The letter *D* denotes coded data, *a* is the address of the node *u*, and *b* is the address of another node, *v*. We say that *u* "points" to *v*, or that *u* and *v* are "linked." Pictorially, the linkage of nodes is described by means of an arrow from node *u* to node *v*, as shown in the list given in figure 2.8.9. Note that the last node has in its pointer cell a special symbol,⊣,

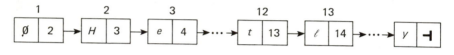

Figure 2.8.9. A list

which denotes the end of the list; i.e., this particular node is not linked to any other nodes. Note also that the empty set symbol, ∅, in the first node denotes the fact that that data cell is empty.

To illustrate the usefulness of the list representation, let us consider the simple example of editing a text, in which a line has the misspelling:

$$\emptyset \text{ He is a lit le boy.}$$
$$\uparrow$$
$$13$$

An extra *t* must be inserted in the 13th position (counting separation spaces) of the line. If the information is stored in the computer memory as a linear array (characters stored in consecutive memory locations according to their occurrence in the sentence), the insertion requires the shifting to successive locations of all the characters following the 12th one in the sentence. On the other hand, if each character of the sentence is stored in a node, as shown in figure 2.8.10, then *t* can be inserted by assigning *t* to a free (or spare) node *k* and linking the 12th node to this spare node, which in turn is linked to the 13th node, as shown in figure 2.8.10.

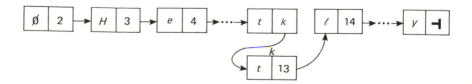

Figure 2.8.10. Insertion of a node in a list

Figure 2.8.9 is an example of a *linear list*, i.e., a list in which each node, with exception of the node whose pointer cell has ⊣, points to a unique node. If some node points to a previous node, then the list becomes a *circular list*. Thus the structure of a list is nothing but a simple path (or circuit). On the other hand, we can use lists to represent more complex structures by noting that, since anything can be stored in the data cell of a node, pointers, in particular, can be stored there. With this modification, a node acquires branching capability. For example, the directed simple graph in figure 2.8.11 (a) is represented as a list in figure 2.8.11 (b).

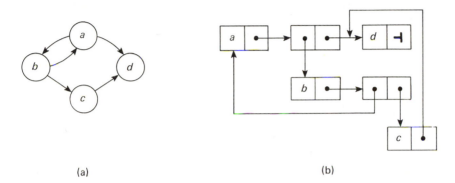

(a)   (b)

Figure 2.8.11. Illustration of using a list to represent a directed simple graph

References

W. C. Gear, *Computer Organization and Programming*, McGraw-Hill, New York, 1969.
D. E. Knuth, *The Art of Computer Programming*, Volume 1, Addison-Wesley, Reading, Mass., 1968.

Note 2.5. Dominating and Independent Sets

A set of vertices *A* in an undirected multigraph *G* is called a *dominating set* if every vertex not in *A* is adjacent to at least one vertex in *A*. Set *A* is called a *minimal dominating set* if it does not properly contain any set which is also a dominating set.

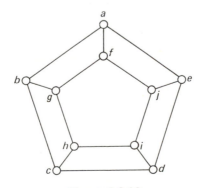

Figure 2.8.12

To illustrate this concept, consider the graph in figure 2.8.12. The set $\{a, g, d, j\}$ is a dominating set. Since each vertex is adjacent to three other vertices and the graph has a total of ten vertices, we conclude that any dominating set must contain at least three elements. Clearly, $\{f, c, j\}$ is a minimal dominating set.

If we think of the graph in figure 2.8.12 as a graphical description of communication links between cities, then the selection of sites for transmitting stations is equivalent to the selection of a minimal dominating set of the corresponding graph.

Considering again the minimal dominating set $A \equiv \{f, c, j\}$ of the graph in figure 2.8.12, we see that the set $\overline{A} \equiv \{a, b, d, e, g, h, i\}$ is also a dominating set since every element in A is adjacent to at least one element in $\overline{A}$. Furthermore, by systematic examination of subsets of $\overline{A}$, we see that $\overline{A}$ contains a minimal dominating set in $\{e, g, h\}$. This fact is indeed a special case of a general result now stated as a theorem.

Theorem 2.8.1. If G is an undirected multigraph without isolated vertices such that each vertex in G has a finite number of incident edges, then for any minimal dominating set in G, there exists another disjoint from it.

As an application of this theorem, let us consider the problem of designing a national defense communication network and its backup system. In order that the backup system can operate independently of the main system during a surprise attack, we are concerned with whether we can build two systems such that no two transmitting stations are located in the same city. Theorem 2.8.1 gives an affirmative answer to this problem, provided that every city is linked to some other cities via some medium of communication, such as telephone lines.

A concept closely related to that of the dominating set is the concept of the independent set. To prepare for this concept, let us consider the problem of the automation of libraries (see Note 1.2). One of the most important tasks in automating a library is the selection of "key words" for indexing documents. If we imagine that documents are represented as vertices of a graph and two vertices are connected by an edge if they are related, i.e., if they share at least one key word, then all documents in the library and their relatedness are represented by an undirected simple graph. One way to extract a minimal set of key words is to find a minimal dominating set such that any two elements in the sets are not related, i.e., not adjacent. This latter property of the set defines the concept of independent set. More specifically, a set of vertices A of a graph is called an *independent set* if no two elements of A are adjacent. An independent set is said to be *maximal* if the addition of any other vertex to the set destroys its characterizing property. The relationship between dominating and independent sets is given as a theorem.

Theorem 2.8.2. An independent set is also a dominating set if and only if it is maximal.

Reference

O. Ore, *Theory of Graphs*, American Mathematical Society, Providence, R.I., 1962.

Note 2.6. Decomposition of a Graph into Layers of Planar Graphs

A systematic procedure to decompose a given graph into planar components (layers) follows. Let G be a graph. Construct a new graph G' such that there is a one-to-one correspondence between edges α in G and vertices $v(\alpha)$ in G'. Furthermore, an edge exists in G' joining vertices $v(\alpha)$ and $v(\beta)$ if and only if α and β intersect in G. By theorem 2.6.1, we know that G' is n-colorable for some n. Coloring the graph G' corresponds to defining an equivalence relation θ on the vertex set of G' such that $v(\alpha)\,\theta\,v(\beta)$ implies that the vertices $v(\alpha)$ and $v(\beta)$ are not adjacent. In terms of the edges of G, this means that α and β do not intersect in G. Therefore, it is possible to decompose G into m planar components.

For illustrative purpose, we again use the utility graph and apply the algorithm to it, as shown in figure 2.8.13. The derived graph has a cycle of length 3 and hence cannot be colored by two colors, according to theorem 2.6.2. It can be colored with three colors, red (R), blue (B), and yellow (Y). The corresponding equivalence relation has three equivalence classes $\{2,5,8\}$, $\{3,6,9\}$, and $\{1,4,7\}$ corresponding to the three colors. We see that with the systematic methods we actually use one more layer than is necessary. This implies that different ways of drawing the graph will affect the number of layers used.

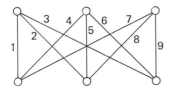

(a)

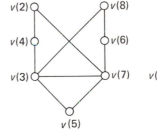

(b)

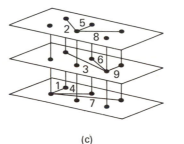

(c)

Figure 2.8.13. Systematic decomposition of the utility graph into layers of planar components

Note 2.7. Every Planar Graph is 5-Colorable

Since the notion of coloration is based on the adjacency of vertices, it is not affected by the multiplicity of edges. Therefore we may restrict ourselves to considering simple graphs.

Theorem 2.8.3. Every planar simple graph is 5-colorable.

Proof. The theorem will be proved by induction on the number of vertices of the graph. We first observe, as the basis of our induction, that the theorem is true for graphs with five or fewer vertices. As an inductive hypothesis, let us assume that the theorem holds for graphs with $(n-1)$ vertices. Let G be a graph with n vertices. We observe that G must contain a vertex v with degree less than or equal to 5, because if the degree of every vertex is larger than 5, then the sum of the degrees of the vertices is at least $6n$, where n is the number of vertices in G. On the other hand, since each edge connects two vertices, the sum of degrees is equal to $2m$, where m is the number of edges in G. The above observations lead to the inequality $2m \geqslant 6n$ or $m \geqslant 3n$. However, by corollary 2.6.5, we must have $m \leqslant 3n - 6$; thus we have a contradiction.

Let G' be the graph obtained from G by the deletion of the vertex v. Since G' is 5-colorable by the inductive hypothesis, we see that G is also 5-colorable if the degree of v is less than 5, because we can assign a color to v which has not been assigned to vertices adjacent to it. Thus we may assume that v is adjacent to five vertices a, b, c, d, e arranged in counterclockwise direction, which are assigned colors α, β, γ, δ, ε, respectively.

Figure 2.8.14 provides help in understanding the rest of the demonstration.

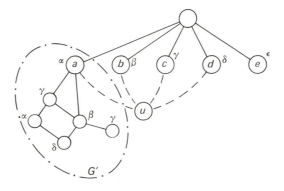

Figure 2.8.14

Let us denote by $G_{\xi\eta}$ the subgraph of G determined by the vertices colored either by ξ or η (ξ and η belong to the set $\{\alpha, \beta, \gamma, \delta, \varepsilon\}$). We then distinguish two cases.

1.   There is no chain between *a* and *c* in the graph $G_{\alpha\gamma}$. It means that vertices *a* and *c* belong to two distinct components. If G'' is the component containing *a*, then we can interchange the assignment of colors α and γ in G''. This results in vertex *a* being assigned the color γ. It follows that the color α can be assigned to vertex *v*; that is, *G* is 5-colorable.

2.   There is a chain between *a* and *c* in the graph $G_{\alpha\gamma}$. We note that there can be no chain between *b* and *d* in $G_{\beta\delta}$. In fact, due to the planarity of *G*, such a chain in $G_{\beta\delta}$ meets the chain in $G_{\alpha\gamma}$ at a vertex *u*, which is assigned two different colors: a contradiction. Thus we argue as in case 1 with respect to the chain between *b* and *d* in $G_{\beta\delta}$. This completes the proof.‖

Note 2.8. Hamilton's Game
The game "Around the world" proposed by Hamilton is graphically represented in figure 2.8.15.

To demonstrate that the graph in figure 2.8.15 contains a hamiltonian circuit,

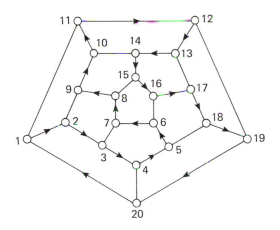

Figure 2.8.15. Graph representation of Hamilton's game

Hamilton noted that when a traveler reaches a vertex he has three choices: (1) taking the edge to his right, an operation denoted by *R*; (2) taking the edge to his left, an operation denoted by *L*; (3) staying where he is, an operation denoted by *I*. We define two sequences of operations to be equivalent if, starting from a given vertex (that is, a vertex with two of its three incident edges chosen as left and right directions), we arrive at the same terminal vertex. Thus a hamiltonian circuit is represented by a sequence of 20 (number of vertices of the graph) symbols from the set $\{L, R\}$, with the following properties.

a) It is equivalent to *I*.
b) None of its proper subsequences is equivalent to *I*.

Such a sequence can be obtained by first observing that the following identities hold.[4]

$$R^5 = L^5 = I$$
$$RL^2R = LRL$$
$$LR^2L = RLR$$
$$RL^3R = L^2$$
$$LR^3L = R^2$$

From the identities above, a sequence is obtained.

$$I = L^5 = L^2L^3 = RL^3RL^3 = (RL^3)^2 = (RL^2L)^2$$
$$= (R(RL^3R)L)^2 = (R^2L^3RL)^2$$
$$= (R^2(RL^3R)LRL)^2 = (R^3L^3RLRL)^2$$
$$= RRRLLLRLRLRRRLLLRLRL$$

A hamiltonian circuit using the sequence above is traced in the graph in figure 2.8.15 (starting from vertex 1).

---

4. It should be noted that identities such as $LR^3 = R$ do not hold because, although the two paths will take a traveler to the same point, adding the *same* edge to the two itineraries would be an operation L in one case, an operation R in the other case.

CHAPTER 3

ALGEBRAIC STRUCTURES

3.1 BINARY OPERATIONS

In the preceding chapters we have seen how the structure of a set A can be enriched by defining a binary relation on A. The role of a binary relation is to express certain properties or relationships existing among the elements of a set. Sometimes binary relations are not powerful enough to entirely represent the structure of a given set. We experienced this insufficiency, however hastily, in Section 1.2, in connection with an example which we now recall. Suppose that the relations "brother of" and "father of" are defined on a set A of people. We noted that from these two relations on A we could *calculate* a new relation on A, "nephew of." How can we formally express this possibility inherent in the structure of A? Clearly a binary relation on A is not adequate for expressing the possibility of combining two elements of a set (here the set of binary relations on a given set A) to obtain a new element of this set. This possibility is realized by the notion introduced in the following general definition.

Definition 3.1.1. *A binary operation* $*$ *on a set* A *is a function* $*: A \times A \to A$. (Usual symbols for binary operations are $*$, $\circ$, $+$, $\cdot$, $\times$, $\oplus$, etc.)

Recall that a function $*: A \times A \to A$ is nothing but a rule which assigns to each element of $A \times A$ (that is, an *ordered* pair in A) a unique element of A.

This definition is of fundamental importance for the development of algebraic structures and deserves some additional comment. As with relations, the adjective "binary" indicates the fact that the operation concerns a pair of elements of A (namely, one element of $A \times A$). With respect to notation, if a_i, a_j, a_k are elements of A, the notation

$$a_i * a_j = a_k$$

will be frequently preferred to the legitimate notations $*: (a_i, a_j) \mapsto a_k$ or $*(a_i, a_j) = a_k$, where a_k is the *result* of the operation $*$ performed on the (*ordered* pair of) *operands* a_i and a_j. When no confusion arises, the operation symbol $*$ may be omitted, and we write $a_i a_j = a_k$. We also note that a binary operation on a set A may be viewed as a ternary relation, that is, as a subset of the three-term cartesian product $A \times A \times A$, where the first, second, and third occurrences of A stand for a choice for the first operand, the second operand, and the result, respectively.[1] This viewpoint will not be further pursued, but it shows how the limitations of a binary relation can be overcome by means of a ternary relation.

Depending on the properties possessed by the operation $*$, different algebraic structures are obtained, as we shall see below. However, one property is possessed unconditionally by binary operations: The result of a binary operation on any two operands in a set always exists and belongs to the same set. This important property, called *closure*, is implicit in the definition of binary operation. Explicit mention is made of it only to stress its importance. As terminological variants, the statements "the operation is *closed* with respect to the set A" and "the set A is closed under the operation $*$" are equivalent to the previous definition.

We must remember that, when dealing with binary relations in Chapter 1, we defined the composition of relations in very much the same way as we defined the function in the paragraph immediately following definition 3.1.1 (that is, as a rule). Why did we not introduce in Chapter 1 the concept of binary operation to encompass "composition of relations"? The answer is that the composition of relations α and β may be performed when the composite $\alpha\beta$ is defined—a circumstance which is not always true. However, when α and β are relations on the same set A, then $\alpha\beta$ is always defined and unique by the rule of composition. Therefore, "composition of binary relations on a set" is a full-fledged binary operation.

The importance of the order of the operands is adequately evidenced by the following example.

Example 3.1.1. Let $A \equiv \{a,b\}$ and $B \equiv \{1,2\}$, and let α be a relation from A to B described by $R_\alpha \equiv \{(a,1), (a,2), (b,1)\}$ and β a relation from B to A described by $R_\beta \equiv \{(1,a), (2,a)\}$. We see that the composition of α and β, $\alpha \circ \beta$ is a relation on A described by $\{(a,a), (b,a)\}$, whereas $\beta \circ \alpha$ is a relation on B described by $\{(1,1),(1,2),(2,1),(2,2)\}$.

Other well-known examples of binary operations are ordinary addition and multiplication, and, somewhat less well-known, the union and intersection of sets (definitions 1.1.5 and 1.1.6). Aside from these somewhat abstract examples, we frequently perform binary operations in our daily lives without being aware of them, as shown by the following example.

Example 3.1.2. When we compare prices of given items, in response to a reasonable desire to economize, we are performing the following operation $*$ defined on

---

1. Note that, when one is considering a binary operation as a ternary relation, in each ordered triple $(a_1, a_2, a_3) \in A \times A \times A$, a_3 is uniquely determined by the ordered pair (a_1, a_2).

the set of positive numbers: $*(a,b) = \text{Min}(a,b)$. Thus $*(3,5) = 3$, $*(5,4) = 4$, $*(6,6) = 6$.

We may observe that when performing a comparison as described in the preceding example, we *order* two items according to some criterion of convenience; we become aware of this order perhaps without really calculating the result of the comparison. In other words, if item A is priced at \$6.95 and item B is priced at \$5.85, we become aware that B is less expensive than A without remembering the actual price of B. However, when we want to compare the prices a, b, c, ... of more than two items A, B, C, ..., the result of the operation must be remembered. If we let $(ab) = \text{min}(a,b)$, the description of the process may be a long expression of the form

$$(\ldots(((ab)c)\ldots)),$$

which states that at first a and b are compared, then $\text{min}(ab)$ is compared with c, and so on. For the sake of simplicity, let us analyze a short expression of the type $(ab)c$. This expression states that the buyer first compares the prices a and b of two items A and B and then compares the smaller one of the two with the price c of another item C. On the other hand, if he were to compare b and c first and then compare the smaller of the two with a, he would have chosen the same term (assuming, of course, that a, b, and c are different). In other words, the following identity holds:

$$(ab)c = a(bc). \tag{3.1.1}$$

The property possessed by a binary operation described by (3.1.1) is not uncommon. Operations such as addition and multiplication of integers and union and intersection of sets also possess this property, which we now formalize.

Definition 3.1.2. A binary operation $*$ on a set A is said to be *associative* if it satisfies identity (3.1.1).

Since associativity means that nested symmetric pairs of parentheses may be inserted arbitrarily in a string of operands, one can simply omit the parentheses and write abc for either $(ab)c$ or $a(bc)$. It is intuitively clear that longer sequences, such as $abcde$ are not ambiguous if the operation is associative.

At this point it is natural for us to inquire whether there exist binary operations which are not associative. Such a binary operation is given in the following example, which is better appreciated with some electrical engineering background, however superficial. If the reader lacks this background, he may omit reading the description of the electrical circuit and continue at the symbol ▶.

Example 3.1.3. A simple circuit is illustrated in figure 3.1.1(a). A voltage waveform v_1 is fed to a circuit called a "differential amplifier." The output of this circuit is a voltage proportional (in our case, equal) to the voltage fed to the

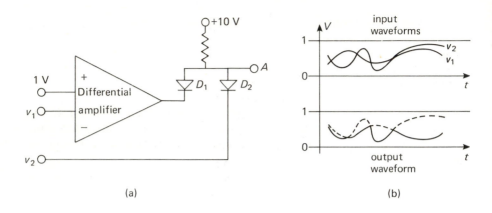

Figure 3.1.1. Circuit illustration of a nonassociative binary operation

input marked $+$ (in our case, a constant voltage equal to 1 volt) minus that fed to the input marked $-$ (the waveform v_1). This output is fed to the cathode of a diode D_1; similarly, another voltage waveform v_2 is fed to the cathode of a diode D_2. The anodes of D_1 and D_2 are connected together and to a terminal A. Since a diode conducts when positive current flows from anode to cathode, and since we may assume that no difference of voltage exists between the anode and the cathode of a conducting diode, the voltage at the terminal A is the smaller of the voltages v_2 and $(1 - v_1)$. The waveforms are sketched in figure 3.1.1(b).

▶ Thus the circuit performs the binary operation $a \circ b = \min(1 - a, b)$ on the set of numbers in the closed interval $[0, 1]$. If we take three numbers, $a = \frac{3}{4}$, $b = \frac{1}{8}$, and $c = \frac{1}{2}$ in $[0, 1]$, we see that $(a \circ b) \circ c = \frac{1}{2}$ and $a \circ (b \circ c) = \frac{1}{4}$; that is, this operation is not associative.

We return to the discussion concerning comparing prices of merchandise. Since it really doesn't matter in which order we select the items to be compared, this means that the operation $*$ defined in example 3.1.2 also has the following property:

$$ab = ba. \tag{3.1.2}$$

Again, operations such as multiplication and addition of integers and union and intersection of sets do satisfy (3.1.2). Hence we formalize this property in the following definition.

Definition 3.1.3. A binary operation $*$ on a set A is said to be *commutative* if it satisfies the identity (3.1.2).

Although many familiar operations are commutative, the operation described in example 3.1.3 is not, as can be seen by the fact that $\frac{1}{8} \circ \frac{1}{2} = \frac{1}{2}$ and $\frac{1}{2} \circ \frac{1}{8} = \frac{1}{8}$.

If a binary operation $*$ is defined on a *finite* set of n elements, it can conveniently be represented by an $n \times n$ array, T_*, called the *operation table* of $*$, by entering the value $*(a, b)$ in the position corresponding to the row labeled by a and the column labeled by b.

Example 3.1.4. Consider the binary operation $\oplus$ defined on the set $A \equiv \{0, 1, 2, 3, 4\}$ such that $i \oplus j = (i + j)(\bmod 5)$, for $i, j \in A$. [This means that $i \oplus j$ is the remainder of the division of the integer $(i + j)$ by 5.] The operation table of $\oplus$ is given in figure 3.1.2.

$\oplus$	0	1	2	3	4
0	0	1	2	3	4
1	1	2	3	4	0
2	2	3	4	0	1
3	3	4	0	1	2
4	4	0	1	2	3

Figure 3.1.2. Table of the operation $\oplus$ in example 3.1.4

Note that if a binary operation is commutative, as is $\oplus$ in the example above, its operation table is symmetric with respect to the main diagonal. In other words, for a commutative binary operation $*$, we have the array identity:

$$T_* = T_*'. \tag{3.1.3}$$

- -

Exercises 3.1

1. a) Which of the following tables of binary operations on the set $\{1, 2\}$ is associative?

	1	2
1	1	1
2	1	2

	1	2
1	1	1
2	2	2

	1	2
1	1	2
2	1	1

	1	2
1	1	2
2	2	2

	1	2
1	2	1
2	1	2

   b) Which of these operations is commutative?

2. Compute the missing entries in the table below so that it defines an associative binary operation $*$ on the set $\{1, 2, 3, 4\}$.

$*$	1	2	3	4
1	1	2	3	4
2	2	1	3	4
3	3	4	3	4
4				

3. Determine whether or not each of the following definitions of $*$ gives a binary operation on the specified set.
   a) On the set of integers, define $*$ by the rule: $a * b = a/b$.
   b) On the set of nonnegative integers, define $*$ by the rule: $a * b = a - b$.
   c) On the set of integers, define $*$ by the rule: $a * b = a + b - ab$.

4. For each of the following binary operations $*$, determine whether it is associative or commutative.
   a) On the integers, $a * b = a - b$.
   b) On the positive integers, $a * b = a^b$.
   c) On the integers, $a * b = ab + 5$.

5. Let $*$ be a binary operation on a set A. For subsets A_i and A_j of A, define $A_i \circledast A_j$, or simply $A_i A_j$, as follows:

$$A_i A_j \equiv \{a * b \mid a \in A_i \text{ and } b \in A_j\}.$$

Show that if $*$ is associative or commutative on A, then $\circledast$ has the same properties on subsets of A.

6. Let $*$ be an associative binary operation defined on a set A. For an element $a \in A$, denote by a^n the element

$$\underbrace{a * a * \cdots * a}_{n \text{ times}}.$$

Show that $a^n * a^m = a^{n+m}$ (where $+$ denotes arithmetic addition).

3.2 SEMIGROUPS AND MONOIDS

We learned in the preceding section that a set acquires a very interesting structure when a binary operation is introduced on it. The objective of this section and several sections that follow is to investigate the properties of these new structures, characterized by the following definition.

Definition 3.2.1. A *semigroup* G consists of a nonempty set S and an associative binary operation $*$ defined on S. The symbolic representation of G is $[S, *]$.

From the discussion in the preceding section, we see that integers under addition, $[I, +]$, and under multiplication $[I, \times]$ are semigroups; furthermore, every finite semigroup is uniquely specified by the operation table of $*$. An interesting case of a semigroup is given in the following example.

Example 3.2.1. Let $A \equiv \{a_1, a_2, \ldots, a_n\}$ be a given finite set, and let A^+ denote the set of all finite sequences[2] of elements of A. Define a binary operation o, called *catenation*,[3] on A^+ by the rule that if x and y are sequences in A^+, then the catena-

---

2. An analysis of the algebraic structure of sequences will be given in Notes 3.6 and 3.7 at the end of the chapter.

3. Also called *concatenation* by some authors.

tion of y to x is the sequence obtained by juxtaposing x on the left and y on the right. For instance, if $x = a_1 a_2 a_3$ and $y = a_2 a_3 a_1$, then $x \circ y = a_1 a_2 a_3 a_2 a_3 a_1$. For brevity we shall write xy instead of $x \circ y$. Clearly, if the sequences x, y, and z are elements of A^+, it really makes no difference as to whether x is catenated with y and the resulting sequence is catenated with z, or y is catenated with z and the resulting sequence is catenated with x on the left. Thus we have the identity

$$(xy)z = x(yz),$$

which shows that catenation $\circ$ is associative (definition 3.1.2), whence $[A^+, \circ]$ is a semigroup, usually called the *free semigroup*[4] generated by the set A. The adjective "free" is used to indicate that $[A^+, \circ]$ is free of any identity among the elements of A^+ except those expressing the associative property.

Example 3.2.2. Let $B(S)$ denote the set of the binary relations on a set S. We know that $B(S)$ is closed under the operation of composition of relations. By theorem 1.2.1, composition is associative; that is, $B(S)$ is a semigroup.

Let us now examine some more familiar examples of semigroups and see whether they will lead us to discover some general properties of semigroups. Considering the examples $[I, +]$ and $[I, \times]$, we see that there is a special element in each of these semigroups, namely, 1 in $[I, \times]$ and 0 in $[I, +]$, such that for each $a \in I$, $a \times 1 = a$ and $a + 0 = a$. This motivates the following definition.

Definition 3.2.2. An element e of a semigroup $G = [S, *]$ is called a *left identity* if $ea = a$ for each a in G. Similarly, f is called a *right identity* if $af = a$ for every a in G. If h is both a left and a right identity in G, then h is said to be a *two-sided identity*.[5]

If we consider the set of positive integers P, we see that 1 is a two-sided identity for the semigroup $[P, \times]$, since for $i \in P$, $i \times 1 = 1 \times i = i$, whereas the semigroup $[P, +]$ has no identities (left, right, or two-sided); we know that the additive identity for positive integers is 0, but 0 does not belong to P, the set of positive integers. On the other hand, a semigroup may have several left or right identities, as demonstrated by the following example.

Example 3.2.3. Let $G = [S, *]$, where $S \equiv \{a, b, c\}$ and where $*$ is given by its operation table in figure 3.2.1.

We see that both elements a and c are right identities; there is no left identity in G.

---

4. The concept of free algebraic structure is discussed in detail in Note 3.6 at the end of the chapter.

5. Note that the word "identity" is used in two connotations: as synonymous with "equality for all choices of the elements" (such as $a + b = b + a$) and as denoting special elements of semigroups. The context always makes the connotation clear.

*	a	b	c
a	a	b	a
b	b	b	b
c	c	b	c

Figure 3.2.1. Operation table of *

We note that in example 3.2.3 the given semigroup has only one-sided identities. Semigroups such as $[I,+]$ and $[I,\times]$, which have a two-sided identity, do not exhibit other identities (one-sided or two-sided). The fact that a two-sided identity is the unique identity of a semigroup is a general property, as demonstrated by the following theorem.

Theorem 3.2.1. If a semigroup G has both a right identity and a left identity, then the two coincide with the same unique two-sided identity.

Proof. Let e be a left identity and f be a right identity in G. We have $ef = f$, since e is a left identity, and $ef = e$, since f is a right identity. It follows that $e = ef = f$; that is, $e = f$ is a two-sided identity. Since any other two-sided identity e' is also a left and a right identity, it coincides with $e = f$; that is, a two-sided identity is unique.‖

Since any two-sided identity in a semigroup is unique, we use the symbol 1 to denote it. Furthermore, we refer to it as *the identity* of the semigroup.

It is always possible to introduce the identity into a semigroup without identity. For instance, consider the semigroup $[A^+, \text{o}]$ discussed in example 3.2.1. We denote by Λ the sequence which consists of no symbol: Λ is called the *empty sequence*. If we adjoin Λ to A^+ to form the set $A^* = A^+ \cup \{\Lambda\}$ and extend the catenation operation o to A^*, so that, for each $x \in A^+$

$$x\Lambda = \Lambda x = x,$$

then the structure $[A^*, \text{o}]$ is a semigroup with the identity Λ. The resulting structure is formalized by the following definition.

Definition 3.2.3. A *monoid* is a semigroup $G = [S, *]$ with the identity.

There are some interesting relationships between semigroups and their substructures. For example, $[I, +]$ is a monoid; however, the set of the even positive integers, which is also closed under addition, is only a semigroup. This motivates the following definition.

Definition 3.2.4. A semigroup (monoid) $G = [S, *]$ is said to be a *subsemigroup* (*submonoid*) of another semigroup (monoid) $G' = [S', \text{o}]$ if S is a subset of S' and * is the restriction of o to S. Furthermore, if G and G' are monoids, their identities coincide.

The reason for insisting that the identity of a submonoid be the same as the identity of the monoid will become clear in Section 3.8.

Example 3.2.4. Let $\mathcal{M} = [S, I, Z, \delta, \lambda]$ be a finite state machine (see definition 1.7.1). $\mathcal{M}$ is described by its flow table (see note 2.1) in figure 3.2.2. If we disregard the output and concentrate on the state behavior of $\mathcal{M}$ we see that each input se-

	0	1
s_1	s_1, 0	s_2, 1
s_2	s_2, 0	s_2, 1

Figure 3.2.2. Flow table of a finite state machine

quence takes a state into a state. For example, sequence 0010 takes s_1 to s_2 and s_2 to s_2. Hence each input sequence $x \in I^*$ induces a function $\delta_x : S \to S$. Since S is finite, there is only a finite number of functions of the form δ_x for $x \in I^*$. Clearly, the set of these functions is closed under composition, composition of functions is associative, and there is at least one sequence (the empty sequence) which induces the identity function on S. We conclude that the set of functions $\{\delta_x | x \in I^*\}$ forms a monoid under composition called the *monoid associated with* $\mathcal{M}$. The monoid associated with $\mathcal{M}$ in figure 3.2.2 consists of the following two elements:

	f_1
s_1	s_1
s_2	s_2

	f_2
s_1	s_2
s_2	s_2

More discussion on the monoids associated with finite state machines will be given in Note 3.1.

Example 3.2.5. An interesting example of subsemigroup is the set G' of all the distinct powers of some element a in a semigroup G (note that, since "multiplication" is assumed to be the operation of G, the "power" a^{j+1} is defined recursively as $a^{j+1} = a^j a$, with $a^1 = a$). Thus elements of G' are of the form a^i for integer i. It is easily shown (for a similar proof, see lemma 2.2.1) that $a^i a^j = a^{i+j}$, which proves closure; moreover, $(a^i a^j)a^k = a^i(a^j a^k) = a^{i+j+k}$, which proves associativity. Therefore G' is a semigroup and is called the *cyclic semigroup* generated[6] by a. If G is finite—that is, $\#(S) = n$ —we have the further property that among the powers $a, a^2, a^3, \ldots, a^{n+1}$ of a, at least two must be equal. Let a^j be the smallest

---

6. The notion of *generator* of an algebraic structure will be analyzed in Note 3.6.

power of a equal to a previous power of a, say $a^i(j>i)$. If we let $m=j-i$, then from $a^i=a^j$ we have $a^i=a^j=a^{i+m}=a^i\,a^m=a^{i+m}\,a^m=\cdots=a^{i+km}$ for any $k\geqslant 0$. Since for any $p\geqslant i$, one has $p=i+s$ and $s=qm+r$, with unique q and r such that $q\geqslant 0$ and $0\leqslant r<m$, we obtain

$$a^p=a^{i+qm+r}=a^{i+r}.$$

Hence it follows that the elements

$$a, a^2, \ldots, a^{i-1}, a^i, a^{i+1}, \ldots, a^{i+m-1}$$

are the distinct elements of the cyclic semigroup G' generated by a. We also note that cyclic semigroups are commutative (definition 3.1.3) since $a^i\,a^j=a^j\,a^i=a^{i+j}$.

Finally, we may consider the operation of multiplying an element $a^i\in G'$ by the generator a as a function on G' ("successor function"). The previous discussion demonstrates that the graph of the successor function is as shown in figure 3.2.3.

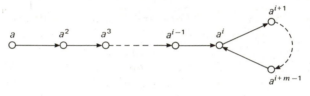

Figure 3.2.3

Examining $[I,+]$ again, we note that the identity 0 in this monoid also identifies pairs of elements $(a,-a)$ in I whose sum is 0. In conventional terminology $-a$ is said to be the "opposite" of a; however, since semigroups and monoids draw their terminology from the operation of multiplication (although the operation $*$ may bear no resemblance to multiplication), we have the following definition.

Definition 3.2.5. An element b of a monoid $G=[S,*]$ is said to be a *left inverse* of a if $ba=1$. Similarly, b is called a *right inverse* of a if $ab=1$. Finally, if b is both a left and a right inverse of a, then it is called a *two-sided inverse* of a.

We now give a theorem about inverses which closely parallels theorem 3.2.1 about identities.

Theorem 3.2.2. If a and c are, respectively, left and right inverses of an element b in a monoid $G=[S,*]$, then a and c coincide with the unique two-sided inverse of b.

Proof. From the fact that $ab=bc=1$ and associativity, we must have

$$a=a1=a(bc)=(ab)c=1c=c.\|$$

Since, by this theorem, a two-sided inverse of an element a of a monoid is unique, we shall, in the sequel, denote this element by the symbol a^{-1} and simply call it *the inverse* of a.

A monoid in which each element has an inverse has additional properties, one of which is the so-called *cancellation property*. (The cancellation property is indeed the rule we resort to in ordinary algebra when we "cancel" identical factors from the two sides of an equality; that is, $ab = ac$ is rewritten as $b = c$). The cancellation property states that $ab = ac$ or $ba = ca$ implies $b = c$. This, of course, is a consequence of the fact that $ab = ac$ implies that $a^{-1}ab = a^{-1}ac$, whence $1 \cdot b = 1 \cdot c$ and $b = c$. However, even a semigroup without identity can have the cancellation property. Such a semigroup will be referred to as a *cancellative semigroup*. A cancellative semigroup without identity is described in the following example.

Example 3.2.6. The semigroup $[A^+, \circ]$ consisting of all finite sequences of elements of $A \equiv \{a_1, \ldots, a_n\}$ under the catenation operation is a cancellative semigroup without identity. To see that it satisfies the cancellation property, we let x and y be two sequences of A^+ such that $x = x_1 z$ and $y = y_1 z$. Now, if $x = x_1 z = y_1 z = y$, then the two sequences x and y must be of identical length, and the symbol in corresponding positions must be identical. This, of course, implies that every initial subsequence in x must be identical to the initial subsequence of the same length in y. In particular, $x_1 = y_1$.

Semigroups of binary relations on a set under composition are usually not cancellative. A specific such semigroup is given in the following example for illustrative purposes.

Example 3.2.7. Let $S \equiv \{a, b, c\}$ and α, β, and γ be relations on S defined by the tables given in figure 3.2.4. It is quite clear that $\alpha \neq \beta$. However, $\alpha\gamma = \beta\gamma$, as shown in figure 3.2.4(d).

α	a	b	c
a	0	1	0
b	0	0	1
c	1	0	0

(a)

β	a	b	c
a	1	0	0
b	0	0	1
c	1	0	0

(b)

γ	a	b	c
a	1	1	0
b	1	1	0
c	0	0	1

(c)

$\alpha\gamma = \beta\gamma$	a	b	c
a	1	1	0
b	0	0	1
c	1	1	0

(d)

Figure 3.2.4. Tabular representation of four relations on S in example 3.2.7

EXERCISES 3.2

1. a) Verify that the following are operation tables of semigroups.

*	1 2		*	1 2		*	1 2		*	1 2		*	1 2		*	1 2
1	1 1		1	1 1		1	1 1		1	1 2		1	1 2		1	2 2
2	1 1		2	1 2		2	2 2		2	1 2		2	2 2		2	2 2

   b) Is there any other semigroup of the form $[\{1,2\},*]$ not included in the preceding list?

   c) Which of the semigroups described in (a) has right identities? Which one has left identities? Which one is a monoid?

2. In how many ways can you complete the following table of the binary operation $*$ on $\{1,2,3\}$ so that $[\{1,2,3\},*]$ is a monoid?

3. For each of the following cases, determine whether the set under the binary operation $*$ forms a semigroup or a monoid. If it is a monoid, identify its identity.

   a) The integers and $a * b = ab$.

   b) The integers and $a * b = a - b$.

   c) Oriented rooted trees. If T_1 and T_2 are two such trees, $T_1 * T_2$ is the result of placing T_1 to the left of T_2 and joining the roots of T_1 and T_2 to a new vertex, which becomes the root of $T_1 * T_2$.

   d) The finite sequences of 0's and 1's containing an even number of 1's; $*$ is catenation.

4. An *idempotent* element a in a semigroup is one such that $aa = a$. Show that the set of all idempotent elements in a commutative semigroup G forms a subsemigroup of G.

5. Show that a cancellative semigroup can contain at most one idempotent element, and if it exists, it is an identity element (see exercise 3.2.4).

6. An element a in a semigroup G is called a *left zero* if $ab = a$ for every b. Similarly, we define the concept of *right zero* and *two-sided* zero in G.

   a) Can a semigroup have more than one left or right zero?

   b) Show that if a semigroup has a left zero and a right zero, then the two coincide with the same unique two-sided zero.

   c) Which of the semigroups given in exercise 3.2.1 has a left zero, which a right zero, which a two-sided zero?

7. A nonempty set A of a semigroup $G = [S, *]$ is called a set of *generators* of G if G is the smallest subsemigroup of G containing A. Give two distinct sets of generators of the semigroup $[I, +]$.

3.3 ISOMORPHISM AND HOMOMORPHISM OF SEMIGROUPS

Consider the semigroups of the integers I under addition, $G_1 = [I,+]$, and of the even integers under addition, $G_2 = [E,+]$. The two sets I and E are in a one-to-one correspondence through the function $\varphi : I \to E$, such that for $i \in I$, $\varphi(i) = 2i$. We note, however, that φ realizes more than a one-to-one correspondence because, for $i, j \in I$,

$$\varphi(i+j) = 2(i+j) = 2i + 2j = \varphi(i) + \varphi(j).$$

That is, the correspondence between I and E through φ also "preserves" an operation, since the image of a sum coincides with the sum of the images.

Definition 3.3.1. Given two semigroups $G_1 = [S,\circ]$ and $G_2 = [T,*]$, an invertible function $\varphi : S \to T$ is said to be an *isomorphism*[7] between G_1 and G_2 if, for every a and b in S,

$$\varphi(a \circ b) = \varphi(a) * \varphi(b); \qquad (3.3.1)$$

G_1 and G_2 are said to be *isomorphic*.

It follows from the definition of isomorphism that isomorphic semigroups differ only in the labeling of their elements and of their respective operations. The meaning of the identity (3.3.1) can be expressed by the operation diagram for

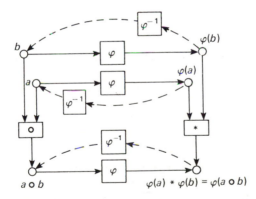

Figure 3.3.1. Operation diagram of the isomorphism φ

isomorphisms given in figure 3.3.1. The boxes $\boxed{\circ}$, $\boxed{*}$, $\boxed{\varphi}$, and $\boxed{\varphi^{-1}}$ may be thought of as kinds of processors (such as adders or multipliers in computers), the elements a, b and $\varphi(a)$, $\varphi(b)$ may be thought of as data fed into the respective processors $\boxed{\circ}$ and $\boxed{*}$, and $a \circ b$ and $\varphi(a) * \varphi(b)$ may be thought of as the processed results.

---

7. This word comes from the Greek words ἴσος (isos), "identical," and μορφή (morphé), "form."

We remark here that in order to show that two semigroups $G_1 = [S_1, \circ]$ and $G_2 = [S_2, *]$ are isomorphic, we can use the following procedure.

1. Define a function $\varphi : S_1 \to S_2$.

2. Show that φ is one-to-one (definition 1.3.3).

3. Show that φ is onto (definition 1.3.2).

4. Show that for each pair of elements s and s' in S_1, $\varphi(s \circ s') = \varphi(s) * \varphi(s')$.

Example 3.3.1. We will illustrate the given procedure with the classical example that the semigroup of the positive real numbers under multiplication is isomorphic to the semigroup of the real numbers under addition.

1. Let the function φ be $\varphi(x) = \log_{10} x$ for each positive real number x.

2. Since $\log_{10} x = y$ means that $10^y = x$, we see that $\varphi(x) = \varphi(y) = z$ implies that $x = 10^z = y$. Hence, φ is one-to-one.

3. Since for each real number x, $\log_{10}(10^x) = x$, we see that φ is onto.

4. For any two arbitrary positive real numbers x and y, we have

$$\varphi(xy) = \log_{10}(xy) = \log_{10} x + \log_{10} y = \varphi(x) + \varphi(y).$$

Hence φ is an isomorphism.

If the semigroups are finite, testing their isomorphism can be accomplished by examining their respective operation tables. The following example illustrates how the operation table of a semigroup G_1 can be obtained from the operation table of a semigroup G_2 isomorphic to G_1 by rearranging the table and renaming the elements.

Example 3.3.2. For simplicity of notation we introduce a slight modification of our standard representation of functions. In figure 1.3.2 a function $f : A \to B$ is illustrated by means of a vertical function table, consisting of a list of pairs, $(a, f(a))$ for each $a \in A$. We now simply transpose the table; that is, the horizontal table of figure 3.3.2(b) replaces the vertical table of figure 3.3.2(a).

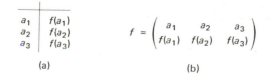

(a)   (b)

Figure 3.3.2. Equivalent tabular representations of a function f

Let $G_1 = [S,*]$ and $G_2 = [T,\text{o}]$ be two semigroups specified by their respective operation tables given in figures 3.3.3(a) and 3.3.3(b). If we define

$$\varphi = \begin{pmatrix} a & b & c \\ 3 & 1 & 2 \end{pmatrix},$$

we see that φ is an invertible function. The function φ is an isomorphism because if we permute the rows in figure 3.3.3(b) cyclically, i.e., row 1 → row 2, row 2 → row 3, and row 3 → row 1, and permute the columns in the same fashion, we obtain figure 3.3.3(c), which is exactly the same as figure 3.3.3(a) if we call elements 3, 1, and 2 by the names a, b, and c, respectively.

*	a	b	c
a	a	b	c
b	b	b	c
c	c	b	c

(a)

o	1	2	3
1	1	2	1
2	1	2	2
3	1	2	3

(b)

o	3	1	2
3	3	1	2
1	1	1	2
2	2	1	2

(c)

Figure 3.3.3. Operation tables of G_1 and G_2 in example 3.3.2

We now recall that in examples 3.2.2 and 3.2.7 we illustrated important properties of semigroups by referring to semigroups of binary relations on a set. We may now ask whether any loss of generality is incurred by restricting ourselves to semigroups (monoids) of functions on a set. We begin with an example and then ask whether general principles can be abstracted from it.

Example 3.3.3. Let us consider the simple case of the set of all functions on a set A consisting of only two elements $\{a,b\}$. There are altogether four functions given in figure 3.3.4, forming a set $G(A)$. On $G(A)$ we may define the binary operation

$$f_1 = \begin{pmatrix} a & b \\ a & b \end{pmatrix}, \quad f_2 = \begin{pmatrix} a & b \\ b & a \end{pmatrix}, \quad f_3 = \begin{pmatrix} a & b \\ a & a \end{pmatrix}, \quad f_4 = \begin{pmatrix} a & b \\ b & b \end{pmatrix}.$$

(a)   (b)   (c)   (d)

o	1	2	3	4
1	1	2	3	4
2	2	1	3	4
3	3	4	3	4
4	4	3	3	4

(e)

Figure 3.3.4. Elements and operation table of the semigroup consisting of functions on a set with two elements

of function composition, whose operation table is shown in figure 3.3.4(e), where the integer i is used to denote the function f_i for simplicity.

It follows from figure 3.3.4 that $G(A)$ is a monoid with f_1 as its identity. The result expressed by this example with regard to a set A of two elements can now be generalized. Indeed, the composite of functions on a set is again a function; composition is associative; i.e., the set of all functions on a set A is a semigroup. Furthermore, the identity function e_A (see Section 1.4) is the identity of the semi-group, and hence the set of all functions on an arbitrary set A is always a monoid. The converse is also true, as illustrated by the following remarkable and beautiful "representation" theorem. This theorem is a slight generalization of a very famous fundamental theorem, derived by Cayley, which states the analogous properties for groups, to be given later as theorem 3.5.3.

Theorem 3.3.1. Every finite monoid $G = [A, *]$ is isomorphic to a monoid con-sisting of a set of functions on a set.

Proof. Let G be a given monoid. We want to construct a monoid of functions to which G is isomorphic. Consider the set of functions $F \equiv \{f_a | a \in G\}$ defined by the rule that for each $b \in G, f_a(b) = ba$. Since multiplication is defined in G, each function f_a is defined. Indeed, if G is finite and given by its operation table, f_a is the column of that table corresponding to the element a. We claim that F under composition of functions is a monoid; that is, the composition of two columns coincides with a column of the operation table of G. In fact, let f_a and f_b be two arbitrary elements of F, and let $f_a f_b = f$. Then, for any $c \in G$

$$f(c) = f_a f_b(c) = f_b(f_a(c)) = f_b(ca) = cab = f_{ab}(c),$$

where the first identity follows from the definition of f, the second from the definition of composition, the third, fourth, and fifth identities from the definition of the set of functions F. Thus, since ab is a member of $G, f = f_{ab}$ is a member of F; that is, F is closed. Since composition of functions is associative, F is also a semi-group. Finally, the identity in G is an element e whose corresponding column in the operation table is identical to the column of the row headings in the same table; i.e., the function f_e is the identity function. Thus F is a monoid.

To show that G is isomorphic to F, we define the mapping $\varphi : G \to F$ such that $\varphi(a) = f_a$ for each $a \in G$. The mapping φ is clearly onto, since each column of the operation table is associated with an element of G; φ is also one-to-one because $a \neq b$ implies that

$$f_a(1) = 1a = a \neq b = 1b = f_b(1);$$

that is, $f_a \neq f_b$ (no two columns of the operation table are identical). Finally,

$$\varphi(ab) = f_{ab} = f_a f_b = \varphi(a) \varphi(b).$$

Hence φ is an isomorphism.$\|$

In many cases, two structures are related but not necessarily isomorphic. Take, for example, the semigroup of the sequences of 0's and 1's under catenation $\circ$, $G_1 = [\{0,1\}^+, \circ]$ and the semigroup $G_2 = [\{0,1\}, \oplus]$, where the operation $\oplus$ called "modulo 2 addition" is defined by the following table.

$\oplus$	0	1
0	0	1
1	1	0

Consider the function $\varphi : \{0,1\}^+ \to \{0,1\}$ such that for $x \in \{0,1\}^+$

$$\varphi(x) = \begin{cases} 1, & \text{if } x \text{ contains an odd number of 1's} \\ 0, & \text{if } x \text{ contains an even number of 1's.} \end{cases}$$

It is easy to see that for $x, y \in \{0,1\}^+$,

$$\varphi(x \circ y) = \varphi(x) \oplus \varphi(y).$$

In other words, φ preserves the operation even though it is not invertible. This gives motivation for the following definition.

Definition 3.3.2. Given two semigroups $G_1 = [S, \circ]$ and $G_2 = [T, *]$, an onto function $\varphi : S \to T$ is said to be a *homomorphism*[8] from G_1 to G_2 if, for every a and b in S,

$$\varphi(a \circ b) = \varphi(a) * \varphi(b);$$

G_2 is said to be the *homomorphic image* of G_1 under φ and is denoted by $\varphi(G_1)$.

It is instructive to contrast this definition with that of an isomorphism (definition 3.3.1). When we consider a homomorphism $\varphi : S \to T$, for s in S, $\varphi(s)$ is still a unique element of T; however, for t in T, $\varphi^{-1}(t)$ is now a nonempty subset of S. When $\varphi^{-1}(t)$ consists of a single element, then φ is also one-to-one; that is, the homomorphism specializes into an isomorphism. The operation diagram of a homomorphism is illustrated in figure 3.3.5.

Example 3.3.4. Let $[P_n, \oplus]$ be the monoid consisting of the set $P_n \equiv \{0, 1, \ldots, n-1\}$, and let the operation $\oplus$ be defined so that $i_1 \oplus i_2 = i_1 + i_2 \pmod{n}$. We will

---

8. This word comes from the Greek words ὁμός (homós), "common", and μορφή (morphé), "form."

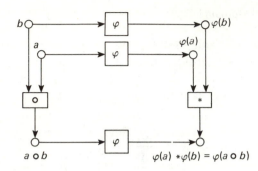

Figure 3.3.5. Operation diagram of a homomorphism

show that $[P_n, \oplus]$ is a homomorphic image of $G = [N,+]$, where N is the set of nonnegative integers, by applying the following procedure.

1. Let φ be a function from N to P_n such that $\varphi(i)$ is the remainder of the division of i by n. Clearly φ is onto.

2. Show that $\varphi(i_1 + i_2) = \varphi(i_1) \oplus \varphi(i_2)$. Indeed, if $i_1 = q_1 n + r_1$ and $i_2 = q_2 n + r_2$, with $0 \leqslant r_1 < n$ and $0 \leqslant r_2 < n$, we see that $\varphi(i_1) = r_1$ and $\varphi(i_2) = r_2$. Thus

$$\varphi(i_1) \oplus \varphi(i_2) = r_1 + r_2 \pmod{n}.$$

Now, if $r_1 + r_2 = q_3 n + r_3$, $0 \leqslant r_3 < n$, then $\varphi(i_1) \oplus \varphi(i_2) = r_3$. But since $i_1 + i_2 = (q_1 + q_2)n + r_1 + r_2 = (q_1 + q_2 + q_3)n + r_3$ and $0 \leqslant r_3 < n$, we have $\varphi(i_1 + i_2) = r_3$. We conclude that

$$\varphi(i_1) \oplus \varphi(i_2) = r_3 = \varphi(i_1 + i_2)$$

We note that in the preceding example $\varphi(0) = 0$. Indeed, if φ is a homomorphism from a monoid G to another monoid G', we have

$$\varphi(1 \circ a) = \varphi(a) = \varphi(1) * \varphi(a). \tag{3.3.2}$$

Equality (3.3.2) implies that the identity 1 of G is mapped by the homomorphism to the identity $1'$ of G'. On the other hand, if a and b belong to $\varphi^{-1}(1')$, then

$$\varphi(a \circ b) = \varphi(a) * \varphi(b) = 1' * 1' = 1',$$

which implies that $a \circ b \in \varphi^{-1}(1')$. These results are summarized in the following lemma.

Lemma 3.3.2. Let φ be a homomorphism from a monoid G to another monoid G'. Then $\varphi(1) = 1'$, and $\varphi^{-1}(1')$ is a submonoid of G.

We now give a theorem which explicitly demonstrates the structure-preserving power of a homomorphism.

Theorem 3.3.3. Let φ be a homomorphism from a semigroup G_1 to another semigroup G_2, and let G_1' be a subsemigroup of G_1 and G_2' be a subsemigroup of G_2. Then $\varphi(G_1')$ and $\varphi^{-1}(G_2')$ are subsemigroups of G_2 and G_1 respectively.,

Proof. Let G_1' be a subsemigroup of $G_1 = [S,\circ]$ and φ a homomorphism from G_1 to $G_2 = [T,*]$. Let t_1, t_2 be two elements in $\varphi(G_1')$. There exist s_1, s_2 in G_1' such that $\varphi(s_1) = t_1$ and $\varphi(s_2) = t_2$. Since φ is a homomorphism, and since G_1' is a semigroup, $s_1 \circ s_2 \in G_1'$ and $\varphi(s_1 \circ s_2) = \varphi(s_1) * \varphi(s_2) = t_1 * t_2 \in \varphi(G_1')$. Hence $\varphi(G_1')$ is closed under the operation $*$. Since associativity holds for subsets of G_2, $\varphi(G_1')$ is a subsemigroup of G_2.

Conversely, assume that G_2' is a subsemigroup of G_2. Then for t_1 and t_2 in G_2', since φ is onto, $\varphi^{-1}(t_1)$ and $\varphi^{-1}(t_2)$ are nonempty subsets of S. Let s_1 and s_2 be two elements of S such that $s_1 \in \varphi^{-1}(t_1)$ and $s_2 \in \varphi^{-1}(t_2)$. Clearly $\varphi^{-1}(t_1) \subseteq \varphi^{-1}(G_2')$ and $\varphi^{-1}(t_2) \subseteq \varphi^{-1}(G_2')$. Since G_2' is a semigroup, $t_1 * t_2 \in G_2'$ (closure). The fact that φ is a homomorphism implies

$$\varphi(s_1 \circ s_2) = \varphi(s_1) * \varphi(s_2) = t_1 * t_2;$$

that is, $s_1 \circ s_2 \in \varphi^{-1}(t_1 * t_2) \subseteq \varphi^{-1}(G_2')$, which shows the closure of the operation $\circ$ in the set $\varphi^{-1}(G_2')$. Thus $\varphi^{-1}(G_2')$ is a subsemigroup of G_1.$\|$

EXERCISES 3.3

1. Determine the truth or falsehood of each of the following statements.

    a) The semigroup of nonzero real numbers under multiplication is isomorphic to the semigroup of real numbers under addition.

    b) Any semigroups of two elements are isomorphic.

    c) Every isomorphism is a one-to-one function, and vice versa.

    d) Every homomorphism is an onto function.

2. Determine a homomorphism from $[P_6, \oplus]$ to $[P_3, \oplus]$ (see example 3.3.4 for a definition of P_i).

3. Show that if φ is a homomorphism from G_1 to G_2 and ψ is a homomorphism from G_2 to G_3, then $\varphi\psi$ is a homomorphism from G_1 to G_3.

4. A homomorphism from a semigroup G to itself is called an *endomorphism* of G. Show that the set of all endomorphisms of G is a monoid.

5. Let $G_1 = [S,*]$ and $G_2 = [T,\circ]$ be two semigroups. Define the *direct product* of G_1 and G_2, denoted by $G_1 \times G_2$, as the system $[S \times T, \times]$, where

$$(s_1, t_1) \times (s_2, t_2) = (s_1 * s_2, t_1 \circ t_2).$$

Show that

a) $G_1 \times G_2$ is a semigroup;

b) G_1 and G_2 are homomorphic images of $G_1 \times G_2$;

c) if G_1 and G_2 are monoids, then G_1 and G_2 are isomorphic to subsemigroups of $G_1 \times G_2$.

3.4 CONGRUENCE RELATIONS ON SEMIGROUPS

In this section we investigate very important structural properties induced by homomorphisms. We shall begin with an informal appreciation of these properties before analyzing them formally.

In the preceding section we illustrated the homomorphism φ from the semigroup $[\{0,1\}^+, \mathrm{o}]$ of the binary sequences under catenation to the semigroup $[\{0,1\}, \oplus]$ of the integers 0 and 1 under modulo 2 addition. We observe that the image $\varphi(x)$ of a binary sequence x does not provide any information about x but the parity of the number of 1's that x contains. Thus, if two sequences x and y map to 0, for example, regarding φ as a relation, we may write $x \varphi 0$ and $y \varphi 0$. Introducing the composite relation $\varphi\varphi^{-1}$, we have $x(\varphi\varphi^{-1})y$. Since φ is a function, we know from theorem 1.5.7 that $\varphi\varphi^{-1}$ is an equivalence relation. We denote it by the symbol γ. It is now natural to inquire whether the fact that A^+ is a semigroup (i.e., there is an associative binary operation defined in A^+) confers special properties to the relation γ. We start with a definition.

Definition 3.4.1. Given a semigroup $G = [S, *]$, an equivalence relation γ on S is called a *congruence relation* if, for arbitrary elements a, b, c, and d in S,

$$a \gamma b \text{ and } c \gamma d \text{ imply } (a * c) \gamma (b * d). \tag{3.4.1}$$

The relation γ is said to have the *substitution property with respect to* the operation $*$ in S.

The choice of the phrase "substitution property" is due to the fact that in $a \gamma b$ we may "substitute" $(a * c)$ for a and $(b * d)$ for b, provided that $c \gamma d$. Furthermore, if γ is interpreted as "=", then an instance of (3.4.1) is "$a = b$ and $c = c$ imply $a * c = b * c$"; that is, we obtain the substitution principle (see Section 0.3).

We observe that the concept of the substitution property is really a special case of the isomorphism concept, as illustrated by the diagram in figure 3.4.1 (the difference between figures 3.3.1 and 3.4.1 is that G_1 in the latter coincides with G_2).

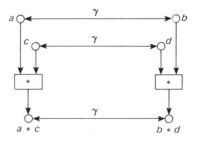

Figure 3.4.1. Graphical illustration of the substitution property concept

Returning to our example, let x, y, x', and y' be four sequences of 0's and 1's such that $x \gamma y$ and $x' \gamma y'$. This means that x and y will either both contain an even number of 1's or both contain an odd number of 1's. Similarly, x' and y', being equivalent with respect to γ, have the same property. But clearly xx' contains an even (odd) number of 1's if and only if yy' contains an even (odd) number of 1's; that is, $xx' \gamma yy'$. We discovered that the relation $\gamma = \varphi\varphi^{-1}$ is indeed a congruence relation.

Lemma 3.4.1. Every homomorphism φ from a semigroup $G_1 = [S,*]$ to a semigroup $G_2 = [T,\circ]$ induces a congruence relation $\varphi\varphi^{-1}$ on G_1.

Proof. Let us define a relation $\gamma = \varphi\varphi^{-1}$ on G_1. Since φ is a function, by theorem 1.5.7 (the characterization theorem of an equivalence relation), γ is an equivalence relation on S. To show that γ satisfies the substitution property, consider the elements a, b, c, and d in S such that $a \gamma b$ and $c \gamma d$. By definition of γ, there exist two elements, say x and y, in T such that $\varphi(a) = \varphi(b) = x$ and $\varphi(c) = \varphi(d) = y$. Since φ is a homomorphism, we must have

$$\varphi(a * c) = \varphi(a) \circ \varphi(c) = x \circ y = \varphi(b) \circ \varphi(d) = \varphi(b * d);$$

therefore, $(a * c) \gamma (b * d)$.‖

Example 3.4.1. Consider the homomorphism φ mapping $[N, +]$ to $[P_n, \oplus]$, as given in example 3.3.4. There are n equivalence classes $E_0, E_1, \ldots, E_{n-1}$ corresponding to the equivalence relation $\gamma = \varphi\varphi^{-1}$, where for $0 \leqslant i < n$, $E_i \equiv \{p \mid p = nl + i, l = 0, 1, \ldots\}$. Let p and p' be elements in E_i, and let q and q' be elements in E_j, namely, $p \gamma p'$ and $q \gamma q'$. We have the following identities:

$$\left. \begin{array}{l} p = nl + i,\, p' = nl' + i,\, \text{for some } l \text{ and } l' \\ q = nm + j,\, q' = nm' + j,\, \text{for some } m \text{ and } m' \end{array} \right\} \tag{3.4.2}$$

If $i + j = k < n$, then we obtain from (3.4.2)

$$p + q = n(l + m) + k \text{ and } p' + q' = n(l' + m') + k,$$

whence both $(p + q)$ and $(p' + q')$ are in E_k; that is, $(p + q) \gamma (p' + q')$.

On the other hand, if $i + j \geqslant n$, then

$$i + j = ns + k', \text{ for some } 0 \leqslant k' < n.$$

In this case, we see that

$$p + q = n(l + m + s) + k' \text{ and } p' + q' = n(l + m' + s) + k',$$

and we conclude that $p + q$ and $p' + q'$ are both in the class $E_{k'}$; again, $(p + q) \gamma (p' + q')$.

It is natural for us to inquire at this point whether the converse of lemma 3.4.1 is also true, namely, whether any congruence relation on a semigroup $G = [S, *]$

always induces a homomorphism from G to some other semigroup. Clearly, the validity of this statement must be shown constructively by exhibiting a homomorphism which in turn induces the given congruence relation on G. To this end we need to construct two items: a new semigroup G' and a homomorphism from G to G'. A semigroup G', in turn, requires a set. Let us choose as this set the partition C_γ induced by γ on G, that is, the set of the equivalence classes of γ (definition 1.5.7). Let us use the symbol $E(a)$ to denote the unique equivalence class of C_γ containing the element $a \in S$. We observe that this notation is independent of the choice of a because it represents the set of all elements which are related to a via γ. Thus, if $a\,\gamma\,b$, then $E(a) = E(b)$.

With the set C_γ given, what is a meaningful binary operation on C_γ so that the resulting semigroup will be a homomorphic image of G? By the fact that γ has the substitution property with respect to $*$, we see that $a\,\gamma\,b$ and $c\,\gamma\,d$ imply $(a * c)\,\gamma\,(b * d)$, which, in turn, implies that all the elements that are the products of elements of two equivalence classes must be contained in the same equivalence class. In other words, there exists a *unique* equivalence class $E(c)$ corresponding to each pair of equivalence classes $E(a)$ and $E(b)$ such that the following set inclusion holds:

$$\{a' b' | a' \in E(a) \text{ and } b' \in E(b)\} \subseteq E(c). \tag{3.4.3}$$

Using example 3.4.1 as an illustration, we see that

$$\{a + b | a \in E_i \text{ and } b \in E_j\} \subseteq E_{i+j(\bmod n)}.$$

The preceding discussion indicates that the natural way to define a binary operation $\circ$ on the set C_γ is as follows:

$$E(a) \circ E(b) = E(a * b).$$

We see that $\circ$ is indeed a binary operation because $E(a * b)$ is the unique class satisfying the relationship above. This leads to the following definition.

Definition 3.4.2. Let γ be a congruence relation on a semigroup $G = [S,*]$. The *quotient* semigroup of G modulo γ, denoted by G/γ, is defined to be the system $[C_\gamma, \circ]$ such that

$$E(a) \circ E(b) = E(a * b), \tag{3.4.4}$$

where $E(a)$ is the unique equivalence class under γ containing $a \in S$.

To verify that G/γ is a semigroup, it is necessary to show that the binary operation $\circ$ has the associative property. But this follows immediately from (3.4.4) since

$$(E(a) \circ E(b)) \circ E(c) = E(a * b) \circ E(c) = E((a * b) * c)$$
$$= E(a * (b * c)) = E(a) \circ E(b * c) = E(a) \circ (E(b) \circ E(c)).$$

Example 3.4.2. Consider γ defined in example 3.4.1. We see that for $n = 5$, the equivalence classes are $E_i(i = 0, 1, \ldots, 4)$.

$$E_0 = \{0, 5, 10, 15, \ldots\}$$
$$E_1 = \{1, 6, 11, 16, \ldots\}$$
$$E_2 = \{2, 7, 12, 17, \ldots\}$$
$$E_3 = \{3, 8, 13, 18, \ldots\}$$
$$E_4 = \{4, 9, 14, 19, \ldots\}$$

The operation on G/γ is given by its table in figure 3.4.2.

$\circ$	E_0	E_1	E_2	E_3	E_4
E_0	E_0	E_1	E_2	E_3	E_4
E_1	E_1	E_2	E_3	E_4	E_0
E_2	E_2	E_3	E_4	E_0	E_1
E_3	E_3	E_4	E_0	E_1	E_2
E_4	E_4	E_0	E_1	E_2	E_3

Figure 3.4.2. Operation table of the semigroup P_5

Continuing our discussion of the converse of lemma 3.4.1, we see that any congruence relation γ on a semigroup G unveils another semigroup G/γ. We next question whether there is a homomorphism from G to G/γ. Since elements of G/γ are equivalence classes in G, a natural choice for a function φ_γ from G to G/γ is the "membership function" $E(a)$ [membership of a in its equivalence class $E(a)$], that is,

$$\varphi_\gamma(a) = E(a). \tag{3.4.5}$$

Since, for any $a \in S$, $E(a)$ is unique, φ_γ is a function. Furthermore, from (3.4.4) and (3.4.5)

$$\varphi_\gamma(a * b) = E(a * b) = E(a) \circ E(b) = \varphi_\gamma(a) \circ \varphi_\gamma(b)$$

by definition of the operation $\circ$ on G/γ. This means φ_γ is indeed a homomorphism from G to G/γ. Thus we have the following result.

Lemma 3.4.2. Every congruence relation γ on a semigroup G induces a homomorphism φ_γ from G to the quotient semigroup G/γ.

The function φ_γ is usually referred to as the *natural homomorphism* from G to G/γ.

Lemmas 3.4.1 and 3.4.2 give us a characterization of congruence relations, which we summarize in the following theorem.

Theorem 3.4.3. (Characterization theorem of congruence relations) An equivalence relation γ on a set S is a congruence relation on the semigroup $G = [S, *]$ if and only if there exists another semigroup G' and a homomorphism φ from G to G' such that $\gamma = \varphi\varphi^{-1}$.

We remark here that theorem 3.4.3 is very similar in its formulation to theorem 1.5.7. The additional feature of the substitution property is reflected by the fact that the function φ is a homomorphism.

We now recall theorem 3.3.3, which states that the homomorphic image of a subsemigroup is itself a semigroup. This result together with the characterization theorem of congruence relations gives us the following fundamental result.

Theorem 3.4.4. (Fundamental theorem of homomorphism of semigroups) If φ is a homomorphism from a semigroup $G = [S, *]$ to another semigroup $G' = [T, \cdot]$ and γ is the congruence relation $\varphi\varphi^{-1}$, then G' is isomorphic to the quotient semigroup $G/\gamma = [C_\gamma, \circ]$.

Proof. We must show that we can construct a function

$$\psi : G/\gamma \to G'$$

which satisfies the requirements of definition 3.3.1. We define

$$\psi(E(a)) = \varphi(a); \tag{3.4.6}$$

that is, ψ maps the equivalence class $E(a)$ containing a to the same element of G' to which φ maps a. First of all, ψ is onto because φ is so. Second, ψ is one-to-one since $E(a) \neq E(b)$ means that a and b are not equivalent under γ; it follows that $\varphi(a) \neq \varphi(b)$ ($\varphi(a) = \varphi(b)$ would imply $a\,\gamma\,b$) which is equivalent to $\psi(E(a)) \neq \psi(E(b))$. Thus ψ is invertible. Finally we have

$$\psi(E(a) \circ E(b)) = \psi(E(a * b)) = \varphi(a * b) = \varphi(a) \cdot \varphi(b) = \psi(E(a)) \cdot \psi(E(b))$$

where the first identity follows from (3.4.4), the second from (3.4.6), the third from (3.3.1), and the fourth from (3.4.6). This completes the proof that ψ is indeed an isomorphism.‖

The relationship between various items mentioned in theorem 3.4.4 can be illustrated by the diagram in figure 3.4.3.

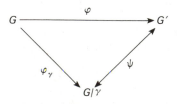

Figure 3.4.3. Diagrammatic illustration of the fundamental theorem of homomorphism of semigroups

---

EXERCISES 3.4

1. Determine which of the following relations are congruence relations, and for each congruence relation found, describe the resulting set of congruence classes.

   a) On the semigroup $[I,+]$ of integers under addition, define γ as follows:

      i) $a\,\gamma\,b$ if and only if $a - b$ is a prime
      ii) $a\,\gamma\,b$ if and only if $a - b$ is even
      iii) $a\,\gamma\,b$ if and only if $a - b$ is odd
      iv) $a\,\gamma\,b$ if and only if $a - b$ is a multiple of 5

   b) Let $A \equiv \{0,1\}$ and A^+ be the semigroup consisting of all finite sequences of elements of A under the catenation operation. Define γ as follows:

      i) $x\,\gamma\,y$ if and only if x and y contain the same number of 1's
      ii) $x\,\gamma\,y$ if and only if both x and y contain either an even or an odd number of 1's
      iii) $x\,\gamma\,y$ if and only if x contains twice as many 1's as y does

2. Let α and β be two congruence relations on a semigroup. Show that $\alpha \cap \beta$ is also a congruence relation.

3. An equivalence relation γ on a monoid $G = [S,\mathrm{o}]$ is said to be *right invariant* if for all a, b, and c in G,

$$a\,\gamma\,b \text{ implies } a\,\mathrm{o}\,c\,\gamma\,b\,\mathrm{o}\,c.$$

   Similarly, γ is *left invariant* if

$$a\,\gamma\,b \text{ implies } c\,\mathrm{o}\,a\,\gamma\,c\,\mathrm{o}\,b.$$

   Show that a congruence relation γ is both a left invariant and a right invariant equivalence relation. Is the converse also true?

*4. Let θ and ρ be two congruence relations on a semigroup G such that $R_\theta \subseteq R_\rho$. Show that there is a unique homomorphism φ from G/θ to G/ρ. (*Hint*: $\varphi(E_a^{(\theta)}) = E_a^{(\rho)}$, where $E_a^{(\theta)}$ is the θ-congruence class containing a.)

5. Show that if G' is a subsemigroup of G and θ is a congruence relation on G, then θ restricted to G' is a congruence relation on G'.

6. Let $\varphi:G_1 \to G_2$ and $\psi:G_2 \to G_3$ be two given homomorphisms; $\varphi\psi$ is a homomorphism from G_1 to G_3. What is the relation between the congruence relations $\varphi\varphi^{-1}$ and $(\varphi\psi)(\varphi\psi)^{-1}$?

3.5 GROUPS

A very familiar problem in the algebra of real numbers is the solution of equations. The simplest kind of problem involves equations of the form $ax = b$ or $a + x = b$, where a and b are known quantities and x is an unknown quantity that makes an identity of the equation. In the process of obtaining the solution of the equation $ax = b$, we assume, at each step, that a, b, and x are elements of an algebraic structure G for which the equality holds, and we apply transformations to both sides which are allowed by the properties of G so that, eventually, we shall be able to express the unknown x in terms of the knowns a and b. This is what we call "solving the equation."

It would be very desirable to be able to solve all equations of the form $ax = b$ involving a binary operation. However, this is not always possible, as demonstrated by the binary operation "composition" of functions on the set $\{f_1, f_2, f_3, f_4\}$ in example 3.3.3. The equation is solvable if we restrict a, b, and x to belonging to the subset $\{f_1, f_2\}$ rather than to the entire set. We observe that this subset is a monoid and that each element has an inverse. Are these conditions sufficient for the existence of a solution to the equation $ax = b$? Let us look at one more example by considering the equation $3x = 4$. Each step in the process of obtaining the solution of this equation is illustrated in the following sequence.

$$3x = 4 \qquad \text{given equation}$$
$$\tfrac{1}{3}(3x) = \tfrac{1}{3}4 \qquad \text{by multiplying both sides by the same quantity } \tfrac{1}{3}$$
$$(\tfrac{1}{3}3)x = \tfrac{1}{3}4 \qquad \text{by associativity}$$
$$1x = \tfrac{1}{3}4 \qquad \text{by the property of inverses}$$
$$x = \tfrac{4}{3} \qquad \text{by the property of the identity 1}$$

The example above calls for the existence of identity and inverse. We now define some extremely important algebraic structures.

Definition 3.5.1. A *group* G is a monoid $[S, *]$ such that each element of S has a unique inverse with respect to $*$ in S. If S is finite, we say that G is a *finite* group and call $\#(S)$ the *order* of G.

We are now ready to demonstrate the existence and the uniqueness in groups of the solution of equations of the form $ax = b$ and $ya = b$.

Theorem 3.5.1. If a, b are elements of a group $G = [S, *]$, then the equations $ax = b$ and $ya = b$ have unique solutions in $x = a^{-1}b$ and $y = ba^{-1}$.

Proof. Since $a(a^{-1}b) = (aa^{-1})b = 1b = b$, we conclude that $x = a^{-1}b$ is a solution of $ax = b$. On the other hand, from $ax = b$ we obtain $x = 1x = (a^{-1}a)x = a^{-1}(ax) = a^{-1}b$; that is, $a^{-1}b$ is the unique solutions of $ax = b$. Similarly, $y = ba^{-1}$ is the unique solution of $ya = b$.$\|$

A result concerning unique inverses for elements of a monoid translates into the following property of groups.

Theorem 3.5.2. Elements in a group $G = [S, *]$ satisfy the left and right cancellation laws; that is, for a, b, and c in S, $ab = ac$ implies $b = c$, and $ba = ca$ implies $b = c$.

Proof. If $ab = ac$, then by multiplying both sides on the left by the element a^{-1}, we have $b = 1b = (a^{-1}a)b = a^{-1}(ab) = a^{-1}(ac) = (a^{-1}a)c = 1c = c$. Hence the left cancellative law holds. Similarly, one can deduce $b = c$ from $ba = ca$ by multiplying both sides on the right by a^{-1}.$\|$

Example 3.5.1. Consider the symmetries of an equilateral triangle, as shown in figure 3.5.1. All symmetries of this triangle arise from two types of operations, namely, rotations by a multiple of 120° and reflections with respect to the lines L_1, L_2, and L_3. Since either a rotation or a reflection along one of the lines

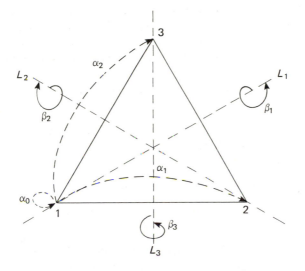

Figure 3.5.1. Symmetries of an equilateral triangle

represents a change of position for the three points 1, 2, and 3, we may denote rotations (counterclockwise) and reflections by the functional notations illustrating the images of the three points 1, 2, and 3 as the result of the transformation. So we have: $\alpha_0 = \binom{123}{123}$, no transformation; $\alpha_1 = \binom{123}{231}$, a rotation of 120°; $\alpha_2 = \binom{123}{312}$, a rotation of 240°; $\beta_1 = \binom{123}{132}$, a reflection with respect to L_1; $\beta_2 = \binom{123}{321}$, a reflection with respect to L_2; $\beta_3 = \binom{123}{213}$, a reflection with respect to L_3. It is easily seen that all symmetries of the triangle are included in the six given above by composition. Furthermore, the six elements form a group S_3 under

∘	α_0	α_1	α_2	β_1	β_2	β_3
α_0	α_0	α_1	α_2	β_1	β_2	β_3
α_1	α_1	α_2	α_0	β_2	β_3	β_1
α_2	α_2	α_0	α_1	β_3	β_1	β_2
β_1	β_1	β_3	β_2	α_0	α_2	α_1
β_2	β_2	β_1	β_3	α_1	α_0	α_2
β_3	β_3	β_2	β_1	α_2	α_1	α_0

Figure 3.5.2. Operation table of the group S_3

composition (of functions); the operation table is given in figure 3.5.2. This group, S_3, is called the *symmetric group* of degree 3 (see Note 3.2).

The attentive reader may have noticed already that the elements of S_3 are exactly the six permutations on the three integers $\{1,2,3\}$. Is this accidental? Certainly not. Indeed, not only do the permutations of a finite set form a finite group, as shown by the preceding example, but every finite group is isomorphic to a group of permutations. We have now arrived at the Cayley representation theorem of groups.

Theorem 3.5.3. (Cayley). Every finite group is isomorphic to a group of permutations.

Proof. The proof is almost identical to the representation theorem of monoids (theorem 3.3.1). Let $G = [S, *]$ be a group with identity 1. Consider the set of functions $F \equiv \{f_a | a \in S\}$ defined on S by the rule that for each $b \in S$, $f_a(b) = ba$. It follows from theorem 3.3.1 that $[F, o]$ is a monoid with the identity e_S, where o is the operation of composition of functions. To show that each element f_a in F has an inverse, we note that

$$e_S = f_1 = f_{aa^{-1}} = f_a f_{a^{-1}},$$

where the first and third equalities follow from theorem 3.3.1 and the second follows from $aa^{-1} = 1$. Hence $f_{a^{-1}}$ is an inverse of f_a in F, and since inverses are unique in a monoid (theorem 3.2.2), $f_{a^{-1}}$ is the inverse of f_a, and F is a group. By defining the mapping $\varphi : G \to F$ such that $\varphi(a) = f_a$, again, by theorem 3.3.1, φ is an isomorphism. It remains to show that the members of F are permutations, i.e., that an arbitrary column of the operation table of G contains no two identical elements. Arguing by contradiction, assume that, for a, b, and c in S, $(b \neq c)$, $f_a(b) = f_a(c)$; by definition of f_a, this implies $ba = ca$, whence, by cancellation, $b = c$, a contradiction.‖

Given the group $G = [S, *]$, the group $[F, o]$, with $F \equiv \{f_a | a \in S\}$ and $f_a(b) = ba$ for $b \in S$, is called the *right regular representation* of G; the adjective "right" refers to the fact that $f_a(b)$ is defined by multiplying b on the right by the fixed a.

$*$	1	a	b
1	1	a	b
a	a	b	1
b	b	1	a

(a)

o	f_1	f_a	f_b
f_1	f_1	f_a	f_b
f_a	f_a	f_b	f_1
f_b	f_b	f_1	f_a

(b)

Figure 3.5.3. Operation tables of a group and its regular right representation

Example 3.5.2. We compute the right regular representation of the group specified by its operation table in figure 3.5.3(a). The functions f_1, f_a, and f_b are as follows:

$$f_1 = \begin{pmatrix} 1\,a\,b \\ 1\,a\,b \end{pmatrix}, \qquad f_a = \begin{pmatrix} 1\,a\,b \\ a\,b\,1 \end{pmatrix}, \qquad f_b = \begin{pmatrix} 1\,a\,b \\ b\,1\,a \end{pmatrix}$$

From the operation table of the composition operation in figure 3.5.3(b) it is easily seen that this group is isomorphic to the given one.

Again the reader may have noticed that there are groups contained in larger groups, such as the group $[I,+]$, which is contained in the group $[R,+]$ where I and R are the sets of integers and of rational numbers, respectively. The concept of *subgroups* will now be given precisely.

Definition 3.5.2. A group $G = [S,*]$ is said to be a *subgroup* of another group $G' = [T,\circ]$ if S is a subset of T and $*$ is the restriction of $\circ$ to S.

It follows from this definition that any subset T' of T which is closed under the operation $\circ$ is a subgroup of $G' = [T,\circ]$. We also note that every group G has two *trivial subgroups* in itself and in the subgroup consisting of identity element alone. All other subgroups of G are called *proper subgroups*.

Example 3.5.3. Let a be an element of a group G. The *cyclic subgroup* G' of G generated by a consists of all the integral powers, a^i, of a, and for $i = 0$, a^0 is defined as the identity element of G. To show that G' is a group, we observe that $a^i a^j = a^{i+j}$; hence G' is a submonoid of G. Furthermore, for each $a^i \in G'$, we have $a^i a^{-i} = a^0 = 1$. So $(a^i)^{-1} = a^{-i}$, and we conclude that G' is indeed a subgroup of G.

It is convenient to have a criterion for determining whether or not a subset of a group G is a subgroup of G. The following theorem gives such a criterion.

Theorem 3.5.4. A nonempty subset T of a group G is a subgroup of G if and only if, for every pair of elements a and b in T, ab^{-1} is also in T.

Proof. If T is a subgroup of G, then clearly $a, b \in T$ implies that $b^{-1} \in T$ and hence $ab^{-1} \in T$.

Suppose now that, for any a and b in T, we know that $ab^{-1} \in T$. Letting $b = a$, the condition $ab^{-1} \in T$ becomes $aa^{-1} \in T$. But $aa^{-1} = 1$; that is, T contains the identity element 1. It follows that, for each element $a \in T$, $1a^{-1} = a^{-1}$ is also in T. Finally, we show that T is closed under the operation of G; indeed, for any two elements a and b in T, also b^{-1} is in T; but $(b^{-1})^{-1} = b$ (because of the uniqueness of inverses), whence $a(b^{-1})^{-1} = ab \in T.\|$

It follows directly from this theorem that if G' is a subgroup of a group G, the identities of G' and G coincide, and for each element g in G', its inverses in G' and G coincide.

Going back to example 3.5.1 on the symmetries of an equilateral triangle, we see that if we restrict ourselves to rotations, the performance of any sequence of rotations on the triangle is also a rotation. This fact is formally expressed by saying that $\{\alpha_0, \alpha_1, \alpha_2\}$—the set of rotations—is a subgroup of the symmetric group of degree 3. Moreover, if we perform an arbitrary rotation followed by a fixed reflection, say β_2, the resulting transformation of the triangle is a reflection. Specifically, in the operation table in figure 3.5.2, the set $\{\beta_1, \beta_2, \beta_3\}$ of the reflections, which is not a subgroup, is obtained by multiplying the elements of the subgroup $\{\alpha_0, \alpha_1, \alpha_2\}$ by an arbitrary element β_j ($j = 1, 2, 3$): in other words, the sets $\{\alpha_0, \alpha_1, \alpha_2\}$ and $\{\beta_j\alpha_0, \beta_j\alpha_1, \beta_j\alpha_2\}$ are a partition of $\{\alpha_0, \alpha_1, \alpha_2, \beta_1, \beta_2, \beta_3\}$. Is this phenomenon accidental or is it a general property of groups? The answer to this interesting structural question is provided by the following lemma. From the viewpoint of notation, for a given subgroup G_1 of G and an element $g \in G$, we let $G_1 g$ denote the subset $\{ag \mid a \in G_1\}$ and $g G_1$ denote the subset $\{ga \mid a \in G_1\}$.

Lemma 3.5.5. Let G_1 be a subgroup of a group G. Then the family of all distinct subsets of G of the form $G_1 g$, $g \in G$, is a partition of G.

Proof. Since $1 \in G_1$ and $G_1 = G_1 1$, we see that every element g in G belongs to some set $G_1 g$. We need to show that if $G_1 g$ and $G_1 h$ have one element in common, they are identical. Suppose that g' is common to both $G_1 g$ and $G_1 h$. This implies that there exist g_1 and g_2 in G_1 such that $g' = g_1 g = g_2 h$. This in turn implies that $g = g_1^{-1} g_2 h$. Now let g'' be any element in $G_1 g$; by definition of $G_1 g$, for some $g_3 \in G_1$ we have $g'' = g_3 g = g_3 g_1^{-1} g_2 h$. Since g_3, g_1, and g_2 are elements of G_1, which is a group, $g_3 g_1^{-1} g_2 \in G_1$. But this implies that $g'' \in G_1 h$. Hence we conclude that $G_1 g \subseteq G_1 h$. Similarly, we can also conclude that $G_1 h \subseteq G_1 g$, and hence $G_1 h = G_1 g$. This demonstrates that the sets $\{G_1 g \mid g \in G\}$ indeed form a partition of G.||

The subset $G_1 g$ in the lemma above is called a *right coset* of G_1. We observe that the lemma is also true if we replace $G_1 g$ by $g G_1$, $g \in G$; the set $g G_1$ is called a *left coset* of G_1. We also observe that cosets of G_1 have the same number of elements as G_1. In fact, $\#(G_1 g) \leqslant \#(G_1)$; if $\#(G_1 g) < \#(G_1)$, we must have $g_1 g = g_2 g$ for some $g \in G$ and two distinct elements g_1 and g_2 in G_1. But this implies that $g_1 = g_2$ by cancellation, a contradiction.

Example 3.5.4. Consider the group P_6 specified by its operation table in figure 3.5.4. The subset $G_1 \equiv \{0, 3\}$ forms a subgroup of P_6. The cosets of G_1 are sets $\{2, 5\} = G_1 2 = G_1 5$ and $\{1, 4\} = G_1 1 = G_1 4$, as shown in the table of figure 3.5.4.

The preceding lemma, when applied to finite groups, gives rise to the following important decomposition theorem discovered by Lagrange.

Theorem 3.5.6. If G is a finite group and H a subgroup of G, the order of H divides the order of G.

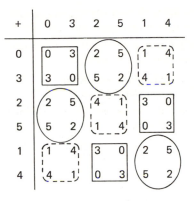

Figure 3.5.4. Operation table of P_6

Proof. Let the orders of G and H be n and m, respectively. Suppose that there are k right cosets of H. Since each right coset has the same number of elements, we conclude by lemma 3.5.5 that $n = km$. Hence m divides n. ‖

Referring to example 3.5.4, we see that the order of the group is 6 and the order of the subgroup G_1 is 2. Clearly 2 divides 6.

An important application of the partition of a group generated by one of its subgroups consists of the so-called *linear codes*, which will be discussed in Note 3.3.

_____

EXERCISES 3.5

1. Construct all the distinct tables of the binary operations on the set {1,2}. (There are 16 of them.)
   a) Select those tables which define semigroups, monoids, groups (see exercises 3.1.1 and 3.2.1).
   b) How many groups are there? Are they isomorphic?
2. Repeat exercise 3.2.3 after replacing the phrase "semigroup or monoid" by the word "group."
3. Determine the truth or falsehood of the following statements.
   a) A group is a monoid.
   b) A group with only three elements is commutative.
   c) The empty set ∅ can be considered a group.
   d) A group may have more than one identity.
   e) A group is a cancellative semigroup.
   f) Every group has at least one proper subgroup.
   g) Every subgroup of a subgroup of a group G is a subgroup of G.
   h) Every subset S' of a group G is a subgroup of G under the operation of G restricted to S'.

4. Find the group generated by the symmetries of a square by following the reasoning used in example 3.5.1.

5. Show that the identity of a group is its only idempotent element (see exercise 3.2.4 for the definition of an idempotent element).

6. Show that if in a semigroup G the equations $ax = b$ and $ya = b$ always have a (common) solution in G, then G is a group.

7. Show that if in a group G the equation $x^2 = 1$ holds for every element x in G, then G is commutative.

3.6 HOMOMORPHISM OF GROUPS

In this section we consider the interesting problem of the homomorphism of groups. Although the development is largely parallel to that of Section 3.3 concerning semigroups and monoids, the structure of groups gives rise to additional interesting properties.

Let φ be a homomorphism from a group G to another group G', and let $1'$ be the identity of G'. What are the properties of $\varphi^{-1}(1')$, the inverse homomorphic image of the identity $1'$? By lemma 3.3.2, we see that $\varphi^{-1}(1')$ is a submonoid of G. To show that it is a subgroup, we observe that $a \in \varphi^{-1}(1')$ implies that $a^{-1} \in \varphi^{-1}(1')$ because

$$\varphi(a^{-1}) = 1' \, \varphi(a^{-1}) = \varphi(a) \, \varphi(a^{-1}) = \varphi(aa^{-1}) = \varphi(1) = 1',$$

where the first equality follows from the fact that $1'$ is the identity in G', the second follows from $\varphi(a) = 1'$, the third is due to the homomorphism, the fourth follows from $aa^{-1} = 1$ and the fifth follows from lemma 3.3.2. The subgroup $\varphi^{-1}(1')$ is called the *kernel* of the homomorphism φ. We also observe that $\varphi^{-1}(1')$ has the special property that each left coset $g\varphi^{-1}(1')$ coincides with the right coset $\varphi^{-1}(1')g$. This is due to the fact that all elements in both cosets $g\varphi^{-1}(1')$ and $\varphi^{-1}(1')g$ have identical homomorphic image $\varphi(g)$ since

$$\varphi(g\varphi^{-1}(1')) = \varphi(g) \, \varphi(\varphi^{-1}(1')) = \varphi(g) \, 1' = \varphi(g) = \varphi(\varphi^{-1}(1')g).$$

Definition 3.6.1. A subgroup G' of a group G is called a *normal subgroup* of G if $gG' = G'g$ for every g in G.

It follows from this definition that if G' is a normal subgroup of G, then for any element g in G, there is a pair of elements g' and g'' in G' such that $gg' = g''g$. This fact is illustrated in example 3.5.4, in which G_1 is a normal subgroup of G. Example 3.5.4 also illustrates that all elements in the same coset are transformed into elements of the same coset after multiplication on the right or left by an arbitrary element of G. In other words, the equivalence relation γ defined by the partition $\{\{0,3\}, \{2,5\}, \{1,4\}\}$ satisfies the substitution property. Indeed, $a \, \gamma \, b$ means that $a = a'c$ and $b = b'c$ for some a', b' in G', and c in G. Since for any element d in G, $ad = a'(cd)$ and $bd = b'(cd)$, we see that $ad \, \gamma \, bd$. Similarly, $da \, \gamma \, db$.

Therefore, γ is indeed a congruence relation on G. Since G is certainly a monoid, G/γ is a monoid because γ is a congruence relation. Is it a group? We recall that the operation defined on G/γ as a monoid is such that for any two elements $G'g_1$ and $G'g_2$ in G/γ, $(G'g_1)(G'g_2) = G'(g_1g_2)$. Hence $(G'g_1)(G'g_1^{-1}) = G'$, which is the identity of G/γ. Therefore we conclude that G/γ is indeed a group, which is called the *quotient group of G modulo* γ.

The preceding discussion is summarized in the following characterization theorem of normal subgroups.

Theorem 3.6.1. Let G be a group. The following statements are equivalent.

1. G' is a normal subgroup of G.

2. The right cosets of G', $\{G'g \,|\, g \in G\}$, define a congruence relation γ on G such that G' coincides with the congruence class containing 1.

3. G' is the kernel of some homomorphism defined on G.

Proof. We will prove the theorem using cyclic implications. To prove that statement 1 implies statement 2, we define the relation γ on G such that $g \gamma g'$ if and only if g and g' are in the same right coset of G'. Clearly, γ is an equivalence relation. To show that it is a congruence relation, we observe that $a \gamma b$ means that $a = a'c$ and $b = b'c$ for some a', b' in G' and c in G. The fact that G' is normal implies that there exist a'' and b'' in G' such that $a = a'c = ca''$ and $b = b'c = cb''$. Thus, if d is any element in G, then

$$da = dca'' \quad \text{and} \quad db = dcb''.$$

Again, by the fact that G' is normal, there exist a''' and b''' in G' such that

$$da = dca'' = a'''(dc), \quad \text{and} \quad db = dcb'' = b'''(dc).$$

But then both da and db are elements of the right coset $G'(dc)$, and hence $da \gamma db$. Similarly, $a \gamma b$ implies $ad \gamma bd$ for any d in G. Hence γ is indeed a congruence relation. Since $1 \in G' = G'1$, we see that G' coincides with the congruence class containing 1.

Assume now that statement 2 holds. Let $\varphi_\gamma : G \to G/\gamma$ be the natural homomorphism from G to G/γ, defined by

$$\varphi_\gamma(g) = G'g.$$

Since $\varphi_\gamma(1) = G'1 = G'$, which is the identity in G/γ, we see that G' is the kernel of φ_γ.

Finally, assuming that statement 3 holds, let φ be a homomorphism from G to another group G'' such that $G' = \varphi^{-1}(1'')$. By lemma 3.3.2, G' is a submonoid of G. Since for any $a, b \in G'$, $\varphi(ab^{-1}) = \varphi(a)\varphi(b)^{-1} = 1''1'' = 1''$, so $ab^{-1} \in G'$, and hence G' is a subgroup of G'' (theorem 3.5.4). Furthermore, it is a normal subgroup because, for any $g \in G$,

$$\varphi(G'g) = \varphi(gG') = \varphi(g). \|$$

This theorem induces the following characterization of congruence relations on groups.

Corollary 3.6.2. A relation γ is a congruence relation on a group G if and only if there is a homomorphism φ from G to another group G' such that $\gamma = \varphi\varphi^{-1}$.

Example 3.6.1. Consider the mapping φ from the set I of the integers to the set $P_n \equiv \{0, 1, \ldots, n-1\}$ defined as follows: $\varphi(i)$ is the remainder of the division of i by n (that is, $i = qn + r$ with $r < n$, and $\varphi(i) = r$ is the *residue of i modulo n*). The reader will convince himself that both $[I, +]$ and $[P_n, \oplus]$ (where $\oplus$ denotes "modulo n addition") are groups. Thus, by an argument identical to the one developed in example 3.3.4, we conclude that φ is a group homomorphism from $[I, +]$ to $[P_n, \oplus]$.

We conclude this section with the following important theorem.

Theorem 3.6.3. (Fundamental theorem of homomorphism of groups.) Let φ be a homomorphism from a group G to a group G', and let γ be the congruence relation $\varphi\varphi^{-1}$. Then $\varphi(G)$ is a group, and there is an isomorphism ψ between G/γ and $\varphi(G)$.

Proof. If we show that both G/γ and $\varphi(G)$ are groups, the results follows from theorem 3.4.4, the fundamental theorem of homomorphism of semigroups. We know already that G/γ is a group. To prove that $\varphi(G)$ is a group, we must show that for each $a' = \varphi(a) \in \varphi(G)$, also $(a')^{-1}$ is in $\varphi(G)$. To this end, we observe that, by theorem 3.3.3, since G is a monoid, $\varphi(G)$ is also a monoid. Furthermore,

$$1' = \varphi(1) = \varphi(aa^{-1}) = \varphi(a)\varphi(a^{-1});$$

that is, $\varphi(a^{-1})$ is the unique inverse of $\varphi(a)$ since $\varphi(G)$ is a monoid (theorem 3.3.3). It follows that $\varphi(a)^{-1} = \varphi(a^{-1}) \in \varphi(G)$; that is, $\varphi(G)$ is a group.$\|$

EXERCISES 3.6

1. Let G be the group of integers under addition. Give two normal subgroups of G.

2. An isomorphism between a semigroup and itself is called an *automorphism*. Show that the set of all automorphisms of a semigroup forms a group under composition.

3. Show that subgroups of a cyclic group are also cyclic groups and, furthermore, that each such subgroup is normal.

4. Show that any subgroup of a commutative group is normal.

5. Show that both the homomorphic and the inverse homomorphic images of a group are groups (see theorem 3.3.3).

6. Show that the right cosets of a subgroup G' of a group G are equivalence classes of a right invariant equivalence relation (see exercise 3.4.3) on G.

7. Consider the cyclic group G of order 15.
   a) Show that there exists a normal subgroup of order 5. Find the quotient group of G with respect to this subgroup.
   b) Find all subgroups and all other normal subgroups of this group.
8. Let G be the direct product of two cyclic groups of order 4. Determine all homomorphisms of G to cyclic subgroups of the cyclic group of order 8. (See exercise 3.3.5.)

3.7 RINGS

The structures we have been discussing so far in this chapter are sets with one binary operation. If we now take, for example, the set of the even integers, we see that two structures are actually recognizable: a commutative group under addition and a semigroup under multiplication. These two structures are related by the fact that multiplication distributes over addition; that is, $a(b + c) = ab + ac$. Note that the distributive property is important, because otherwise, the additive and multiplicative structures would become totally unrelated. The structure of even integers under both addition and multiplication is called a *ring*.

Definition 3.7.1. A *ring* R is a system $[S, +, \circ]$, where the pair $[S, +]$ is a commutative group and the pair $[S, \circ]$ is a semigroup such that, for arbitrary elements a, b, and c in S, the following two distributive laws hold:

$$a(b + c) = ab + ac \qquad (b + c)a = ba + ca$$

For convenience, in this section we will refer to $[S, +]$ and $[S, \circ]$ as the *additive* and *multiplicative* structures of the ring R. The same distinction of terminology will carry over to the notions of identities and inverses. Thus 0—the *additive* identity—is the identity in the group $[S, +]$, and 1—the *multiplicative* identity—if it exists, is the identity in the semigroup $[S, \circ]$. Similarly, for $a \in S$, $(-a)$ and a^{-1} denote the inverses (if they exist) of a in $[S, +]$ and in $[S, \circ]$, respectively. Note, however, that the relationship between the ring operations and conventional addition and multiplication does not extend, in general, beyond the terminology.

In the ring of even integers, the multiplication operation is also commutative, and we say the ring is a *commutative ring*. A noncommutative ring is presented in the following example.

Example 3.7.1. Let I be the set of integers, and let M denote the set of all 2×2 arrays with entries in N. We define two operations, $*$ and $\circ$, on M as follows:

$$\begin{bmatrix} a & c \\ b & d \end{bmatrix} * \begin{bmatrix} e & g \\ f & h \end{bmatrix} = \begin{bmatrix} a+e & c+g \\ b+f & d+h \end{bmatrix}$$

$$\begin{bmatrix} a & c \\ b & d \end{bmatrix} \circ \begin{bmatrix} e & g \\ f & h \end{bmatrix} = \begin{bmatrix} ae+cf & ag+ch \\ be+df & bg+dh \end{bmatrix}$$

It is easily seen that $[M, *]$ is a commutative group with the identity $[\begin{smallmatrix} 0 & 0 \\ 0 & 0 \end{smallmatrix}]$, and that $[M, \circ]$ is a semigroup. Distributive laws hold in $[M, *, \circ]$ since they hold in $[\,I, +, \cdot\,]$. Finally, M is not a commutative ring since, for example,

$$\begin{bmatrix} 2 & 2 \\ 2 & 2 \end{bmatrix} \circ \begin{bmatrix} 3 & 3 \\ 4 & 2 \end{bmatrix} = \begin{bmatrix} 14 & 10 \\ 14 & 10 \end{bmatrix} \neq \begin{bmatrix} 12 & 12 \\ 14 & 14 \end{bmatrix} = \begin{bmatrix} 3 & 3 \\ 4 & 2 \end{bmatrix} \circ \begin{bmatrix} 2 & 2 \\ 2 & 2 \end{bmatrix}$$

It follows from the definition of a ring that many of the more elementary properties of a ring $R = [S, +, \circ]$ are already known through its group and semigroup structures. For example, the equation $a + x = b$ has the unique solution $x = (-a) + b$. Some less trivial properties of rings are given by the following lemma.

Lemma 3.7.1. If $R = [S, +, \circ]$ is a ring with additive identity 0, then for any $a, b \in R$, we have

1) $0a = a0 = 0$
2) $(-a)b = a(-b) = -ab$
3) $(-a)(-b) = ab$

Proof.

1) Consider the sequence of equalities

$$a0 = a(0 + 0) = a0 + a0,$$

of which the first follows from $0 = 0 + 0$ and the second from the distributive property of R. From $a0 = a0 + a0$ we obtain $a0 + a0 = a0 + 0$, whence, by the cancellative property of the additive group $[S, +]$, $a0 = 0$. Likewise, we can conclude that $0a = 0$.

2) Since $(-a) + a = 0$, by definition of inverse, we have

$$(-a)b + ab = ((-a) + a)b = 0b = 0,$$

where use has been made of the distributive property and of result (1). It follows that $(-a)b$ is the additive inverse of ab; that is, $(-a)b = (-ab)$. Similarly, $a(-b) = (-ab)$.

3) Using (2) and the distributive property of R, we have

$$(-a)(-b) + (-ab) = (-a)(-b) + (-a)b = (-a)(-b + b) = (-a)0 = 0.$$

Therefore $(-a)(-b)$ is the additive inverse of $(-ab)$; that is, $(-a)(-b) = ab.\|$

Since a ring has two binary operations, the concept of substructure as well as of structure-preserving maps must be reconsidered. We begin by identifying substructures of a ring.

Definition 3.7.2. A subset T of a ring $R = [S, +, \circ]$ is called a *subring* of R if T is a ring with respect to the additive and multiplicative operations of R.

It follows from this definition that for a subset T of a ring to be a subring, T must be an additive group. That is, for a, $b \in T$, $(a + b)$ must also be in T. Also, T needs to be a multiplicative semigroup. Since associativity is automatic for subsets of a ring, T needs only to satisfy the closure property. This result is stated as a theorem.

Theorem 3.7.2. A subset T of a ring R is a subring of R if and only if $(a + b)$ and ab are in T for every a and b in T.

We have noted for semigroups that a homomorphism is a structure-preserving function. A natural extension is that a ring homomorphism must preserve both the additive and multiplicative structures of the ring as well as distributivity.

Definition 3.7.3. A function φ is said to be a (*ring*) *homomorphism* from a ring R onto another ring R' if for each a, b in R, we have

$$\varphi(a + b) - \varphi(a) + \varphi(b), \qquad \text{and} \qquad \varphi(ab) = \varphi(a)\varphi(b).$$

This definition indicates that a ring homomorphism is both a group homomorphism and a semigroup homomorphism. Note, however, that the preservation of the distributive property is a consequence of definition 3.7.3; in fact,

$$\varphi(a(b + c)) = \varphi(a)\varphi(b + c) = \varphi(a)(\varphi(b) + \varphi(c))$$
$$= \varphi(a)\varphi(b) + \varphi(a)\varphi(c) = \varphi(ab + ac).$$

Thus a ring homomorphism also preserves the distributive property. This together with theorem 3.3.3 and exercise 3.6.5 gives the following result.

Lemma 3.7.3. The homomorphic image and the inverse homomorphic image of a ring are also rings.

Example 3.7.2. We consider again the mapping φ from I to P_n discussed in example 3.6.1 (mapping from the integers to the residue classes modulo n). We know that the mapping φ from $[I, +]$ to $[P_n, \oplus]$ is a group homomorphism. Let $\cdot$ denote the "multiplication of integers" and let $*$ (multiplication modulo n) be defined as the remainder of the division of $i \cdot j$ by n, for i, j in P_n. Clearly, both $[I, \cdot]$ and $[P_n, *]$ are monoids. We claim that φ is a monoid homomorphism from $[I, \cdot]$ to $[P_n, *]$. Indeed, let $i_1 = q_1 n + r_1$ and $i_2 = q_2 n + r_2$; clearly, $\varphi(i_1) = r_1$ and $\varphi(i_2) = r_2$. We now have

$$\varphi(i_1 i_2) = \varphi[(q_1 n + r_1)(q_2 n + r_2)] = \varphi[n(nq_1 q_2 + q_1 r_2 + q_2 r_1) + r_1 r_2] = \varphi(r_1 r_2).$$

Letting $r_1 r_2 = qn + r$, $0 \leqslant r < n$, we have

$$\varphi(i_1 i_2) = \varphi(r_1 r_2) = \varphi(qn + r) = r;$$

on the other hand,

$$\varphi(i_1) * \varphi(i_2) = \varphi(q_1 n + r_1) * \varphi(q_2 n + r_2) = r_1 * r_2 = r.$$

Hence φ is indeed a ring homomorphism.

Although a ring has two structures, the group structure is the dominating one, in general. We saw in the preceding section that the kernel of a group homomorphism also has the group structure. Let φ be a ring homomorphism from R_1 to R_2, and let 0_j be the additive identity of $R_j (j = 1, 2)$. We know that the kernel of φ, $\varphi^{-1}(0_2)$, is an additive subgroup of R_1. Furthermore, for a, $b \in \varphi^{-1}(0_2)$, $\varphi(ab) = \varphi(a)\varphi(b) = 0_2$, and hence $\varphi^{-1}(0_2)$ is indeed a subring of R_1. Moreover, $\varphi^{-1}(0_2)$ satisfies the additional property that for each element $a \in R_1$ and $b \in \varphi^{-1}(0_2)$,

$$\varphi(ab) = \varphi(a)\varphi(b) = \varphi(a)0_2 = 0_2 = 0_2\varphi(a) = \varphi(ba);$$

that is, $a\varphi^{-1}(0_2) = \varphi^{-1}(0_2)a = \varphi^{-1}(0_2)$. A subring with this additional property is referred to as an *ideal*.

Definition 3.7.4. A subring R' of a ring R satisfying the properties $ar \in R'$ and $ra \in R'$ for all a in R and r in R' is called an *ideal* of R.

Does the ring structure cover many of the mathematical structures we are familiar with? Consider the set of the integers under addition and multiplication. One of its most important properties is that the product of two integers can be zero only if at least one of the factors is zero. This fact certainly does not hold for rings in general. For example, consider the ring $R = [S, \oplus, \circ]$, where $S \equiv \{0, 1, 2, 3, 4, 5\}$, $i \oplus j = i + j$ (mod 6) and $i \circ j = i \cdot j$ (mod 6). We see that $2 \circ 3 = 0$ in this ring. In general, if $ab = 0$ in a ring, a and b are called, respectively, the *left* and *right divisors of zero*. With a little reflection, we realize that the characteristic condition for the absence of divisors of zero is that the multiplicative semigroup has the cancellation property. In fact, by the cancellation property, the equalities

$$ab = 0 = a0$$

imply that $b = 0$ and similarly that $a = 0$, a contradiction when both factors are nonzero. Therefore, the integers form a ring whose multiplicative semigroup is, in reality, a cancellative commutative monoid. In other words, it has more structure than a semigroup but not enough to make it a group, since multiplicative inverses do not exist for nonzero integers, except for the identity 1. Certainly, if the multiplicative semigroup becomes a commutative group,[9] it will be an even more powerful structure than strictly required for ruling out divisors of zero. This is indeed true of the rational or real numbers, which belong to the extremely important class of algebraic structures encompassed by the following definition.

Definition 3.7.5. A *field* is a ring in which the nonzero elements form a commutative group under the multiplicative operation.

It is clear that rationals or real numbers under addition and multiplication are fields.

---

9. Very frequently the phrase "commutative group" is replaced by the phrase "abelian group," after the mathematician G. Abel. We shall not use the latter phrase in this text.

_____

EXERCISES 3.7

1. Determine whether each of the following structures is a ring, a field, or neither.

   a) Integers I under addition and multiplication.

   b) Positive integers under addition and multiplication.

   c) Real numbers under addition and multiplication.

2. Let R be a ring. A *polynomial $f(x)$ with coefficients in R* is an infinite sum

$$\sum_{i=0}^{\infty} a_i x^i = a_0 + a_1 x + \cdots + a_k x^k + \cdots,$$

where $a_i \in R$ and $a_i = 0$ for all but a finite number of values of i. If

$$f(x) = \sum_{i=0}^{\infty} a_i x^i \quad \text{and} \quad g(x) = \sum_{i=0}^{\infty} b_i x^i$$

are two polynomials, then we can define two operations, addition $(+)$ and multiplication $(\cdot)$, on $R[x]$, the set of all polynomials with coefficient in R, as follows:

$$f(x) + g(x) = \sum_{i=0}^{\infty} (a_i + b_i) x^i;$$

$$f(x) \cdot g(x) = \sum_{n=0}^{\infty} \left(\sum_{i=0}^{\infty} a_i b_{n-i} \right) x^i.$$

   Show that $R[x]$ is a ring under $+$ and $\cdot$.

3. An element u in a ring R is called a *unit* if it has a multiplicative inverse, i.e., if there exists $a \in R$ such that $ua = au = 1$.

   a) Show that if U is the set of all units in $R = [S, +, \cdot]$, then $[U, \cdot]$ is a group usually referred to as the *unit group* of R.

   b) What is the unit group of the ring of integers?

4. Let $\varphi : R \to R'$ be a ring homomorphism. If 0 and 1 are respectively the zero and identity elements of R, show that $\varphi(0)$ and $\varphi(1)$ are respectively the zero and identity elements of R'.

5. Determine all automorphisms of the ring of integers (see exercise 3.6.2).

6. Show that the cancellation laws for multiplication (see Section 3.2) hold in a ring R if and only if R has no left or right divisors of zero.

_____

*3.8 UNIVERSAL ALGEBRAS

In this book we have considered various algebraic structures. We started with sets which have no structure in Chapter 1, and then in Chapter 2 we augmented their structure by introducing binary relations. In this chapter we have seen that the most familiar structures in number theory are sets with one or two binary operations. Of course, we can continue building up structures indefinitely by adding operations. On the other hand, in the preceding sections we have given evidence of the parallelism between the structures of groups and those of rings, especially the similarity between normal subgroups and ideals. It is therefore expedient to construct a unified theory from which results concerning different structures, such as groups and rings, can be derived as special cases. The purpose of this section is to investigate general algebraic structures with an arbitrary number of operations, not necessarily of the same kind.

First of all, we want to generalize the notion expressed by definition 3.1.1. In the same way as a binary operation on a set A is defined as a function $f: A \times A \to A$, an n-ary operation may be defined. We introduce the notion of n-term cartesian product

$$\underbrace{A \times A \times \cdots \times A}_{n \text{ times}}$$

of the set A as the collection of the ordered n-tuples of the elements of A.

Definition 3.8.1. An *n-ary operation on a set A* is a function

$$f: \overbrace{A \times A \times \cdots \times A}^{n \text{ times}} \to A.$$

In other words, for every ordered n-tuple $(a_1, a_2, \ldots, a_n)$ of elements of A there is a unique element $f(a_1, a_2, \ldots, a_n) \in A$ which is the result of applying f to this n-tuple. For $n = 2$, we again have our familiar concept of binary operation. For $n = 1$, we have a *unary operation* which is a function $f: A \to A$, that is, an ordinary function on a set. For $n = 0$, we have the so-called *nullary* or *0-ary operation* which may be seen, conventionally, as a map from the empty set to a fixed element A or equivalently, as the selection of an element of A independent of the selection of any elements in A. For example, in choosing the identity element of a group, we are making use of a nullary operation. The notion of n-ary operation is a prerequisite for the following definition.

Definition 3.8.2. A *universal algebra* (or simply an *algebra*) is a system $[S, f_1, f_2, \ldots, f_k]$ which consists of a set S and $k(k \geqslant 0)$ operations $f_1, f_2, \ldots, f_k$ on S such that each $f_i (1 \leqslant i \leqslant k)$ is an n_i-ary operation on S for some $n_i \geqslant 0$.

We will now illustrate how the previously defined algebraic structures can fit into this general framework of universal algebras.

Example 3.8.1. A semigroup $G = [S, *]$ is clearly an algebra with a single binary operation. On the other hand, if G is finite, then we may regard G as an algebra with a finite number of unary operations f_a, $(a \in G)$ such that for an arbitrary element b in G, $f_a(b) = ba$.

Thus, for the semigroup given by its table in figure 3.8.1, we have the following corresponding unary algebra[10] $[S, f_a, f_b, f_c]$, where $S \equiv \{a, b, c\}$ and

$$f_a = \begin{pmatrix} a\,b\,c \\ a\,b\,c \end{pmatrix}, \quad f_b = \begin{pmatrix} a\,b\,c \\ b\,b\,a \end{pmatrix}, \quad \text{and} \quad f_c = \begin{pmatrix} a\,b\,c \\ a\,b\,c \end{pmatrix}.$$

*	a	b	c
a	a	b	a
b	b	b	b
c	c	a	c

Figure 3.8.1. Operation table of a semigroup

The preceding example demonstrated that an algebraic structure may be represented differently in the framework of universal algebras. This is due to the fact that universal algebras are formulated in the most general fashion.

Example 3.8.2. The set of positive integers P is certainly a semigroup under multiplication and hence is an algebra $[P, \times]$. However, it is also a monoid, which, represented in the framework of universal algebras, is the algebra $[P, *, 1]$, where 1 is the nullary operation of selecting the special element 1 as the identity. This is the reason identities of a monoid and its submonoids coincide (see definition 3.2.4).

Example 3.8.3. The set of real numbers, R, under multiplication is a group and hence can be thought of as an algebra with one binary operation (multiplication), one unary operation (inverse), and one nullary operation (selecting the identity). On the other hand, since equations $ax = b$ and $ya = b$ have unique solutions in groups, this group may also be thought of as the algebra $[R, f_1, f_2, f_3]$, where each $f_i (i = 1, 2, 3)$ is a binary operation such that the following statements hold for a, b in R.

$$f_1(a, b) = ab \qquad \text{(multiplication)}$$
$$f_2(a, b) = a/b = ab^{-1} \qquad \text{(right division)}$$
$$f_3(a, b) = b/a = a^{-1}b \qquad \text{(left division)}$$

Again, in the preceding example, we have an instance of the fact that an algebraic structure may be viewed as two distinct universal algebras. At this point we recall that we have always compared only structures of the same type via the concept of homomorphism. We do not, for example, consider homomorphism from a group to a ring. How should we compare two universal algebras? Will

---

10. A *unary algebra* is an algebra whose operations are unary and, possibly, nullary.

different representations in the framework of universal algebras cause two structures of the same type, say groups, to be incomparable? Let us first consider what algebras are comparable. Since structures, as we know them from our experience so far, are characterized in terms of operations defined on a set, it is natural to say that two structures are comparable if they have the same set of operations. Of course, this means that if we choose to represent a group as two algebras with distinct sets of operations, as in example 3.8.3, we are not able to compare a group with itself. This is one of the disadvantages of the generalization.

Definition 3.8.3. Two universal algebras $A = [S, f_1, \ldots, f_k]$ and $A' = [T, g_1, \ldots, g_l]$ are said to be *similar* if there is a one-to-one correspondence between the sets of operations $\{f_1, \ldots, f_k\}$ and $\{g_1, \ldots, g_l\}$ such that for each operation f_i in A and g_j corresponding to it in A', f_i and g_j are n-ary operations with one and the same n.

We note, by this definition, that if A and A' are similar, then $k = l$. For the sake of simplicity, we shall assume that f_i always corresponds to the operation g_i in A' with the same index. Thus, by the definition of similarity of algebras, semigroups and monoids need not be similar. We are now ready to define the concepts of homomorphism and isomorphism between similar algebras.

Definition 3.8.4. Let $A = [S, f_1, \ldots, f_k]$ and $A' = [T, g_1, \ldots, g_k]$ be two similar algebras. An onto function $\varphi : S \to T$ is called a *homomorphism* from A to A' if for each n-ary operation f_i in A and elements $s_1, \ldots, s_n$ in S we have

$$\varphi(f_i(s_1, \ldots, s_n)) = g_i(\varphi(s_1), \ldots, \varphi(s_n)) \tag{3.8.1}$$

We call A' a *homomorphic* image of A, and we call φ an *isomorphism* between A and A' if, in addition, φ is one-to-one. In this case, A and A' are said to be *isomorphic* (in symbols, $A \cong A'$).

We recall that a function φ is a homomorphism from a semigroup $G = [S, *]$ to another semigroup $G' = [T, \circ]$ (definition 3.3.2) if

$$\varphi(s_1 * s_2) = \varphi(s_1) \circ \varphi(s_2),$$

which can also be written in the following form [see (3.8.1)]:

$$\varphi(*(s_1, s_2)) = \circ (\varphi(s_1), \varphi(s_2)).$$

Similarly, the fact that group homomorphisms must map identity into identity (lemma 3.3.2) is a consequence of the result that homomorphism of algebras preserve nullary operations. In other words, when two groups G and G' are treated as universal algebras $A = [S, f_1, f_2]$ and $A' = [T, g_1, g_2]$, where f_2 and g_2 are corresponding nullary operations of taking the identity elements, then by (3.8.1) we must have

$$\varphi(f_2) = g_2.$$

Lemma 3.8.1. If φ and ψ are algebraic homomorphisms from A to A' and from A' to A'' respectively, then their composite, $\varphi\psi$, is a homomorphism from A to A''.

Proof. Let f_i be an arbitrary n-ary operation of A and $s_1, \ldots, s_n$ arbitrary elements of A. We have

$$(\varphi\psi)(f_i(s_1,\ldots,s_n)) = \psi(f'_i(\varphi(s_1),\ldots,\varphi(s_n)))$$
$$= f''_i(\psi(\varphi(s_1)),\ldots,\psi(\varphi(s_n)))$$
$$= f''_i(\varphi\psi(s_1),\ldots,\varphi\psi(s_n)).$$

Since the composite of two functions is also a function, $\varphi\psi$ is indeed a homomorphism from A to A''.$\|$

In order to demonstrate the structure-preserving capabilities of a homomorphism, the following definition is needed.

Definition 3.8.5. An algebra $A' = [S', f'_1, \ldots, f'_k]$ is said to be a *subalgebra* of another algebra $A = [S, f_1, \ldots, f_k]$ if $S' \subseteq S$ and for each $1 \leqslant i \leqslant k$, f'_i is the restriction of f_i to S'.

Theorem 3.8.2. The homomorphic and inverse homomorphic images of subalgebras are also subalgebras.

Proof. Let φ be a homomorphism from $A = [S, f_1, \ldots, f_k]$ to $B = [T, g_1, \ldots, g_k]$, and let A' and B' be subalgebras of A and B, respectively. By (3.8.1) we can conclude immediately that $\varphi(A')$ is a subalgebra of B. To show that $\varphi^{-1}(B')$ is a subalgebra of A, for each n-ary operation f_i in A and elements $s_1, \ldots, s_n$ in $\varphi^{-1}(B')$, we must show closure of f_i in $\varphi^{-1}(B')$, that is,

$$f_i(s_1, s_2, \ldots, s_n) \in \varphi^{-1}(B').$$

Note that, since φ is a homomorphism,

$$\varphi(f_i(s_1,\ldots,s_n)) = g_i(\varphi(s_1),\ldots,\varphi(s_n));$$

but $s_k \in \varphi^{-1}(B')$ means $\varphi(s_k) \in B'$ $(k = 1, 2, \ldots, n)$, and since B' is a subalgebra, $g_i(\varphi(s_1), \ldots, \varphi(s_n)) \in B'$; but this implies that

$$f_i(s_1, \ldots, s_n) \in \varphi^{-1}(g_i(\varphi(s_1), \ldots, \varphi(s_n))) \in \varphi^{-1}(B'),$$

which completes the proof.$\|$

We recall from previous sections that the concept of homomorphism is intimately related to that of a congruence relation, and that a congruence relation on a semigroup $G = [S, *]$ is defined to be an equivalence relation α such that $a \, \alpha \, b$ and $a' \, \alpha \, b'$ implies $aa' \, \alpha \, bb'$. This concept is extended to the general case in the following definition.

Definition 3.8.6. Let $A = [S, f_1, \ldots, f_k]$ be a given algebra. An equivalence relation α on S is called a *congruence relation* on A if for an arbitrary n-ary operation f_j

in A and arbitrary elements a_i, a_i', $i = 1, 2, \ldots, n$ in S, it follows from $a_i \, \alpha \, a_i'$, $i = 1, 2, \ldots, n$, that

$$f_j(a_1, \ldots, a_n) \, \alpha \, f_j(a_1', \ldots, a_n').$$

This definition enables us to define a quotient structure (the structure of the equivalence classes) since, if $A_1, A_2, \ldots, A_n$ are equivalence classes under α (briefly, α-classes), then the class B containing the elements $f_j(a_1, \ldots, a_n)(a_i \in A_i, i = 1, \ldots, n)$ does not depend on how a_i is chosen in $A_i (i = 1, \ldots, n)$. Therefore we can define an n-ary operation $\overline{f_j}$ on C_α, the set of the α-classes, by putting

$$\overline{f_j}(A_1, A_2, \ldots, A_n) = B. \tag{3.8.2}$$

Since (3.8.2) holds for all $f_j(j = 1, \ldots, k)$ in A, this converts C_α into a universal algebra $[C_\alpha, \overline{f_1}, \ldots, \overline{f_k}]$ which is similar to A. We will call this newly formed algebra the *quotient algebra* of A modulo α and denote it by A/α.

We are now ready to give the characterization theorem of congruence relations on universal algebras.

Theorem 3.8.3. An equivalence relation γ on an algebra $A = [S, f_1, \ldots, f_k]$ is a congruence relation on A if and only if there is a similar algebra $B = [T, g_1, \ldots, g_k]$ and a homomorphism $\varphi : A \to B$ such that $\gamma = \varphi \varphi^{-1}$.

Proof. If γ is a congruence relation on A, then we define a function $\varphi_\gamma : A \to A/\gamma$ as follows:

$$\varphi_\gamma(a) = E_\gamma(a), \text{ for any } a \in S.$$

Clearly, φ_γ is a function. Furthermore, for any n-ary operation f_j and arbitrary elements $a_1, \ldots, a_n$ in A, we have

$$\begin{aligned}
\varphi_\gamma(f_j(a_1, \ldots, a_n)) &= E_\gamma(f_j(a_1, \ldots, a_n)) \\
&= \overline{f_j}(E_\gamma(a_1), \ldots, E_\gamma(a_n)). \\
&= \overline{f_j}(\varphi_\gamma(a_1), \ldots, \varphi_\gamma(a_n)).
\end{aligned}$$

Hence φ_γ is a homomorphism and $\gamma = \varphi_\gamma \varphi_\gamma^{-1}$. Conversely, let γ be an equivalence relation on A. Let f_j be any n-ary operation on A and $a_i \gamma a_i'$ $(i = 1, \ldots, n)$. Since $a_i \, \gamma \, a_i'$ means that $\varphi(a_i) = \varphi(a_i')$ by the definition of γ, we have

$$\begin{aligned}
\varphi(f_j(a_1, \ldots, a_n)) &= \overline{f_j}(\varphi(a_1), \ldots, \varphi(a_n)) \\
&= \overline{f_j}(\varphi(a_1'), \ldots, \varphi(a_n')) \\
&= \varphi(f_j(a_1', \ldots, a_n')).
\end{aligned}$$

We conclude that $f_j(a_1, \ldots, a_n) \, \gamma \, f_j(a_1', \ldots, a_n')$, and hence γ is a congruence relation on A. $\|$

In accordance with the terminology we used in previous sections, the function φ_γ in the theorem above is called the *natural homomorphism* from G to G/γ. The following result is an immediate consequence of theorem 3.8.3. Since the proof is similar to that of theorem 3.4.4, it is omitted here.

Corollary 3.8.4. If φ is a homomorphism from an algebra A to another algebra B and γ is the congruence relation $\varphi\varphi^{-1}$, then there exists an isomorphism ψ between B and A/γ such that the composite $\varphi\psi$ coincides with the natural homomorphism φ_γ from A onto A/γ.

Although we have obtained some common properties of various algebraic systems utilizing the concept of universal algebra, the concept of similar algebras is still far too general to distinguish classes of algebraic systems. For example, consider the class of algebras with one binary operation, one unary operation, and one nullary operation. Clearly, not every algebra in this class is a group although it is similar to a group. Groups can be separated from other algebras in this class by the fact that they all satisfy the following identical relations for arbitrary elements x, y, and z:

$$(xy)z = x(yz), \qquad xx^{-1} = 1, \qquad x1 = 1x = x. \tag{3.8.3}$$

Indeed, the concept of identical relations or "characteristic identities" of an algebra is the tool by which similar algebraic structures are further separated. However, their consideration is beyond the scope of this book and will not be pursued further.

_____

EXERCISES 3.8

1. Describe the structures of rings and fields in the framework of universal algebras.
2. Let A be a given algebra and B a subalgebra of A. Let γ be a congruence relation of A and $\varphi : A \rightarrow A'$ a homomorphism. Then the restriction of γ to B is a congruence of B and $\varphi | B$ is a homomorphism of B to a subalgebra of A'.
3. Show that if α and β are congruence relations of an algebra, then so is $\alpha \cap \beta$. (See exercise 1.5.9.)
4.   a) Show that the set of all endomorphisms of an algebra A form a semigroup called the *endomorphism semigroup* of A. (See exercise 3.3.4.)
     *b) Show that every finite monoid is isomorphic to the endomorphism semigroup of some algebra.
5.   a) Show that the set of all automorphisms of an algebra A form a group called the *automorphism group* of A. (See exercise 3.6.2.)
     *b) Show that every finite group is isomorphic to the automorphism group of some algebra.

_____

3.9 BIBLIOGRAPHY AND HISTORICAL NOTES

The modern study of algebraic structures has its root in the classical theory of equations developed, principally by G. Abel and E. Galois, at the beginning of the nineteenth century. Abstract group theory arose in the mid-nineteenth century (Cayley), originally for describing symmetries of geometric figures. Since the beginning of the twentieth century, the theory of modern algebra has undergone explosive and far-reaching development.

Although the literature on the subject is extensive, only a few references are given below. Three valuable books cover a fairly broad spectrum of algebraic structures:

> J. B. Fraleigh, *A First Course In Abstract Algebra*, Addison-Wesley, Reading, Mass., 1969.

> A. G. Kurosh, *Lectures on General Algebra*, Chelsea, New York, 1963.

> G. Birkhoff and T. C. Bartee, *Modern Applied Algebra*, McGraw-Hill, New York, 1970.

A very readable book devoted only to semigroups is

> E. S. Ljapin, *Semigroups* [in Russian], Moscow, 1960. English translation, American Mathematical Society, Providence, R.I., 1964.

A very clear and pleasant treatment of the theory of rings is

> N. H. McCoy, *Rings and Ideals*, Mathematical Association of America (distributed by Wiley), 1948.

A beautifully written book on group theory is

> A. G. Kurosh, *Theory of Groups*, Vol. 1 (translated and edited by K. A. Hirsch), Chelsea, New York, 1960.

Finally, an excellent book on universal algebra is

> G. Grätzer, *Universal Algebra*, Van Nostrand, Princeton, 1968.

3.10 NOTES AND APPLICATIONS

Note 3.1. The Associated Monoid of a Finite State Machine

As usual, a finite state machine $\mathcal{M}$ is a system $[S, I, Z, \delta, \lambda]$, and the reader is referred to definition 1.7.1 for the meaning of the symbols S, I, Z, δ, and λ. In the rest of this note we shall ignore the output function λ of $\mathcal{M}$.

For any input sequence $x = i^{(1)} i^{(2)} \cdots i^{(n)} \in I^*$, and for a fixed (initial) state $s_0 \in S$, the sequence x induces a state sequence $s^{(1)} s^{(2)} \cdots s^{(n+1)}$, where $s^{(1)} = s_0$ and for $0 < t < n$, $s^{(t+1)}$ is the unique state $\delta(s^{(t)}, i^{(t)})$. Therefore, every input sequence will cause a unique transition from a state to a state. This transition can be defined formally by extending the domain of the function δ to $S \times I^*$ as follows:

> i) $\delta(s, \Lambda) = s$, for all $s \in S$.

> ii) If $x \in I^*$ and $i \in I$, then $\delta(s, xi) = \delta(\delta(s, x), i)$, for all $s \in S$.   (3.10.1)

Condition (i) of (3.10.1) is a formalization of the intuitive notion that when the machine receives no input (recall that Λ is the empty sequence), it will not make any transition. Condition (ii) says that if the machine is in state s, then the resulting state, $\delta(s, xi)$, after the sequence xi is applied as an input to the machine, is the same as the state reached by first applying the input sequence x [to reach the state $\delta(s,x)$] and then the input i [to reach the state $\delta(\delta(s,x),i)$]. Indeed, it is clear that the transition caused by a sequence $x_1 x_2$ is the same as that caused by just applying first x_1 and then x_2. The same reasoning is now applied in the proof of the following lemma.

Lemma 3.10.1. Let $\mathscr{M} = [S, I, Z, \delta, \lambda]$ be a given finite state machine. For arbitrary input sequences x and y in I^*, and $s \in S$, we have

$$\delta(s, xy) = \delta(\delta(s,x), y). \tag{3.10.2}$$

Proof. We will prove (3.10.2) using induction on the length of y. Clearly, when $y = \Lambda$, by condition (i) of (3.10.1), we have

$$\delta(\delta(s,x), \Lambda) = \delta(s, x) = \delta(s, x\Lambda).$$

Let us assume now that (3.10.2) holds for any y such that y contains k symbols for some $k \geqslant 0$. Let z be a sequence of length $k + 1$. Then $z = iy$, for some sequence y of length k and $i \in I$.
We have

$$\begin{aligned}
\delta(s, xz) &= \delta(s, xiy) = \delta(\delta(s, xi), y) \\
&= \delta(\delta(\delta(s, x), i), y) \\
&= \delta(\delta(s, x), iy) = \delta(\delta(s, x), z),
\end{aligned}$$

where the second and fourth equalities follow from the inductive hypothesis, and the third follows from condition (ii) of (3.10.1). Hence the lemma is proved.‖

The two conditions in (3.10.1) state that each input sequence induces a transformation on the state set. Since it is natural to call two input sequences "equivalent" if they induce the same transformation, we define a relation γ on I^* such that

$$R_\gamma \equiv \{(x, y)\,|\, \text{for all } s \in S,\ \delta(s, x) = \delta(s, y)\}, \tag{3.10.3}$$

and thus γ is clearly an equivalence relation. Furthermore, γ is a congruence relation (definition 3.4.1) because, if $x_1\, \gamma\, y_1$ and $x_2\, \gamma\, y_2$, then for any $s \in S$, we have

$$\begin{aligned}
\delta(s, x_1 x_2) &= \delta(\delta(s, x_1), x_2) = \delta(\delta(s, x_1), y_2) \\
&= \delta(\delta(s, y_1), y_2) = \delta(s, y_1 y_2),
\end{aligned}$$

where the first and fourth equalities follow from (3.10.2), and the second and third equalities follow from $x_1\, \gamma\, y_1$ and $x_2\, \gamma\, y_2$. It follows that $x_1 x_2\, \gamma\, y_1 y_2$, whence γ is indeed a congruence relation on the monoid $[I^*, \cdot]$. Therefore $[I^*, \cdot]/\gamma$ is a monoid referred to as the *associated monoid of $\mathscr{M}$*.

Since a congruence class under γ contains all input sequences which induce the same transformation on S, there is a one-to-one correspondence between congruence classes under γ and a set of functions on S. Indeed, let us define, for each $x \in I^*$, a function $\delta_x : S \to S$ such that for $s \in S$,

$$\delta_x(s) = \delta(s, x). \tag{3.10.4}$$

Let $S(\mathscr{M})$ denote the class of all functions of the form δ_x, as x ranges over I^*. We see that $S(\mathcal{M})$ is a semigroup under the operation of function composition $\circ$. Furthermore, $[S(\mathcal{M}),\circ]$ is a monoid with the identity δ_A. This follows from the property

$$\delta_x \circ \delta_y = \delta_{xy}.$$

Hence

$$\delta_A \circ \delta_y = \delta_{Ay} = \delta_{yA} = \delta_y = \delta_y \circ \delta_A.$$

By the one-to-one correspondence between $S(\mathcal{M})$ and γ-congruence classes, we see that the two monoids $[I^*, \cdot]$ and $[S(\mathcal{M}),\circ]$ are isomorphic.

It is natural to ask whether every monoid is the associated monoid (up to isomorphism) of a finite state machine. The answer is affirmative, as stated in the following theorem.

Theorem 3.10.2. Every finite monoid is isomorphic to the associated monoid of some finite state machine.

Proof. Let $G = [S,\circ]$ be a given finite monoid. Define a finite state machine $\mathcal{M}_G = [S, S, S, \delta, \text{arbitrary } \lambda]$ such that for s_1, s_2 in S

$$\delta(s_1, s_2) = s_1 \circ s_2.$$

Let γ be the congruence relation defined on S^* by (3.10.3), and let $[x]_\gamma$ denote the congruence class under γ containing x. We define a function $\varphi : S \to S^*/\gamma$ such that for each $s \in S$,

$$\varphi(s) = [s]_\gamma.$$

We will leave as an exercise the task of showing that φ is indeed an isomorphism between G and S^*/γ.

Reference

R. McNaughton and S. Papert, *Counter-free Automata*, M.I.T. Press, Cambridge, Mass., 1971.

Note 3.2. Permutations

We have already considered permutations from two main viewpoints. First we looked at a permutation as an invertible function on a set, and then we examined the class of permutations of the elements of a set as an algebraic structure and found it to be a group. In this note we develop some additional notions concerning these very important mathematical concepts. Some of these notions will be useful in Chapter 6.

First of all, we reconsider permutations as functions on a set. Specifically, let $p : S \to S$ be a permutation of a set $S \equiv \{s_1, s_2, ..., s_n\}$; that is, p is a rearrangement of the objects s_1, s_2, ..., s_n. To simplify the notation, we may consider these objects as the integers $1, 2, ..., n$ and refer to p as a permutation of n integers (also called a permutation of *degree n*). Since permutations of n integers are all and only the *invertible* functions on $\{1,2,...,n\}$, the graph of p (see Section 1.2) is characterized by the following properties:

  i) each vertex has exactly one outgoing arc (because p is a function) ;

  ii) each vertex has exactly one incoming arc (because p is an invertible function) ;

it follows that the graph of p must consist of circuits called the *cycles* of p (and their set is called the *cycle set* of p). For example, given

$$p = \begin{pmatrix} 1\,2\,3\,4\,5\,6 \\ 4\,1\,6\,2\,5\,3 \end{pmatrix},$$

the graph of p is

by which we display the set of the cycles of p. We realize that we need not draw the diagram of a cycle to represent it. In fact, since arcs are directed, we need only list the sequence of vertices as $(s_i, p(s_i), p[p(s_i)], ...)$; here we adopt the convention that there is an arc from the j-th to the $(j+1)$-th integers of this sequence and an arc from the last integer to the first integer. Thus the cycles above are represented by the sequences $(1,4,2)$, $(3,6)$, and (5). The notation can be further simplified by not listing cycles consisting of a single vertex, since such cycles do not operate any rearrangement of $\{1,2,...,n\}$. Thus a permutation is completely specified by its cycles of length greater than 1 ; furthermore, we may omit commas both between elements of a cycle and between cycles. For the permutation p given above we obtain

$$p = (142)\,(36). \tag{3.10.5}$$

Here the symbol "$=$" must be temporarily interpreted as "is represented as." This is standard notation for permutations.

Consider now the permutation

$$p_1 = \begin{pmatrix} 1\,2\,3\,4\,5\,6 \\ 4\,1\,3\,2\,5\,6 \end{pmatrix}.$$

Since p_1 contains only one cycle of length greater than 1, we write $p = (142)$. Similarly, we can represent the permutation

$$p_2 = \begin{pmatrix} 1\,2\,3\,4\,5\,6 \\ 1\,2\,6\,4\,5\,3 \end{pmatrix}$$

as $p_2 = (36)$. The product $p_1 p_2$ is a rearrangement of $\{1,2,...,n\}$ obtained by performing *first* p_2 and *then* p_1; clearly $p_1 p_2$ is another permutation, since the permutations of n integers form a group (theorem 3.5.3). Moreover, we recognize that $p_1 p_2 = p_2 p_1$, since the two cycles (142) and (36) have no common element, whereas for arbitrary p_1 and p_2, $p_1 p_2 \neq p_2 p_1$. Thus we can interpret (3.10.5) as a true product of single cycle permutations and $=$ becomes the equality symbol in the permutation group. This is an example of the following straightforward theorem (where the word "cycle" must now be interpreted as "permutation consisting of a single cycle").

Theorem 3.10.3. Any permutation is a product of disjoint cycles.

The simplest nontrivial cycles have length 2. They are called *transpositions*, since their effect is to interchange, or transpose, two members of the set $\{1, 2, \ldots, n\}$. Note that the cycle (132) can be obtained as the product of (12) (23), as illustrated by the following diagrams.

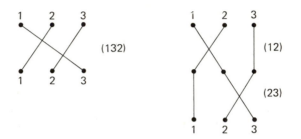

In other words, we obtain a permutation (132) as the product of two transpositions. This is an instance of the following general theorem.

Theorem 3.10.4. Every permutation of n integers can be written as a product of transpositions.

Proof. The theorem is trivial for $n = 2$. Assume that the theorem is true for a permutation of $(n - 1)$ integers and consider the permutation p:

$$\begin{pmatrix} 1 & 2 & \cdots n \\ p(1) & p(2) & \cdots p(n) \end{pmatrix}.$$

There is exactly one i for which $p(i) = n$. If we perform the transposition (i, n), the rightmost element of the resulting arrangement is correct, and the leftmost $(n - 1)$ elements must still be rearranged. Thus we have a permutation of $(n - 1)$ elements, which we know to be obtainable as a sequence of transpositions.‖

Note that any transposition (i, j) with $(i < j)$ can be obtained as the product of transpositions of consecutive elements:

$$(i, j) = (i, i + 1) \cdots (j - 2, j - 1)(j - 1, j) \cdots (i + 1, i + 2)(i, i + 1)$$

$\underbrace{\qquad\qquad\qquad\qquad}_{(j - i - 1) \text{ transpositions}}$ $\underbrace{\qquad\qquad\qquad\qquad}_{(j - i) \text{ transpositions}}$

$\longleftarrow$

Order in which transpositions are performed

The rightmost $(j - i)$ transpositions have the effect of bringing i to the place of j, while displacing all the other elements one position to the left. The leftmost $(j - i - 1)$ transpositions bring j to the place of i, while restoring all the other elements to their original position. Combining this result with the preceding theorem, we can say that every permutation of degree n can be expressed as an appropriate product of the $(n-1)$ transpositions $(i, i+1)$ for $i = 1, 2, \ldots, n-1$. This is concisely expressed by saying that the group of the permutations of degree n— also called S_n, *the symmetric group of degree n*—is generated by the transpositions $\{(i, i+1) | i = 1, 2, \ldots, n-1\}$. Is this the minimum number of generators of S_n? No; the permutations $p_1 = (12)$ and $p_2 = (23 \cdots n)$ generate S_n. As proof, it is sufficient to show that any transposition $(i, i+1)$ is expressible as a product containing only p_1 and p_2 as factors. In fact, starting from the arrangement $(1, 2, \ldots, n)$,

after applying	we obtain
p_2 $(n-i+1)$ times,	$(1, i, i+1, \ldots, i-1)$;
p_1 once,	$(i, 1, i+1, \ldots, i-1)$;
p_2 $(n-2)$ times,	$(i, i+1, i+2, \ldots, i-1, 1)$;
p_1 once,	$(i+1, i, i+2, \ldots, i-1, 1)$;
p_2 once,	$(i+1, 1, i, i+2, \ldots, i-1)$;
p_1 once,	$(1, i+1, i, i+2, \ldots, i-1)$;
p_2 $(i-2)$ times,	$(1, 2, \ldots, i+1, i, \ldots, n)$.

This sequence describes the transposition $(i, i+1)$, thus proving our claim. It can also be shown that two is the minimum number of generators of S_n, for $n > 2$.

Reference

C. Berge, *Principles of Combinatorics,* Academic Press, New York, 1971.

Note 3.3. Communication and Error Correcting Codes

Although communication is so basic to human beings, we seldom pause to consider its nature and mechanisms. If we now undertake this task, at least superficially, we notice that every form of communication is aimed at transferring to a "destination" some "meaning" originating within a "source." Here, source and destination are generic designations for specific pairs, such as speaker and listener, writer and reader, or two interconnected computers. We prefer not to enter into a discussion here of "meaning" ; rather, we stop at the intuitive notion that the source wants to convey to the destination some message (a statement, a question) which is to be "understood" by the latter. The basic goal underlying communication is that the meaning of the message as understood by the destination coincides, for all practical purposes, with the meaning conceived by the source. Normally, this transfer of information is accomplished by representing the message in a form suitable for communication. For written communication in English, for example, this form will be a sequence of characters from the Latin alphabet (plus blanks and punctuation marks) ; for spoken communication, it will be a sequence of phonemes of some natural language ; for telegraphy it will be a sequence of dots, dashes, and spaces ; and so on. For all purposes, the form will be a sequence of symbols from some set, usually called—in this context—an *alphabet*. This representation is called a coding of the message.

In all practical instances, communication is disturbed by a variety of random causes present in the medium through which transmission occurs. These disturbances are usually classed as "noise" (for example, running water in a room where two persons talk impairs the intelligibility of their words). We are all instinctively aware of one method of combating the noise of the environment when we talk: we raise our voices; that is, we utter our sounds (the "signal") in such a way that the noise power becomes negligible by comparison. This approach, known as increasing the transmitter power, is applicable to most communication systems; however, transmitter power is not unlimited. Fortunately, there is an alternative and more subtle approach to combating noise: redundancy in the message. This is not a sophisticated technical invention; living organisms make constant use of the technique, and engineers are only trying to apply it systematically and economically to communication systems. For example, we all understand the string of symbols "col_er w_th ch_nce o_ lig_t sho_rs," notwithstanding the fact that seven letters have been erased. We understand because the letters of the alphabet are not used independently in natural languages. On the other hand, in the string of symbols "This suit costs \$8_.55" there is no way to fill in the missing digit. Redundancy in messages makes them resistant to disturbance (errors), such as alterations and deletions. The essence of redundancy techniques is to select for representation of messages a *proper* subset of the constructible strings of symbols; this proper subset is usually referred to as the *code*. If the destination receives a string which does not belong to the code, it is alerted to the fact that some error has occurred (error detection), and it may use the received string to guess the most likely transmitted string (error correction).

Let us now see how we can construct a code suitable for error correction. We assume that our *physical* transmission medium is a *binary channel*, that is, a "box" which accepts a binary symbol as input and delivers a binary symbol as output (figure 3.10.1). The noisiness is modeled by specifying a nonzero probability of receiving a 1 upon transmission of a 0, and conversely. If messages are encoded as sequences of n symbols from the alphabet $\{0,1\}$ (binary sequences), and a given sequence is transmitted, *any* other sequence may be received. If it is just as likely that a 0 will be changed to a 1 as that a 1 will be changed to a 0, the channel is called *symmetric*. The effect of the channel may then be thought of as the "addition" of an *error sequence* to the transmitted sequence. Sequence addition, $\oplus$, is a binary operation on the set S_n of the binary sequences of n symbols, defined as follows: Let $\mathbf{s}^{(1)} = (a_1^{(1)} a_2^{(1)} \cdots a_n^{(1)})$ and $\mathbf{s}^{(2)} = (a_1^{(2)} a_2^{(2)} \cdots a_n^{(2)})$ be two sequences of S_n; $\mathbf{s} = \mathbf{s}^{(1)} \oplus \mathbf{s}^{(2)}$ is a sequence $(a_1 a_2 \cdots a_n)$ such that $a_j = 0$ if $a^{(1)}$ and $a_j^{(2)}$ are equal and $a_j = 1$ if $a_j^{(1)}$ and $a_j^{(2)}$ are different. For example, for $n = 6$,

$$\mathbf{s}^{(1)} = 010111$$
$$\mathbf{s}^{(2)} = 110010$$
$$\mathbf{s}^{(1)} \oplus \mathbf{s}^{(2)} = 100101.$$

Thus $\mathbf{e} = (e_1 e_2 \cdots e_n)$ is an error sequence, and if $e_j = 1$, the j-th symbol of the transmitted sequence is received altered (that is, a 0 is changed to a 1, or vice versa). The reader can verify that the operation $\oplus$ is commutative and associative, that $(00 \cdots 0)$ is the additive identity, and that every string is its own inverse with

Figure 3.10.1. Illustration of a binary channel

respect to $\oplus$ (that is, $\mathbf{s} \oplus \mathbf{s} = (00\cdots0)$). Thus S_n is a group $[S_n, \oplus]$ of order 2^n (definition 3.5.1) since, as we shall see in Section 6.1, there are 2^n binary sequences of length n.

Suppose that we select a set $C \equiv \{\mathbf{s}^{(1)}, \mathbf{s}^{(2)}, \ldots, \mathbf{s}^{(m)}\}$ of m sequences ($m < 2^n$) of S_n as the code. The sequences of C are called *code words*. If $\mathbf{r} = (r_1 r_2 \cdots r_n)$ is the received sequence, since $\mathbf{r} = \mathbf{r} \oplus (\mathbf{s}^{(j)} \oplus \mathbf{s}^{(j)}) = (\mathbf{r} \oplus \mathbf{s}^{(j)}) \oplus \mathbf{s}^{(j)}$, $\mathbf{r} \oplus \mathbf{s}^{(j)}$ is the error sequence which could have mapped the hypothetically transmitted sequence $\mathbf{s}^{(j)}$ to the actually received sequence $\mathbf{r}$. Therefore, for a given $\mathbf{r}$, we may compute the m sequences:

$$\mathbf{r} \oplus \mathbf{s}^{(j)} = \mathbf{e}^{(j)}, j = 1, 2, \ldots, m.$$

If we make the reasonable assumption that, of two arbitrary error sequences, the one with the smaller number of 1's is the more likely to occur, clearly the best criterion is to interpret $\mathbf{r}$ (to "decode" $\mathbf{r}$) as that $\mathbf{s}^{(j)}$ for which $\mathbf{e}^{(j)}$ contains a smallest number of 1's, for $j = 1, 2, \ldots, m$. This can be done for every $\mathbf{r} \in S_n$, and a result is a function (*decoding rule*) $\delta : S_n \to C$ which maps the received sequences to the transmitted sequences. We know by theorem 1.5.7 that every function $\delta : S_n \to C$ determines an equivalence relation $\delta\delta^{-1}$ on S_n, that is, a partition of S_n into classes, called *decoding sets*, such that all the members of a given decoding set are decoded as the same code word.

This approach is quite general and apparently simple; however, it is extremely complicated to implement in practical cases. For example, if $n = 31$, for decoding we should store a table of δ containing $2^{31} \simeq 10^9$ entries. It is here that one resorts to algebra, in the hope that structure, i.e., order and regularity, may bring about simplicity of implementation. Specifically, the instrument used here is the theory of groups.

Suppose that the code C is a subgroup of S_n (C is called a *group code*). We know, by the fundamental lemma 3.5.5, that for every $\mathbf{s} \in S_n$ the family of subsets $\{\mathbf{s} \oplus \mathbf{s}' | \mathbf{s}' \in C_n\}$ is a partition of S_n into C and its cosets (whether right or left is immaterial because $\oplus$ has the commutative property). For each coset C' of C, let $\mathbf{m}'$ be a sequence with smallest number of 1's in C' ($\mathbf{m}'$ may not be unique). Using the notation of Section 3.5, C' can be expressed as $\mathbf{m}' \oplus C$. It follows that if $\mathbf{r} \in C'$, $\mathbf{r}$ can be expressed uniquely as

$$\mathbf{r} = \mathbf{m}' \oplus \mathbf{s}^{(j)} \tag{3.10.6}$$

for some $\mathbf{s}^{(j)} \in C$, and clearly $\mathbf{s}^{(j)}$ is the code word which differs from $\mathbf{r}$ in the smallest number of positions; that is, $\delta(\mathbf{r}) = \mathbf{s}^{(j)}$. Since each sequence is its own inverse, (3.10.6) can be rewritten as

$$\mathbf{m}' \oplus \mathbf{r} = (\mathbf{m}' \oplus \mathbf{m}') \oplus \mathbf{s}^{(j)} = \mathbf{s}^{(j)}.$$

This leads to the following decoding algorithm.

1. Given a received r, determine C' such that $r \in C'$.

2. From C', determine $m' \in C'$ such that m' has a smallest number of 1's.

3. Compute $m' \oplus r = s^{(j)}$.

We shall not discuss the details of determining C' from r (*syndrome decoding*). However, to appreciate how the choice of a group code may reduce the complexity of decoding, consider a code C with $n = 31$ and with 2^{21} code words. From Lagrange's theorem (theorem 3.5.6), there are $2^{31}/2^{21} = 2^{10}$ cosets. Therefore, if one can simply determine the coset C' to which r belongs, a table with $2^{10} = 1024$ entries is sufficient for representing the function $C' \to m'$ (compare this with the 2^{31}-entry table representing δ). Of course, other structural properties of groups can be cleverly exploited to further reduce the complexity of implementation. However, their presentation is well beyond our scope.

   This concludes our elementary presentation of group codes and of the principles of algebraic coding theory. Coding theory is concerned with the systematic construction of error-correcting and detecting codes of prescribed capabilities and with the development of algorithms for their encoding and decoding. Coding theory is a well-developed area of applied mathematics.

References

W. W. Peterson, *Error Correcting Codes*, M.I.T. Press, Cambridge, Mass., and Wiley, New York, 1961.
R. G. Gallager, *Information Theory and Reliable Communication*, Wiley, New York, 1968 (Chapter 6).

Note 3.4. The Residue Number System

We now illustrate an interesting application of the notion of congruence to a representation system for the integers which has been the subject of considerable investigation in the area of computer arithmetic.

   In this chapter, several examples (3.3.4, 3.4.1, 3.4.2, 3.6.1, 3.7.2) have illustrated some intermediate results, culminating in the following conclusion. The remainder of the division of an integer i by a fixed integer m (called the *modulus*) is a function r from the ring $[I, +, \cdot]$ of the integers under addition and multiplication to the ring $[P_m, \oplus, *]$ of the residue classes modulo m under addition and multiplication modulo m. Formally, for each integer $i \in I$, $i = qm + r(i)$ holds for a unique $0 \leqslant r(i) < m$. This function r is a ring homomorphism (see example 3.7.2).

   Note that $r(i)$ does not uniquely specify the integer i, but it confines our ignorance to within a residue class modulo m. If we also know that $0 \leqslant i \leqslant m - 1$, then $r(i)$ coincides with i.

   Suppose now that we choose two distinct moduli, m_1 and m_2. Under which circumstances does the pair $[r_1(i), r_2(i)]$ uniquely specify the integer i? The answer is provided by the following theorem.

Theorem 3.10.5. For two integers i and j, $r_1(i) = r_1(j)$ and $r_2(i) = r_2(j)$ if and only if $(i - j)$ is a multiple of the least common multiple of m_1 and m_2 (where $r_1(i)$ and $r_2(i)$ are the residues of i modulo m_1 and m_2, respectively).

Proof. Assume that $r_k(i) = r_k(j)$ for $k = 1, 2$. We have

$$r_k(i) = i - q'm_k = j - q''m_k = r_k(j), \quad (k = 1, 2) ;$$

that is, $(i - j) = (q' - q'')m_k$, which shows that $(i - j)$ is a multiple of m_1 and m_2, that is, of their least common multiple.

Conversely, assume that $(i - j)$ is a multiple of m_k; that is, $(i - j) = a_k m_k$ for some integer a_k. We also have $i - j = (q'm_k + r_k(i)) - (q''m_k + r_k(j))$, that is,

$$i - j = (q' - q'')m_k + (r_k(i) - r_k(j)) \quad (k = 1, 2).$$

Since the left side is a multiple of m_k, the right side must also be a multiple of m_k; this implies that $r_k(i) - r_k(j)$ is a multiple of m_k. However, by the property of the remainder, $0 \leqslant r_k(i) \leqslant m_k - 1$ and $0 \leqslant r_k(j) \leqslant m_k - 1$. Thus

$$-(m_k - 1) \leqslant r_k(i) - r_k(j) \leqslant (m_k - 1),$$

and in this interval only 0 is a multiple of m_k.$\|$

As a consequence of this theorem, we see that if the integers i satisfy the condition $0 \leqslant i \leqslant l.\text{c.m.}(m_1, m_2) - 1$, there is a one-to-one correspondence between i and $[r_1(i), r_2(i)]$; we say that $[r_1(i), r_2(i)]$ is a *residue representation* of i (or a *modular representation* of i). Note also that the argument used in the proof of theorem 3.10.5. clearly holds for an arbitrary number of moduli. Thus, let us assume that $m_1, m_2, \ldots, m_s$ are relatively prime numbers (that is, their greatest common divisor is 1). As a consequence, their least common multiple is the integer $N = m_1 m_2 \cdots m_s$. Therefore, if our integers i satisfy the condition $0 \leqslant i \leqslant N - 1$, the sequence $[r_1(i), r_2(i), \ldots, r_s(i)]$ is a residue representation of the integer i with s moduli. We note that $\{r_1, r_2, \ldots, r_s\}$ is a collection of ring homomorphisms. Specifically, r_i is the homomorphism from the ring $[I, +, \cdot]$ to the ring $[P_{m_i}, \oplus, *]$. It follows that for two integers i and j in I, we have the following operation rules for the set of the sequences $[r_1, r_2, \ldots, r_s]$.

$$[r_1(i), \ldots, r_s(i)] + [r_1(j), \ldots, r_s(j)] = [r_1(i+j), \ldots, r_s(i+j)]$$
$$[r_1(i), \ldots, r_s(i)] \cdot [r_1(j), \ldots, r_s(j)] = [r_1(i \cdot j), \ldots, r_s(i \cdot j)]$$

A very interesting fact is that as long as the results of the operations $(i + j)$ or $(i \cdot j)$ do not exceed $(m_1 \cdots m_s - 1)$, the arithmetic can be carried out entirely on the residues, and the uniqueness of conversion from, say $[r_1(i+j), \ldots, r_s(i+j)]$ to $(i + j)$, is guaranteed.

The residue representation system has excited considerable interest among workers in the arithmetic of digital computers. In fact, one of the most serious problems affecting the speed of execution of arithmetic operations with integers represented in positional notation (such as the decimal notation) is the propagation of carries from the least significant digits to the most significant digits. For example, when performing the addition

$$
\begin{array}{ll}
1\ 1\ 1\ 1 & \text{carries} \\
0\ 0\ 0\ 2 & \\
9\ 9\ 9\ 9 & \\
\hline
1\ 0\ 0\ 0\ 1 &
\end{array}
$$

this carry propagates from the position of the units to that of tens of thousands. This drawback disappears in the residue representation system because there is no interaction whatsoever between residues corresponding to different moduli.

Example 3.10.1. Consider the integers 25 and 17, and let $m_1 = 3, m_2 = 4, m_3 = 5$, $m_4 = 11$; clearly l.c.m. $(m_1, m_2, m_3, m_4) = 660$. We have the representations $25 \leftrightarrow [1, 1, 0, 3], 17 \leftrightarrow [2, 1, 2, 6]$, and the operations are illustrated as follows:

$$[1, 1, 0, 3] + [2, 1, 2, 6] = [3, 2, 2, 9] \leftrightarrow 42,$$
$$[1, 1, 0, 3] \cdot [2, 1, 2, 6] = [2, 1, 0, 7] \leftrightarrow 425.$$

There is an extensive literature on residue number systems and their ramifications. For further study the reader is referred to the following book.

N. S. Szabo and R. I. Tanaka, *Residue Arithmetic and Its Applications to Computer Technology*, McGraw-Hill, New York, 1967.

Note 3.5. Infix, Prefix, and Postfix Notations for Expressions

In this note we consider again the problem of writing algebraic expressions involving binary operations. The conventional way of writing expressions, as discussed in example 2.5.5 in connection with oriented rooted trees, is usually referred to as *infix notation*, since the operation symbol appears *within* its two operands, for example, $(A + B)$. A less trivial example is the following arithmetic expression.

$$(A + B) \times C + (D - A)/E \tag{3.10.7}$$

It was pointed out by the Polish logician Łukasiewicz that algebraic expressions involving only binary operations, such as (3.10.7), can be written in parenthesis-free form without recourse to a precedence rule for the operations (see Section 0.4), simply by placing the operation symbol *immediately before* its operands. In this new notation, $(A + B)$ becomes $+AB$, and the expression (3.10.7) becomes

$$+ \times + ABC/ - DAE \tag{3.10.8}$$

Since operation symbols *precede* their operands, this notation is usually referred to as *prefix* notation, or *Polish* notation in tribute to its inventor. We now discuss a systematic way for effecting the conversion between the infix and the prefix forms of an expression. Both notations obviously describe the same oriented rooted tree (definition 2.5.4). The relationship between the tree and the corresponding expression in infix notation was discussed in example 2.5.5. Therefore, since we know how to transform an expression in infix notation into its corresponding tree (and vice versa), we need to analyze only the relationship between the tree and the expression in prefix notation. For example, the tree representing (3.10.7) is given in figure 3.10.2. Starting from the root, we traverse the binary tree in the following way. At each vertex we traverse in the forward direction the leftmost previously untraversed arc; if either the vertex is terminal or both of its outgoing arcs have been traversed, we proceed in the backward direction along the arc, reaching the vertex; we stop when we reach the rightmost terminal vertex. The prefix notation is obtained by writing down from left to right the sequence of symbols as en-

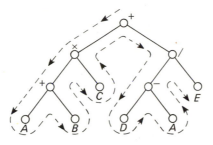

Figure 3.10.2

countered in the forward direction when traversing the tree as specified above. The transformation from the expression in prefix form to the corresponding tree should be obvious.

We may also choose to represent an expression by writing the operation symbol immediately *after* its operands. The resulting parenthesis-free expression, said to be in *postfix* notation, is obtained from the tree by the procedure given above, simply interchanging the words "right" and "left" whenever they occur. Thus expression (3.10.7) becomes

$$AB + C \times DA - E/+$$

in postfix notation.

It is a simple exercise to extend the preceding considerations to expressions involving a mixture of *n*-ary operations. The reader is referred to the next note and Section 5.7 for a discussion of well-formed formulas and propositional calculus.

Note 3.6. The Algebra of Words

In this note we shall examine more closely the monoid of the finite sequences of the elements of a set (see Section 3.2).

We begin with an interesting case. Let $X \equiv \{x_1 \ x_2, \ldots, x_n\}$ be a nonempty set; $x_1, \ldots, x_n$ are called *variables*. A *word over X* is a finite sequence of elements of X. For example, if $X \equiv \{0, 1\}$, then 001 and 11100011 are words over X. Here again, we adopt the convention of using the symbol Λ to denote the word which consists of no symbols, and we call it the *empty word*. We denote by X^* the family of all words, including the empty word, over X. We recall from Section 3.2 that a binary operation, catenation, is defined on X^*, so that if w_1 and w_2 are words over X, then the catenation of w_2 to w_1 is the word $w_1 w_2$ obtained by juxtaposing w_1 on the left and w_2 on the right. We also recall that catenation is associative; that is, X^* under catenation is a monoid with identity Λ.

We can also look at X^* from another point of view. Consider Λ, the symbol for the empty word, as a variable, and consider each $x \in X$ as the name of a unary operation f_x. We define inductively the set X' as follows:

i) $\Lambda \in X'$;
ii) if $x \in X$ and $w \in X'$, then $f_x(w) = wx$ belongs to X'.

Thus, if $X \equiv \{x_1, x_2, ..., x_n\}$, X' will consist of the elements Λ, Λx_1, Λx_2, ..., Λx_n, $\Lambda x_1 x_2$, $\Lambda x_1 x_3$, ... That is, X' is the set $\{\Lambda w | w \in X^*\}$. Clearly, the function

$$\varphi : \Lambda w \rightarrow w, \; w \in X^*$$

is a one-to-one correspondence between the sets X' and X^*. Furthermore, since for each $x \in X$ and $\Lambda w \in X'$,

$$f_x(\Lambda w) = \Lambda w x \in X',$$

the structure $[X', f_{x_1}, ..., f_{x_n}]$ is called a *unary algebra*.[11]

The preceding discussion suggests a couple of very interesting comments. First of all, we note that the same structure, the set of words over X, may be interpreted either as X^* or as X', another instance of the fact that in the context of universal algebras (Section 3.8), a structure may be interpreted as different algebras. Second, the elements of X^* are obtained—one could say "generated"—by combining the elements of X by means of catenation; thus, it would be intuitively appealing to say, "X^* is generated by X." Similarly, any element of X' is obtained by applying an appropriate sequence of the operations f_{x_1}, ..., f_{x_n} to the element Λ; so we could say, "X' is generated by Λ." We now want to make precise the notion of "generators."

Definition 3.10.1. A nonempty subset K of a universal algebra $A = [S, f_1, ..., f_n]$ is called a set of *generators* of a subalgebra $A' = [S', f_1, ..., f_n]$ of A if A' is the smallest subalgebra containing K. If $A' = A$, then K is said to be a set of generators of A.

Intuitively, a set of generators K of an algebra A is one such that all and only elements of A can be generated from K by applying operations of the algebra. For example, all elements of the monoid X^* over the set X can be generated either by $K = X$ in the algebra $[X^*, o]$ or by $K \equiv \{\Lambda\}$ in the algebra $[X', f_{x_1}, ..., f_{x_n}]$.

We now want to explore in some depth the correspondence between the monoid X^* and the unary algebra X'. We need some preparatory definitions.

Definition 3.10.2. A *ranked alphabet* is an ordered pair (Σ, σ), where $\Sigma \equiv \{a_1, ..., a_n\}$ is a set (called *alphabet*) and σ is a function mapping Σ into the set of nonnegative integers N. For each $a \in \Sigma$, $\sigma(a)$ is called the *rank* of a.

An interpretation of the ranked alphabet is that each element $a \in \Sigma$ is the name of a function f_a having $\sigma(a)$ arguments. If $\sigma(a) = 0$, then f_a is referred to as a *constant*. For notational convenience, we will use the symbols Σ_n to denote the set $\{a | a \in \Sigma, \sigma(a) = n\}$.

Definition 3.10.3. Let X be an arbitrary set of variables and (Σ, σ) a ranked alphabet. The set of *words*, denoted by $\Sigma(X)$, generated by (Σ, σ), together with X, is defined inductively as follows:

i) Elements of Σ and X are words;

ii) if $a \in \Sigma_n$ and w_1, w_2, ..., w_n are words, so is $a w_1 w_2 ... w_n$;

iii) nothing is a word except those obtained by a finite number of applications of (i) and (ii).

---

11. See p. 157 (footnote) for a definition of unary algebra.

As an example, consider the *well-formed formulas* (wff) which arise in propositional logic (see Section 5.7), defined inductively as follows: Let the symbols $\lor$ and $\land$ represent binary operations, and let the symbol $\lnot$ represent a unary operation; a set $X \equiv \{x_1 \ldots, x_m\}$ of variables is also given.

i) A variable is a wff;

ii) if α and β are wff's, then so are $\lnot\alpha$, $\alpha \lor \beta$ and $\alpha \land \beta$;

iii) an expression is a wff if and only if its being so follows from finitely many applications of (i) and (ii).

Clearly, the set of wff's is generated by the ranked alphabet $(\{\lnot,\land,\lor\},\sigma)$ and by a set X of variables, with $\sigma(\lnot) = 1$ and $\sigma(\lor) = \sigma(\land) = 2$.

Returning to definition 3.10.3, we see that the set $\Sigma(X)$ of words can be viewed as a universal algebra $[\Sigma(X), f_{a_1}, \ldots, f_{a_n}]$, referred to as a *word algebra*.

An important property of the word algebra $\Sigma(X)$ is that each word in $\Sigma(X)$ has a unique representation. In other words, no nontrivial identical relation of the form

$$w_1 = w_2 \tag{3.10.9}$$

holds in this algebra, where w_1 and w_2 are distinct words in $\Sigma(X)$. This property, known as the *unique factorization property*, is stated as a theorem below. The proof of the theorem (by induction on the length of the word) will be omitted here.

Theorem 3.10.6. If a word w in $\Sigma(X)$ has two representations $fx_1x_2 \cdots x_n$ and $gy_1y_2 \cdots y_m$, then $f = g$, $m = n$, and $x_i = y_i$ for each $1 \leqslant i \leqslant n$ (where x_i and y_j are variables in X).

The unique factorization property certainly does not hold for any arbitrary semigroup. For example, in the semigroup of symmetries of a triangle (example 3.5.1), we have

$$\beta_3 = \alpha_2\alpha_1\alpha_2\beta_2\alpha_1 = \beta_2\alpha_2.$$

Indeed, the problem (known as the *word problem*) of deciding whether or not two words of a semigroup are equivalent is, in general, algorithmically unsolvable. That is, it can be shown that there is no algorithm for solving this problem in general. (The concept of algorithmic unsolvability is discussed in Section 7.6.)

The fact that no nontrivial identical relation of type (3.10.9) holds in $\Sigma(X)$ is expressed by saying that $\Sigma(X)$ is *totally free* (of special identical relations). This introduces us to the notion of "free" algebraic structures, which we shall now make precise. We begin by asking the question: What happens if we allow special identities to hold in $\Sigma(X)$?

Define a binary relation Γ on $\Sigma(X)$ as follows: $w_1 \Gamma w_2$ if $w_1 = w_2$ is a special identity of the type (3.10.9) which holds in $\Sigma(X)$. We now construct a binary relation Γ^* on $\Sigma(X)$, as follows. For w_1 and w_2 in $\Sigma(X)$,

1. if $w_1 \Gamma w_2$, then $w_1 \Gamma^* w_2$;

2. if $w_1' \Gamma^* w_2'$, $w_1 = uw_1'v$ and $w_2 = uw_2'v$, then $w_1 \Gamma^* w_2$;

3. $w_1 \Gamma^* w_2$ if and only if its being so follows from finitely many applications of (1) and (2).

In other words, $w_1 \, \Gamma^* \, w_2$ if either $w_1 = w_2$ is an identity of type (3.10.9) or there is a finite sequence of transformations from w_1 to w_2.

Clearly, Γ^* is an equivalence relation on $\Sigma(X)$. Furthermore, it is a congruence relation in $\Sigma(X)$ because if $w_i \, \Gamma^* \, w_i'(i = 1,2,...,n)$, then obviously the application of an n-ary operation to words $w_1, ..., w_n$ does not affect the transformation rules; i.e., if f is an n-ary operation in $\Sigma(X)$, then

$$f(w_1, ..., w_n) \, \Gamma^* \, f(w_1', ..., w_n').$$

The quotient algebra $\Sigma(X)/\Gamma^*$ is called the *free algebra of type* Γ. The adjectives indicate that it is an algebra which is free of all nontrivial identical relations except those in Γ.

We are now ready to reconsider the monoid X^* generated by X. Consider the word algebra A generated by the ranked alphabet $(\{o,1\}, \sigma)$ and by the set X, with $\sigma(o) = 2$ (the catenation operation) and $\sigma(1) = 0$ (the nullary operation identifying the identity element). Clearly, X^* is obtained from A by imposing on it the identical relation

$$(xy) z = x(yz). \tag{3.10.10}$$

Since (3.10.10) expresses the associative property, which is the characteristic relation of monoids, and since no other special identical relation holds in X^*, we refer to X^* as the *free monoid* generated by X. For example, if we add to (3.10.10) the identical relations $ax = b$ and $ya = b$, for fixed a, b in X and for some x and y in X, we obtain the structure of groups. The reader will also convince himself that the unary algebra $[X', f_{x_1}, ..., f_{x_n}]$ discussed at the beginning of this section is correctly denoted as the *free unary algebra* generated by Λ.

As another example, let us consider the infix and prefix notations for arithmetic expressions presented in Note 3.5. It is clear that an arithmetic expression in infix notation, such as (3.10.7), is an element of the word algebra generated by $(\{\times, /, +, -\}, \sigma)$ and the set $X \equiv \{A,B,C,D,E\}$, where $\sigma(\times) = \sigma(/) = \sigma(+) = \sigma(-) = 2$, and for each $\Delta \in \{\times,/,+,-\}$, $f_\Delta(x,y) = x\Delta y$. The expression (3.10.8) in prefix notation is a word in $\Sigma(X)$, where $\Sigma \equiv \{\times,/,+,-\}$, and, for each $\Delta \in \{\times,/,+,-\}$, $f_\Delta(x,y) = \Delta xy$.

Note 3.7. Phrase-Structure Grammars

Written languages are normally described in terms of expressions, letters, words, sentences, paragraphs, etc. A sentence is accepted as grammatically well formed on the basis of certain grammatical rules. For our purpose, we regard any "formal" language as a set of sequences over a finite alphabet, and the only rule of combining two words is the operation of catenation. The question to be considered in this note is whether there are grammatical rules for generating and classifying formal languages.

Consider the sentence "The man eats the apple" (given in Section 2.5), whose parsing tree is reproduced as figure 3.10.3. To find a set of rules which will generate the given sentence from the parsing tree, we may take the vertex whose label is "sentence" as the starting point and replace it by the two vertices with respective labels "noun phrase" and "verb phrase." These symbols will then be replaced by

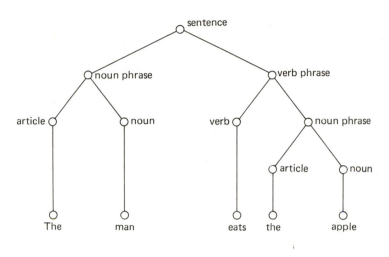

Figure 3.10.3

the labels of the succeeding vertices at the next level until we have reached the terminal vertices. The labels of the terminal vertices from left to right will then form the sentence.

More specifically, let S, NP, VP, A, N, and V represent, respectively: sentence, noun phrase, verb phrase, article, noun, and verb. For words α and β in A^*, let the notation $\alpha \to \beta$ denote the fact that any occurrence of α as a subword in a word α' can be replaced by the word β. Consider the following rules:

$$S \to NP\ VP$$
$$NP \to A\ N$$
$$VP \to V\ NP$$
$$A \to \text{the} \qquad\qquad (3.10.12)$$
$$N \to \text{man}$$
$$N \to \text{apple}$$
$$V \to \text{eats}$$

Beginning with the symbol S, the given sentence can then be derived using rules given in (3.10.12) by a sequence of "substitutions":

$$S \to NP\ VP \to A\ N\ VP \to \text{The N VP} \to \text{The man VP}$$
$$\to \text{The man V NP} \to \text{The man eats NP}$$
$$\to \text{The man eats A N} \to \text{The man eats the N}$$
$$\to \text{The man eats the apple.}$$

Of course, the rules given above generate other sentences, such as "The apple eats the man" and "The apple eats the apple." We note that the symbols S, NP, VP, A, N, V do not appear in the final sentence. They are merely markers used to help the derivation of sentences.

Motivated by the example above, we now define a *formal grammar F* as a quadruple $[V_N, V_T, S, P]$, where $V_N \cup V_T = V$ is called the *alphabet* of the grammar, V_N is called the set of *nonterminals*, V_T is called the set of *terminals*, and $V_N \cap V_T = \varnothing$. The letter S is the *sentence symbol* and is a member of V_N. The set P is a collection of grammatical rules, called *productions* (or *rewriting rules*) of the form $\alpha \rightarrow \beta$, where α, β are elements of V^* and α must be nonempty and contain at least one nonterminal. This type of grammar is usually referred to as a *phrase-structure* grammar for obvious reasons, as illustrated by the preceding example.

In order to define the language generated by a phrase-structure grammar, we need the following two relations: $\Rightarrow$ and $\overset{*}{\Rightarrow}$. Let G be a given phase-structure grammar. For x and y in V^*, we write $x \Rightarrow y$ if there exists z_1, z_2, α, and β such that $x = z_1 \alpha z_2$, $y = z_1 \beta z_2$ and $\alpha \rightarrow \beta \in P$. We write $x \overset{*}{\Rightarrow} y$ if either $x = y$ or there exist $z_0, z_1, \ldots, z_k$ such that $z_0 = x$ and $z_k = y$, and $z_i \Rightarrow z_{i+1}$ for each $1 \leqslant i \leqslant k$. Finally, the language generated by G, denoted by $L(G)$, consists of all words α over V_T such that $S \overset{*}{\Rightarrow} \alpha$. That is, $L(G) \equiv \{\alpha \mid \alpha \in V_T^* \text{ and } S \overset{*}{\Rightarrow} \alpha\}$.

As an example, consider the grammar $G = [V_N, V_T, S, P]$, where $V_N \equiv \{S\}$, $V_T \equiv \{(,)\}$, and P consists of the following productions:

$$S \rightarrow \Lambda, \qquad S \rightarrow SS, \qquad S \rightarrow (S), \qquad (3.10.11)$$

where Λ is the empty word. The given grammar will generate the set of all well-formed strings of parentheses. For illustrative purposes, we now give a typical derivation:

$$S \Rightarrow (S) \Rightarrow (SS) \Rightarrow ((S)S) \Rightarrow ((S)(S))$$
$$\Rightarrow ((S)((S))) \Rightarrow (( )((S))) \Rightarrow (( )(( ))).$$

The productions given in (3.10.11) may be written more compactly by combining all productions having the same left-hand side into one production with this unique left-hand side expression followed by the symbols "::=" and separating the different right-hand sides by vertical bars. Furthermore, in order to avoid ambiguity, elements of V_N are enclosed in angular brackets. Thus the productions given in (3.10.11) can be written compactly as follows:

$$\langle S \rangle ::= \Lambda \mid \langle SS \rangle \mid (\langle S \rangle).$$

The compact notation just described, usually referred to as the *Backus normal form,* is used in programming languages.

As another illustration of the Backus normal form, we give the following productions of the formal language called ALGOL.

1. $\langle \text{identifier} \rangle ::= \langle \text{letter} \rangle \mid \langle \text{identifier} \rangle \langle \text{letter} \rangle \mid \langle \text{identifier} \rangle \langle \text{digit} \rangle$.
2. $\langle \text{unsigned integer} \rangle ::= \langle \text{digit} \rangle \mid \langle \text{unsigned integer} \rangle \langle \text{digit} \rangle$.
3. $\langle \text{label} \rangle ::= \langle \text{identifier} \rangle \mid \langle \text{unsigned integer} \rangle$.

Phrase-structure grammar can be classified into several broad classes by restricting the form of productions. A phrase-structure grammar G is called *context-sensitive* if its productions are of the form

$$\alpha A \beta \rightarrow \alpha \gamma \beta \qquad \text{or} \qquad S \rightarrow \Lambda,$$

where $\alpha, \beta \in V^*$, $A \in V_N$, and $\gamma \in V^* - \{\Lambda\}$. The adjective "context-sensitive" refers to the fact that when one uses a rule of the form $\alpha A \beta \to \alpha \gamma \beta$, A is replaced by γ only in the context of "$\alpha \cdots \beta$." A grammar G is said to be *context-free* if each of its productions is of the form

$$A \to \alpha, A \in V_N, \alpha \in V^*.$$

The adjective "context-free" refers to the fact that the rewriting of A by α is independent of the context. Finally, G is called *regular* (or right linear) if each of its productions has one of the following forms:

$$A \to Ba, A \to a, A \to \Lambda \; (A, B \in V_N, a \in V_T).$$

The relationship between the various classes of grammars mentioned above is illustrated in figure 3.10.4.

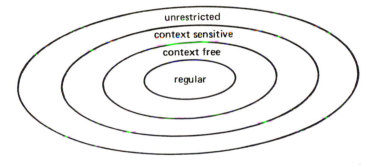

Figure 3.10.4

As another example, we will give a context-free grammar in Backus normal form which generates correct arithmetic expressions.

1. $\langle$digit$\rangle ::= 0|1|2|3|4|5|6|7|8|9$

2. $\langle$variable$\rangle ::= A|B|C|D$

3. $\langle$constant$\rangle ::= \langle$digit$\rangle | \langle$constant$\rangle\langle$digit$\rangle$

4. $\langle$atomic expression$\rangle ::= \langle$variable$\rangle | \langle$constant$\rangle | (\langle$arithmetic expression$\rangle)$

5. $\langle$arithmetic expression$\rangle ::=$

$\langle$atomic expression$\rangle | \langle$arithmetic expression$\rangle + \langle$atomic expression$\rangle |$

$\langle$arithmetic expression$\rangle - \langle$atomic expression$\rangle |$

$\langle$arithmetic expression$\rangle * \langle$atomic expression$\rangle |$

$\langle$arithmetic expression$\rangle / \langle$atomic expression$\rangle$

As an illustration, we give the following derivation for the arithmetic expression $((A + B)/(C * D))$. We shall use the shorthand $\langle$A$\cdot$E$\rangle$ and $\langle$At$\cdot$E$\rangle$ for $\langle$arithmetic expression⟩ and ⟨atomic expression⟩, respectively; the notation $\underset{i}{\Rightarrow}$ indicates that the i-th production has been used in that derivation.

$$\langle A \cdot E \rangle \underset{5}{\Rightarrow} \langle At \cdot E \rangle \underset{4}{\Rightarrow} (\langle A \cdot E \rangle) \underset{5}{\Rightarrow} (\langle A \cdot E \rangle / \langle At \cdot E \rangle) \underset{5}{\Rightarrow} (\langle At \cdot E \rangle / \langle At \cdot E \rangle)$$

$$\underset{4}{\Rightarrow} ((\langle A \cdot E \rangle) / \langle At \cdot E \rangle) \underset{5}{\Rightarrow} ((\langle A \cdot E \rangle + \langle At \cdot E \rangle) / \langle At \cdot E \rangle)$$

$$\underset{4}{\Rightarrow} ((\langle A \cdot E \rangle + \langle At \cdot E \rangle) / (\langle A \cdot E \rangle)) \underset{5}{\Rightarrow} ((\langle At \cdot E \rangle + (At \cdot E \rangle) /$$

$$(\langle A \cdot E \rangle * \langle At \cdot E \rangle))$$

$$\underset{5}{\Rightarrow} ((\langle At \cdot E \rangle + \langle At \cdot E \rangle) / (\langle At \cdot E \rangle * \langle At \cdot E \rangle))$$

$$\underset{4}{\Rightarrow} ((\langle variable \rangle + \langle variable \rangle) / (\langle variable \rangle * \langle variable \rangle))$$

$$\underset{2}{\Rightarrow} ((A + B) / (C * D))$$

CHAPTER 4

LATTICES

4.1 PARTIAL ORDERINGS

The reader is certainly familiar with the usual analysis of a rather complex system in terms of its components, such as an automobile comprising the engine, the body, etc. These components can be further analyzed in terms of smaller components, and so on, until one reaches parts that, for practical purposes, do not admit of further subdivision, usually because each consists of a single physical piece. Let us analyze more carefully the relation between the various constituents we have mentioned. Considering the product line of a large car manufacturer, we find that a certain engine is used in various models of cars, a certain gasoline pump is adopted in different engines, and so on. All these components—engines, gasoline pumps, etc.—are identifiable subassemblies. We can readily realize that this collection of assemblies is organized on the basis of the relation "to be part of" (or "to be contained by"). This relation is clearly reflexive (each subassembly is contained by itself) and transitive (if rubber ring r is used in gasoline pump p and gasoline pump p is used in engine e, then rubber ring r is used in engine e). However, this relation is evidently not symmetric; more strongly, it is antisymmetric. The reader will recall that this kind of relation on a set (reflexive, antisymmetric, and transitive) is called *partial ordering* (see Section 1.4). Actually, the given example is an instance of "set inclusion," which in turn is an instance of partial ordering.

A somewhat more abstract example is offered by the relation of divisibility over the set I of the positive integers. This relation is usually denoted by the symbol "|"; that is, for two positive integers a and b, $a|b$ means "a divides b" or, equivalently, "b is a multiple of a" ($b = ja$, for some positive integer j). The relation | is clearly reflexive since $a = 1 \times a$. It is also transitive since if for some positive integers j_1 and j_2, $b = j_1 a$ and $c = j_2 b$, then $c = (j_1 j_2)a$, and $j_1 j_2$ is a positive integer. Finally, it is antisymmetric since if $a = j_1 b$ and $b = j_2 a$ for positive integers j_1 and j_2, then we have, by substitution, $a = j_1 j_2 a$; that is, $j_1 j_2 = 1$. But this implies that $j_1 = j_2 = 1$, whence $a = b$. (Compare the last result with definition 1.4.3.)

It is worth pointing out that in both of the preceding examples the sets contain elements which are unrelated. With reference to the second example, think of two integers, neither of which divides the other, such as 4 and 10. This motivates the choice of the denotation *partial* ordering, in that *not all* elements are related. Related elements are referred to as *comparable*, whereas unrelated elements are said to be *incomparable*.

We now recall the following definition.

Definition 4.1.1. A *partial ordering* is a reflexive, antisymmetric, and transitive relation on a set.

The usual symbol for a partial ordering is "$\leqslant$" with "$\geqslant$" denoting the inverse relation. Note that, in accordance with theorem 1.4.1 on the preservation of properties through inversion of relations, the relation $\geqslant$ is reflexive, antisymmetric, and transitive. This deserves to be stated formally as a corollary.

Corollary 4.1.1 (of theorem 1.4.1). The inverse relation of a partial ordering is also a partial ordering.

We shall return to this point later.

Example 4.1.1. Let $|$ be the relation of divisibility on the set of the integers $\{1,2,3,4,6,8,12\}$. It is instructive to construct both the tabular and the graphical representations of the relation $|$; they are given in figure 4.1.1(a) and (b), respectively.

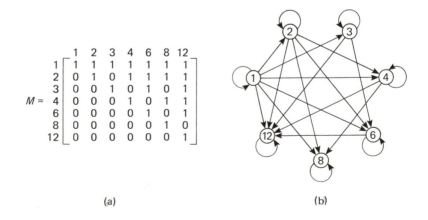

$$
M = \begin{array}{c|ccccccc}
 & 1 & 2 & 3 & 4 & 6 & 8 & 12 \\
\hline
1 & 1 & 1 & 1 & 1 & 1 & 1 & 1 \\
2 & 0 & 1 & 0 & 1 & 1 & 1 & 1 \\
3 & 0 & 0 & 1 & 0 & 1 & 0 & 1 \\
4 & 0 & 0 & 0 & 1 & 0 & 1 & 1 \\
6 & 0 & 0 & 0 & 0 & 1 & 0 & 1 \\
8 & 0 & 0 & 0 & 0 & 0 & 1 & 0 \\
12 & 0 & 0 & 0 & 0 & 0 & 0 & 1 \\
\end{array}
$$

(a)   (b)

Figure 4.1.1. Tabular and graphical representations of

$|$ on $\{1,2,3,4,6,8,12\}$

We expect, because of the antisymmetric property, that the graph of a partial ordering does not contain any symmetric pair (see figure 1.4.2(b)) or, in the terminology introduced in Chapter 2, any circuit of length 2 (see definition 2.1.5). This is true, for example, in figure 4.1.1(b). The antisymmetric property, however, induces an even stronger property on the directed graph of a partial ordering relation, as stated by the following simple but significant theorem.

Theorem 4.1.2. The directed graph of a partial ordering relation contains no circuit of length greater than 1.

Proof. The proof is by contradiction. Indeed, assume that the directed graph of $\leqslant$ contains a circuit of length n. If $n = 2$, a circuit of length 2 directly violates the antisymmetric property, and the theorem holds trivially. For $n > 2$, let $a_1, a_2, \ldots, a_n$ be the vertices of such a circuit consisting of arcs $(\overrightarrow{a_1, a_2})$, ..., $(\overrightarrow{a_{n-1}, a_n})$, $(\overrightarrow{a_n, a_1})$. The existence of arcs (a_1, a_2) and (a_2, a_3) implies, by the transitive property, the existence of (a_1, a_3); by the same argument, (a_1, a_3) and (a_3, a_4) imply the existence of (a_1, a_4), and so on. So we conclude that the graph contains the arc (a_1, a_n). (See figure 4.1.2.) But the assumed circuit contains also the arc $(\overrightarrow{a_n, a_1})$, by definition of circuit; that is, the graph contains the symmetric pair $(\overrightarrow{a_1, a_n})$ and $(\overrightarrow{a_n, a_1})$, a fact which violates the hypothesis that $\leqslant$ is antisymmetric. Therefore the thesis holds.‖

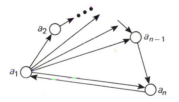

Figure 4.1.2. Aid to the proof of theorem 4.1.1

EXERCISES 4.1

1. Determine which of the following relations is a partial ordering.

   a) The relation "less than or equal to" on the set of integers.

   b) The relation "is a subset of" in a collection of sets.

   c) The relation "is a proper subset of" in a collection of sets.

   d) The lexicographic ordering of the English words.

   e) The relation γ on the set of all real functions which are continuous over the interval $[0, 1]$ defined as follows: $f \gamma g$ if and only if $f(x) \leqslant g(x)$ for all x in the closed interval $[0, 1]$.

   f) The relation γ defined on the set of all subgroups of a group G such that $G_1 \gamma G_2$ if and only if G_1 is a subgroup of G_2.

2. Suggest one or more partial ordering relations on the set $\{1,4,9,16,25,\ldots\}$.

3. Show that the identity relation on a set is the only relation which is both a partial ordering and an equivalence relation.

4. Enumerate all partial orderings on a three-element set.

5. Let $\leqslant$ be a partial ordering on a set A and $B \subseteq A$. Define a relation $\leqslant'$ on B such that for $a, b \in B$, $a \leqslant' b$ if and only if $a \leqslant b$. Show that $\leqslant'$ is a partial ordering on B.

6. Let F denote the set of all partial orderings of a set A. Define a relation $\leqslant$ on F such that for α and β in F, $\alpha \leqslant \beta$ means that for arbitrary elements a,b in A, $a\alpha b$ implies that $a\beta b$. Show that $\leqslant$ is a partial ordering on F.

*7. A *preordering* (or quasi ordering) is a reflexive and transitive relation. For example, in the monoid M_{15} of the integers modulo 15 under multiplication, we say that $a|b$ for a and b in M_{15}, if $ax = b$, for some $x \in M_{15}$. The relation $|$ is reflexive and transitive but not antisymmetric. In fact $2|8$, since $2 \cdot 4 = 8$ and $8|2$ since $8 \cdot 4$ mod $15 = 2$, but $2 \neq 8$. Thus $|$ is a preordering but *not* a partial ordering.

Can you think of some other example of a preordering which is not a partial ordering?

4.2 POSETS

A set on which a partial ordering relation is defined is worth a special denotation.

Definition 4.2.1. A *partly ordered set* (*poset*) consists of a set S and a partial ordering relation $\leqslant$ on S. A poset is usually denoted by the pair $[S, \leqslant]$, which also denotes the graph of the relation $\leqslant$ on S.

An interesting case of posets is offered by the class of *simply* (or *linearly*) *ordered sets* or *chains*, that is, those posets R for which any pair of elements is comparable. A simple instance of such sets is offered, for example, by the integers $\{1,2,3,4,5,6\}$ under the relation of "magnitude comparison." Suppose now that we want to label each element of a finite chain R with distinct successive positive integers. This operation, referred to as *enumeration*, is the assignment of the integer $i(a)$ to the element $a \in R$. It is intuitively clear that we can always enumerate the elements of a finite chain R so that, for a and b in S, we have $a \leqslant b$ if and only if $i(a)$ is less than $i(b)$.

A question which naturally arises at this point is whether something analogous can be done with a generic finite poset $[S, \leqslant]$. In other words, we are asking whether we can find an *enumeration* of the elements of S (note that "enumeration" is synonymous with "simple ordering") which is *consistent* with the partial ordering of S. A consistent enumeration means that, for a and b in S, if $a \leqslant b$, then $i(a)$ must be less than $i(b)$. (Note that the converse is not true, since if $i(a)$ is less than $i(b)$, still a and b may not be comparable. Contrast this remark with the case of simply ordered sets, any two elements of which are comparable.) This question

has an affirmative answer, as we shall shortly show. We first need the following·
definition.

Definition 4.2.2. A *consistent enumeration* of a finite poset $S \equiv \{s_1, s_2, \ldots, s_n\}$ is an
assignment of integers $i(s_j)$ to s_j so that $s_p \leqslant s_q$ implies $i(s_p) < i(s_q)$.[1]

With this nomenclature, we can now state and prove the following theorem.

Theorem 4.2.1. (Theorem of consistent enumeration). Every finite poset $S \equiv \{s_1, s_2, \ldots, s_n\}$ admits of a consistent enumeration.

Proof. The proof is of a constructive nature; that is, we give a procedure for
obtaining the function $i: S \to I$, from the set S to the set of the positive integers,
so that i is a consistent enumeration.

The key to the proof is an induction argument. We note first of all that a
consistent enumeration for the single element s_1 exists trivially; that is, $i(s_1) = 1$.
Then we assume that we have obtained a consistent enumeration for the subset
$\{s_1, s_2, \ldots, s_k\}$ of S and show that we can construct a consistent enumeration for
$\{s_1, s_2, \ldots, s_k, s_{k+1}\}$. Indeed, if s_{k+1} is incomparable with any element of $s_1, \ldots, s_k$,
we choose

$$i(s_{k+1}) = \max_{1 \leqslant j \leqslant k} i(s_j) + 1$$

and obtain a consistent enumeration of $\{s_1, \ldots, s_k, s_{k+1}\}$. Otherwise, consider the
set $P \equiv \{s_j | \text{either } s_j \leqslant s_{k+1} \text{ or } s_{k+1} \leqslant s_j, j \leqslant k\}$, i.e., the set of the elements of
$\{s_1, \ldots, s_k\}$ which are comparable with s_{k+1}. Clearly, P is a subset of $\{s_1, \ldots, s_k\}$,
and let $P \equiv \{s_1', s_2', \ldots, s_h'\}$. Note that P, a subset of a consistently enumerated set,
is also consistently enumerated. Now we compare s_{k+1} with the elements of P, and
we have one of the following three cases.

1. $s_{k+1} \leqslant s_p'$ for every $s_p' \in P$. In this case we set $i(s_{k+1}) = 1$ and increment by 1
    the previous values of $i(s_j)$, for $j = 1, 2, \ldots, k$.

2. $s_p' \leqslant s_{k+1}$ for every $s_p' \in P$. In this case we set

    $$i(s_{k+1}) = \max_j i(s_j) + 1.$$

3. There are two elements s_p' and s_q' in P such that $i(s_p')$ is the largest and $i(s_q')$ is
    the smallest integer in the enumeration of $\{s_1, \ldots, s_k\}$ for which $s_p' \leqslant s_{k+1} \leqslant s_q'$.
    In this case we set $i(s_{k+1}) = i(s_p') + 1$ and increment by 1 the previous values
    of $i(s_j)$, for $i(s_j) > i(s_p')$.

By construction, in all three cases we obtain a consistent enumeration.‖

---

1. Note that in the definition the symbol "$\leqslant$" denotes the partial ordering in S, and the
symbol "$<$" denotes the simple ordering of integers.

Example 4.2.1. Consider the poset $S \equiv \{s_1, s_2, s_3, s_4, s_5\}$, whose graph is given in figure 4.2.1. We shall now apply the constructive procedure embodied in the proof of theorem 4.2.1 to obtain consistent enumerations for the sets $\{s_1\}$, $\{s_1, s_2\}$,

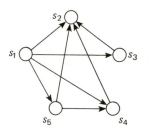

Figure 4.2.1. A poset

$\{s_1, s_2, s_3\}$, $\{s_1, s_2, s_3, s_4\}$, and $\{s_1, s_2, s_3, s_4, s_5\}$. It is convenient to illustrate each step as the insertion of one vertex in a linear array of vertices (figure 4.2.2).

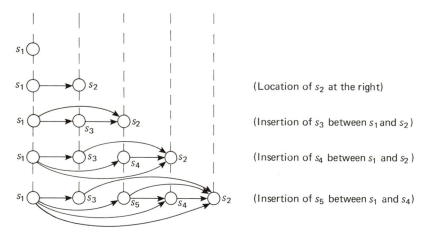

Figure 4.2.2. Sequence of consistent enumerations

We remark that a consistent enumeration of a poset, as produced by the given method is, in general, not unique. Indeed, reexamining the procedure, we note that in step 3, when $i(s'_p)$ and $i(s'_q)$ are not consecutive, we have the choice of inserting s_{k+1} in any "slot" between them. In example 4.2.1, in the progression from the enumeration of $\{s_1, s_2, s_3, s_4\}$ to the one of $\{s_1, s_2, s_3, s_4, s_5\}$, the element s_5 may be inserted either between s_3 and s_4 (as in figure 4.2.2) or between s_1 and s_3.

That this is not true for chains, however, can be seen very simply. The element s_{k+1} is by hypothesis comparable with all the elements $\{s_1, s_2, \ldots, s_k\}$, which themselves form a chain. Hence there are two unique elements s_p and s_q for which

$i(s_q) = i(s_p) + 1$ such that $s_p \leqslant s_{k+1} \leqslant s_q$, and the insertion of s_{k+1} can be performed in only one way; i.e., the consistent enumeration of a chain is unique—as was intuitively clear!

The result of theorem 4.2.1 could be interpreted in the following way. The graph of a poset $[S, \leqslant]$ can always be redrawn, simply by displacing the vertices, so that all the arrows point in the same direction (for example, in figure 4.2.2, all arrows point from left to right). This property holds also if the vertices in the redrawn graph do not form a linear array, as long as their projection on a straight line is a consistent enumeration. This, then, suggests a simplification in drawing the graph of a poset $[S, \leqslant]$ in that we may place the vertices so that the arrows can be omitted. The usual convention is that the horizontal projection of the vertices of $[S, \leqslant]$ on a vertical line yields a consistent enumeration, so that for any edge the arrow is intended to point to the upper extreme of the edge (see figure 4.2.3). Recall that a similar convention was adopted for representing rooted trees (definition 2.5.3; figure 2.5.8). With this convention the graph of figure 4.1.1(b) can be redrawn as in figure 4.2.3 (where loops have been omitted).

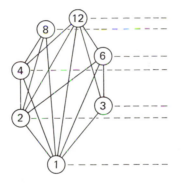

Figure 4.2.3. A simplified version of the diagram of figure 4.1.1(b)

Looking at figure 4.2.3, we notice the presence of pairs of vertices a and b such that $a \leqslant b$ and there is no other vertex c for which $a \leqslant c \leqslant b$. This is true, for example, of vertices 2 and 6 in figure 4.2.3, but not of vertices 2 and 8, since $2 \leqslant 4 \leqslant 8$. Pairs of elements for which this happens are related in a more special way than two generically comparable elements. This special relation is specified by the following definition.

Definition 4.2.3. In a poset, $[S, \leqslant]$, a is the *immediate predecessor* of b if $a \leqslant b$ and there is no other element c in S for which $a \leqslant c \leqslant b$. This relation is denoted by the symbol $<\!\!\cdot$. Its inverse relation is termed the *immediate successor*.

The relation of "immediate predecessor" (which is not reflexive, symmetric, or transitive) is defined on a poset; i.e., it appears to imply an underlying "partial ordering." Indeed, the two relations contain the same information in the sense that if one is specified, we can easily obtain the other. The following discussion gives operational substance to this statement.

Assume that a poset $[S, \leqslant]$ is given by specifying the relation "partial ordering," $\leqslant$. In order to obtain the pairs related under the "immediate predecessor" relation, $<$, in the graph $[S, \leqslant]$, we must remove arcs such as $\overrightarrow{(a,c)}$ from each three-edge configuration, as shown in figure 4.2.4. Note that $\overrightarrow{(a,c)}$ exists if and

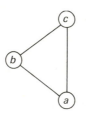

Figure 4.2.4. A transitive three-edge configuration

only if there is a path of *two* arcs from a to c connecting the *three* distinct vertices a, b, and c. Therefore a path of two arcs, one of which is a loop, is excluded. This immediately suggests a method to remove the unwanted arcs from the graph of a "partial ordering" in order to obtain the graph of the corresponding "immediate predecessor." Indeed, we first remove the loops from the graph $[S, \leqslant]$ and thereby ensure that any two-arc path traverses three vertices; let $[S, \alpha]$ be the resulting graph, where α coincides with $\leqslant$ except for reflexivity. Second, we note that to any two-arc path in $[S, \alpha]$ there corresponds a single arc in the graph $[S, \alpha^2]$,[2] that is, the graph of the composite relation α^2. We denote by M, M_α, and $M_{\alpha 2}$ the arrays of the relations $\leqslant$, α, and α^2, respectively (recall that by corollary 2.2.2, $M_{\alpha 2} = M_\alpha^2$). We know from Section 1.4 that, because of transitivity, if an entry of M_α^2 is a 1, so is the corresponding entry of M_α. If we now delete from M_α all the 1's for which the corresponding entries in M_α^2 are 1's, the resulting array describes the relation of immediate predecessor, $<$, associated with the partial ordering, $\leqslant$.

Example 4.2.2. Refer to the poset of Example 4.1.1. We easily compute, by means of the array multiplication rule (definition 1.2.4), the array M_α^2, as shown in figure 4.2.5.

Comparing M_α^2 and M_α (figure 4.1.1(b)), we obtain the array $M_<$ of $<$ (figure 4.2.6), which describes the desired "immediate predecessor" relation.

---

2. By the symbol α^k we denote, as usual, the k-th power of α (definition 2.2.2).

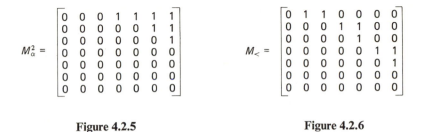

$$M_\alpha^2 = \begin{bmatrix} 0 & 0 & 0 & 1 & 1 & 1 & 1 \\ 0 & 0 & 0 & 0 & 0 & 1 & 1 \\ 0 & 0 & 0 & 0 & 0 & 0 & 1 \\ 0 & 0 & 0 & 0 & 0 & 0 & 0 \\ 0 & 0 & 0 & 0 & 0 & 0 & 0 \\ 0 & 0 & 0 & 0 & 0 & 0 & 0 \\ 0 & 0 & 0 & 0 & 0 & 0 & 0 \end{bmatrix} \qquad M_< = \begin{bmatrix} 0 & 1 & 1 & 0 & 0 & 0 & 0 \\ 0 & 0 & 0 & 1 & 1 & 0 & 0 \\ 0 & 0 & 0 & 0 & 1 & 0 & 0 \\ 0 & 0 & 0 & 0 & 0 & 1 & 1 \\ 0 & 0 & 0 & 0 & 0 & 0 & 1 \\ 0 & 0 & 0 & 0 & 0 & 0 & 0 \\ 0 & 0 & 0 & 0 & 0 & 0 & 0 \end{bmatrix}$$

Figure 4.2.5   **Figure 4.2.6**

Assume now, conversely, that a poset is given by means of the "immediate predecessor" relation $<$ (that is, by means of the graph $[S, <]$) and that we want to obtain the associated partial ordering $\leqslant$. To this end, we must introduce the arcs expressing the reflexive property (i.e., loops) and those expressing the transitive property; namely, in the graph $[S, \leqslant]$ for any pair of arcs such as $\overrightarrow{(a,b)}$ and $\overrightarrow{(b,c)}$ in figure 4.2.4, there must be the arc $\overrightarrow{(a,c)}$. If we compute the composite $<^2$, we recognize that, for any pair $\overrightarrow{(a,b)}$, $\overrightarrow{(b,c)}$ in the graph $[S, <]$, the graph $[S, <^2]$ contains $\overrightarrow{(a,c)}$. However, there may be arcs in $[S, <]$ which disappear in $[S, <^2]$. To ensure that each arc of $[S, <]$ does not disappear through composition, we simply need to add loops to each node of $[S, <]$. Let $[S, \beta]$ be the resulting graph. Since β is reflexive, we know (see Section 1.4) that β^2 covers β. It is still possible, however, that $[S, \beta^2]$ contains configurations of the type of figure 4.2.4 with the arc $\overrightarrow{(a,c)}$ missing; therefore we must iterate the composition of β until for some k, the $(k-1)$-time composite and the k-time composite of β contain the same arcs. This calculation is certain to end. We state without proof—but a little intuition will justify the statement—that if S contains n elements, the relation β must be composed at most $(n-1)$ times (compare with definition 2.2.3). Quite justifiedly, $[S, \leqslant]$ is said to be the transitive and reflexive closure of $[S, <]$.

Example 4.2.3. This example is the converse of example 4.2.2; i.e., we place 1's on the main diagonal of $M_<$, call the resulting array M_β, and form M_β^2, M_β^3, ... (figure 4.2.7); clearly $M_\beta^3 = M_\beta^4$ is the array of the partial ordering relation $\leqslant$ associated with $<$.

The structure of the poset of figure 4.2.3 can be exhibited by means of the graph of $<$ on S (see figure 4.2.8). It is almost superfluous to note that the graphical representation of posets by means of the graph $[S, <]$ is much more economical in terms of drawn arcs than that given by the graph $[S, \leqslant]$. (Indeed, $\leqslant$ covers $<$!) The former is the *standard* representation of posets, called a *Hasse diagram*. A path in a Hasse diagram is normally referred to as a *chain*, although it should be called a path since $[S, \leqslant]$ is a directed graph (compare definitions 2.1.4 and 2.3.2).

We conclude this section with some important nomenclature on posets.

$$M_\beta = \begin{bmatrix} 1 & 1 & 1 & 0 & 0 & 0 & 0 \\ 0 & 1 & 0 & 1 & 1 & 0 & 0 \\ 0 & 0 & 1 & 0 & 1 & 0 & 0 \\ 0 & 0 & 0 & 1 & 0 & 1 & 1 \\ 0 & 0 & 0 & 0 & 1 & 0 & 1 \\ 0 & 0 & 0 & 0 & 0 & 1 & 0 \\ 0 & 0 & 0 & 0 & 0 & 0 & 1 \end{bmatrix}, \quad M_\beta^2 = \begin{bmatrix} 1 & 1 & 1 & 1 & 1 & 0 & 0 \\ 0 & 1 & 0 & 1 & 1 & 1 & 1 \\ 0 & 0 & 1 & 0 & 1 & 0 & 1 \\ 0 & 0 & 0 & 1 & 0 & 1 & 1 \\ 0 & 0 & 0 & 0 & 1 & 0 & 1 \\ 0 & 0 & 0 & 0 & 0 & 1 & 0 \\ 0 & 0 & 0 & 0 & 0 & 0 & 1 \end{bmatrix}$$

$$M_\beta^3 = \begin{bmatrix} 1 & 1 & 1 & 1 & 1 & 1 & 1 \\ 0 & 1 & 0 & 1 & 1 & 1 & 1 \\ 0 & 0 & 1 & 0 & 1 & 0 & 1 \\ 0 & 0 & 0 & 1 & 0 & 1 & 1 \\ 0 & 0 & 0 & 0 & 1 & 0 & 1 \\ 0 & 0 & 0 & 0 & 0 & 1 & 0 \\ 0 & 0 & 0 & 0 & 0 & 0 & 1 \end{bmatrix} = M_\beta^4 = M$$

Figure 4.2.7

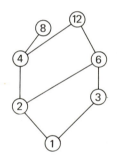

Figure 4.2.8. Hasse diagram of the poset of figure 4.1.1(b), the graph of the associated immediate predecessor relation

Definition 4.2.4. An element $m \in [S, \leqslant]$ is *maximal* when there is no other element a in $[S, \leqslant]$ for which $m \leqslant a$ (a *minimal* element is defined similarly by considering the relation $\geqslant$).

Referring to the Hasse diagram of $[S, \leqslant]$, we note that a maximal (minimal) element is not connected to any element above (below) it; for example, in figure 4.2.8, elements 8 and 12 are maximal and element 1 is minimal.

Definition 4.2.5. An element $I \in [S, \leqslant]$ is the *greatest* when $a \leqslant I$ for every $a \in [S, \leqslant]$ (the *least* element O is defined similarly).

Note the difference between a maximal element and the greatest element: The latter is also maximal but is, in addition, comparable to every other element. This property also leads directly to a demonstration of the uniqueness of the greatest element. Indeed, assume that there are two such elements, I_1 and I_2;

then we have simultaneously $I_1 \leqslant I_2$ and $I_2 \leqslant I_1$; that is, $I_1 = I_2$ (by the very definition of the antisymmetric property, definition 1.4.3). The greatest and least elements, I and O, are also referred to as *upper and lower universal bounds*, and hereafter we shall use the latter terminology.

A finite poset certainly has minimal and maximal elements but not necessarily the universal bounds. For example, the poset in figure 4.2.8 has the lower universal bound (the element 1) but not the upper universal bound.

EXERCISES 4.2

1. Let $S \equiv \{1,2\}$ and let $\mathscr{P}(S)$ denote the set of all subsets of S. Find all consistent enumerations of the poset $[\mathscr{P}(s), \subseteq]$ using the integers 1, 2, 3, 4 for the enumeration.

2. List all the nonisomorphic posets of three elements.

3. Compute and sketch the Hasse diagrams of the posets $[S, <']$ and $[S, <'']$, where $S \equiv \{s_1, s_2, s_3, s_4, s_5\}$ and the partial orderings $<'$ and $<''$ are described by the following arrays.

$$M_{\leqslant'} = \begin{bmatrix} 1 & 0 & 0 & 1 & 0 \\ 1 & 1 & 1 & 1 & 1 \\ 0 & 0 & 1 & 1 & 0 \\ 0 & 0 & 0 & 1 & 0 \\ 0 & 0 & 0 & 1 & 1 \end{bmatrix}, \quad M_{\leqslant''} = \begin{bmatrix} 1 & 0 & 1 & 0 & 1 \\ 0 & 1 & 1 & 0 & 1 \\ 0 & 0 & 1 & 0 & 1 \\ 1 & 1 & 1 & 1 & 1 \\ 0 & 0 & 0 & 0 & 1 \end{bmatrix}$$

4. Draw the Hasse diagram, and list all minimal and maximal elements and the least and greatest elements (if they exist) of each of the following posets.

   a) $[\mathscr{P}(S), \subseteq]$ as given in exercise 4.2.1
   b) $[S, |]$, where $S \equiv \{1, 2, 3, 5, 10, 15, 30\}$ and $|$ is the relation "to be divisible by"

5. A *linear* (or *simple*, or *total*) *ordering* on a set S is a partial ordering $\leqslant$ such that for any two elements a, b in S, either $a \leqslant b$ or $b \leqslant a$.

   a) What does the Hasse diagram of a linearly ordered set look like?
   b) Give an example of a partial ordering which is not a linear ordering.

6. Let $[S, \leqslant]$ be a poset and S' a subset of S. Show that S' together with $\leqslant$ restricted to S' is also a poset.

7. Let S' and S be as given in exercise 4.2.6. Show that if $x \in S'$ is the greatest (least) element of S, then x is the greatest (least) element of S'.

8. Let $[S, \leqslant]$ be a poset, and let a, b be two elements of S such that $a \leqslant b$. The set $\{x | a \leqslant x \leqslant b\}$ is called an *interval* of S, denoted by $[a, b]$. Show that $[a, b]$ is a poset containing a least element and a greatest element. Furthermore, show that $[a, b] = [c, d]$ if and only if $a = c$ and $b = d$.

9. Let $P = [S, \leqslant]$ and $P' = [S', \leqslant']$ be two posets. An onto function $\varphi : S \to S'$ is said to be *order-preserving* if for all $a, b \in S$,

$$a \leqslant b \text{ implies that } \varphi(a) \leqslant' \varphi(b).$$

If φ is also one-to-one, then P and P' are called *order-isomorphic*. Show that the following statements are true.

a) If $\varphi : P_1 \to P_2$ and $\psi : P_2 \to P_3$ are order-preserving functions, so is their composite $\varphi\psi$.

b) Two posets have the same Hasse diagram if and only if they are order-isomorphic.

c) An order-preserving onto function maps the greatest (least) element into the greatest (least) element.

d) An order-preserving onto function maps a maximal (minimal) element into a maximal (minimal) element.

4.3 THE OPERATIONS OF JOIN AND MEET; LATTICES

We are now ready for the introduction of binary operations on partly ordered sets.

We shall consider a company which designs, manufactures, and sells digital computers of various sizes, all built with the same types of components. In other words, the designers have available a basic set of parts—moving parts, a family of printed circuits, etc.—and each new computer model is a new assembly of the same set of parts. When a computer is delivered to a customer, a stock of spare parts must also be delivered for the purpose of preventive and corrective maintenance. We will assume here that the stock for a given computer contains exactly one of each item used in the computer. (This assumption, made to simplify the discussion, is somewhat unrealistic. In practice, spare part stocks are tailored to the observed frequencies of failure, so that some parts may be omitted from the stock, whereas more than one unit of others may be included.) Several computers of different models, manufactured by this company, may be installed at the same customer site. Specifically, suppose that s_i is the part stock for computer c_i, and that computers c_i and c_j are installed at the same site. If we provide the installation with both s_i and s_j, clearly the stock is adequate; but we see that the possibly smaller stock, composed of all and only the parts which are in either s_i or s_j (or both), is also adequate. We recognize that collections of spare parts form a poset under the relation of inclusion. A collection s which includes both s_i and s_j is said to be an "upper bound" of s_i and s_j; if no member can be removed from s without s ceasing to be an upper bound of s_i and s_j, then s is said to be a "least upper bound." This notion is now formalized.

Definition 4.3.1. For a and b in a poset $[S, \leqslant]$ a *least upper bound* (l.u.b) or *join* or *sum* of a and b is an element c of S which satisfies the relationships $a \leqslant c$ and $b \leqslant c$, and there is no other x in S for which $a \leqslant x \leqslant c$ and $b \leqslant x \leqslant c$. If two elements a and b in S have a unique join, the latter is denoted by $(a \vee b)$.

Our definition of *join* rests entirely on the partial ordering ≤ defined on S, and it may assume completely different connotations depending on the nature of ≤. This will become apparent in subsequent examples.

Given two arbitrary elements a and b of a poset $[S, ≤]$, a join of a and b may or may not exist as an element of S. The existence of a join clearly depends on the membership of S and the nature of ≤. For instance, referring to example 4.1.1, where $S ≡ \{1, 2, 3, 4, 6, 8, 12\}$ under divisibility, we recognize that the join of 3 and 8 (the least integer divisible by both 3 and 8, that is, 24) does not belong to the poset. The following theorem offers a necessary and sufficient condition for the existence of a join of a and b in a finite poset $[S, ≤]$.

Theorem 4.3.1. For any two arbitrary elements a and b in a finite poset $[S, ≤]$, a join of a and b exists in $[S, ≤]$ if and only if $[S, ≤]$ has the universal upper bound I.

Proof. Assume that $[S, ≤]$ contains the element I. Note that if a and b in $[S, ≤]$ have an upper bound in $[S, ≤]$, then they also have a least upper bound, i.e., a join, in $[S, ≤]$. But I is an upper bound of a and b, since $a ≤ I$ and $b ≤ I$ by definition of universal upper bound, whence a join of a and b exists in $[S, ≤]$.

Conversely, suppose that $[S, ≤]$ does not contain the element I. Then it must contain at least two maximal elements, say, m_1 and m_2. Assume now that a join m of m_1 and m_2 exists in $[S, ≤]$; then it must satisfy $m_1 ≤ m$ and $m_2 ≤ m$, but this violates the statement that m_1 and m_2 are maximal. Since a contradiction has been reached, m cannot exist in $[S, ≤]$.∥

Given a finite poset $[S, ≤]$, a join e of two arbitrary elements a and b of S can be readily obtained with the aid of the Hasse diagram of the poset. Indeed, $a ≤ e$ means the existence of an ascending chain from a to e (see figure 4.3.1); so does the relationship $b ≤ e$. Therefore e is an element common to ascending chains from a and from b, such that no other element $d ≤ e$ enjoys the same property. This also gives a significance to the term "join". A join of a and b is a vertex where chains from a and b first join together.

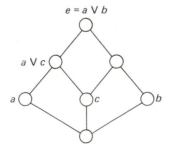

Figure 4.3.1. Determination of joins on the Hasse diagram

Example 4.3.1. In Section 1.5 we introduced the concept of *partition of a set A* as a collection of subsets of A such that each a in A belongs to exactly one of them. From this concept we now derive that of *partition of an integer*, by replacing each subset by its cardinalities; i.e., a partition of a positive integer n is a collection of positive integers whose sum is n. For example $\{1,1,1,1\}$, $\{1,1,2\}$, $\{1,3\}$, $\{2,2\}$, $\{4\}$ are all the partitions of the integer 4. Now, on the finite set $Q_n \equiv \{P_1, P_2, \ldots\}$ of the partitions of n, we consider a partial ordering, as follows: $P_i \leqslant P_j$ if P_j can be obtained from P_i by grouping summands of P_i (for example, $\{1,1,2\} \leqslant \{2,2\}$, since one can group two summands of $\{1,1,2\}$ to obtain a summand of $\{2,2\}$). By so doing, we get the Hasse diagram of $[Q_4, \leqslant]$ shown in figure 4.3.2(a), and we see that any two elements have a *unique* join. Consider now, in an analogous manner, the partitions of the integer 5. They are $Q_5 \equiv \{\{1,1,1,1,1\}, \{1,1,1,2\}, \{1,2,2\}, \{1,1,3\}, \{2,3\}, \{1,4\}, \{5\}\}$, and the Hasse diagram of $[Q_5, \leqslant]$ for the same partial ordering is shown in figure 4.3.2(b). The reader will notice that, although any two elements have a join because $P_7 \equiv \{5\}$ is the universal upper bound of Q_5, P_4 and P_3 do not have a unique join; indeed, both P_5 and P_6 are joins of P_3 and P_4.

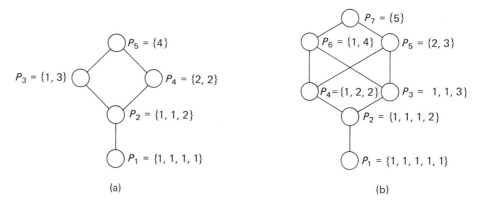

Figure 4.3.2. Hasse diagrams of the partitions of the integers 4 and 5

When for any two elements a and b in $[S, \leqslant]$ the unique join $a \vee b$ exists, then "join" becomes a function $\vee : \{S \times S\} \rightarrow S$; that is, it is a binary operation on the set S (definition 3.1.1). For the sake of brevity, the phrase "a poset with a binary operation" means that the result of the binary operation is a unique element of S for arbitrary operands in S (equivalently, join is closed in S).

At this point we consider a new notion in a poset with a definition that parallels definition 4.3.1.

Definition 4.3.2. For a and b in a poset $[S, \leqslant]$ a *greatest lower bound* (g.l.b) or *meet* or *product* of a and b is an element d of S which satisfies the relationships

$d \leqslant a$ and $d \leqslant b$, and there is no other x in S for which $d \leqslant x \leqslant a$ and $d \leqslant x \leqslant b$. If two elements a and b in S have a unique meet, the latter is denoted by $a \wedge b$.

Comparing the formal definitions of l.u.b. (definition 4.3.1) and g.l.b. (definition 4.3.2), we notice that we may obtain one from the other by interchanging the symbols $(\wedge, \vee)$ and $(\leqslant, \geqslant)$. This indicates that the "meet" coincides with the "join" of the inverse relation, which, by corollary 4.1.1, is also a partial ordering. This has a very interesting interpretation in terms of the Hasse diagrams; indeed, the Hasse diagram of $[S, \geqslant]$ is the Hasse diagram of $[S, \leqslant]$ *turned upside-down*. We also realize that all the previous discussion concerning the join can be repeated exactly for the meet with the following substitutions.

$$\text{join} \leftrightarrow \text{meet}$$
$$\text{l.u.b.} \leftrightarrow \text{g.l.b.}$$
$$\vee \leftrightarrow \wedge$$
$$\leqslant \leftrightarrow \geqslant$$
$$I \leftrightarrow O$$

Therefore any statement proved for join is automatically proved for meet with the given substitutions. This is known as the *duality principle* of posets, and it descends directly from corollary 4.1.1. Accordingly, join and meet are referred to as *dual* operations.

Posets $[S, \leqslant]$ with two binary operations join and meet form an extremely important class, as expressed by the following definition.

Definition 4.3.3. A *lattice* is a poset $[S, \leqslant]$, any two elements of which have unique join and meet. The symbol for a lattice is $[S, \vee, \wedge]$.

It is important to realize at this point that join and meet, hence lattices, have been defined exclusively in terms of l.u.b. and g.l.b., which in turn hinge on the partial ordering. Therefore the properties of join and meet, which we shall now elucidate, are due entirely to the relation $\leqslant$ and the fact that the results of the operations exist and are unique. Because of the duality principle, we need to analyze only one of the two operations, for example, join.

First, note that, for $a \in [S, \leqslant]$, the l.u.b. of a and a is a itself, expressed formally as

$$a \vee a = a. \quad \text{(property of idempotency)} \tag{4.3.1}$$

Observe now that the l.u.b. of a and b does not change when the order in which a and b are selected is changed; that is,

$$a \vee b = b \vee a. \quad \text{(property of commutativity)} \tag{4.3.2}$$

Next, we define the l.u.b. of a set of three elements a, b, $c \in [S, \leqslant]$ as an element d for which $a \leqslant d$, $b \leqslant d$, and $c \leqslant d$ and such that there is no other element

$x \in [S, \leqslant]$ for which $a \leqslant x \leqslant d$, $b \leqslant x \leqslant d$, and $c \leqslant x \leqslant d$. We claim that $d = (a \vee b) \vee c$. Indeed, $a \vee b \leqslant d$. If it were $d \leqslant a \vee b$ and $a \vee b$ and d were distinct, then d rather than $a \vee b$ would be the l.u.b. of a and b. Moreover, suppose that there is an element $e \in [S, \leqslant]$ for which $a \vee b \leqslant e \leqslant d$ and $c \leqslant e \leqslant d$. Then we readily obtain $a \leqslant a \vee b \leqslant e$ (that is, $a \leqslant e \leqslant d$) and $b \leqslant a \vee b \leqslant e$ (that is, $b \leqslant e \leqslant d$), which, combined with $c \leqslant e \leqslant d$, violate the definition of d. By an identical argument, we can show that $d = a \vee (b \vee c)$, whence

$$(a \vee b) \vee c = a \vee (b \vee c). \quad \text{(associative property)} \quad (4.3.3)$$

Naturally, as a consequence of the associative property (see also definition 3.1.2), we may omit the parentheses and write $(a \vee b) \vee c = a \vee b \vee c$. Moreover, consider the element $a \vee (a \wedge b)$. Since $a \leqslant a$ (trivially) and $a \wedge b \leqslant a$ (by definition of meet), we have $a \vee (a \wedge b) \leqslant a$. Clearly, there is no other x for which $a \leqslant x \leqslant a$, whence a is the join of a and $a \wedge b$; that is,

$$a \vee (a \wedge b) = a. \quad \text{(property of absorption)} \quad (4.3.4)$$

Finally, let a and b be two distinct comparable elements in the partial ordering of $[S, \vee, \wedge]$, and assume that $a \leqslant b$. Thus there is a descending chain from b to a in the Hasse disgram of $[S, \vee, \wedge]$, and clearly, $a \wedge b = a$ and $a \vee b = b$. Conversely, if $a \wedge b = a$, since $a \wedge b \leqslant b$, we have, by substitution, $a \leqslant b$; similarly, we show that $a \vee b = b$ implies $a \leqslant b$. We summarize by saying that join, meet, and partial ordering are collectively related by the basic statement

$$a \vee b = b, \ a \wedge b = a, \ a \leqslant b \text{ are equivalent.} \quad \text{(consistency principle)} \quad (4.3.5)$$

We also summarize this as a theorem.

Theorem 4.3.2. In a lattice the binary operations of join and meet are idempotent, commutative, and associative, and they satisfy the absorption property and the consistency principle.

This completes the introduction of lattices as formal structures. They have been obtained by selecting those posets on which join and meet are defined. We now exhibit a few examples. The Hasse diagram is a powerful aid in deciding whether a finite poset is a lattice.

Example 4.3.2. The poset of figure 4.3.2(a) is a lattice; that of figure 4.3.2(b) is not, because joins and meets are not unique.

Example 4.3.3. Consider the poset $[S, \leqslant]$ where $S \equiv \{1, 2, 3, 4, 5, 6, 10, 12, 15, 20, 30, 60\}$ and $\leqslant$ is the relation of divisibility. The Hasse diagram of $[S, \leqslant]$ is given in figure 4.3.3. Since each pair has unique join and meet, $[S, \leqslant]$ is a lattice. Notice that, by the nature of $\leqslant$, join is synonymous with "least common multiple" and meet with "greatest common divisor."

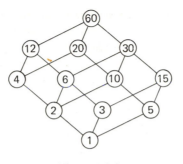

Figure 4.3.3

We know (recall definition 1.1.3) that if we select some elements of a set A, we obtain a new collection called a subset of A. If we perform the same operation on a poset $[S, \leqslant]$, we obtain a subset S' of S which is still partly ordered (if $a \leqslant b$ in S and a and b are assigned to the subset S', we still have $a \leqslant b$). Does the same occur for a lattice $L = [S, \vee, \wedge]$? The answer is, in general, negative, and it is provided by the observation that a lattice is an algebra (definition 3.8.2) with two binary operations. Hence all the results in Section 3.8 hold for lattices. We single out the following definition of sublattice for later convenience.

Definition 4.3.4. A lattice $L = [S', +, \cdot]$ is called a *sublattice* of another lattice $L = [S, \vee, \wedge]$ if S' is a subset of S and $+$ and $\cdot$ are the respective restrictions of $\vee$ and $\wedge$ to S'.

In other words, $+$ and $\cdot$ must be closed in S'. The following example should illustrate the point.

Example 4.3.4. Consider the lattice L of figure 4.3.4(a). The poset $S_1 \equiv \{a, d, e, f\}$ as shown in figure 4.3.4(b), is not a lattice, because it does not contain the meet of

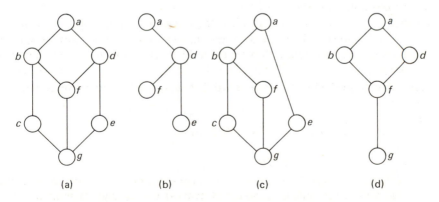

          (a)   (b)   (c)   (d)

Figure 4.3.4. Examples of a lattice and of some of its subsets

f and e. In figure 4.3.4(c) we have the Hasse diagram of the poset $S_2 \equiv \{a,b,c, e,f,g\}$, which is a lattice, because each pair of elements has a unique join and a unique meet. However, this is not a sublattice of L, since $f \vee e = d$ does not belong to S_2. Finally, the poset $S_3 \equiv \{a,b,d,f,g\}$, whose Hasse diagram is shown in figure 4.3.4(d) is not only a lattice but also a sublattice of L, as can be easily verified.

EXERCISES 4.3

1. Determine which of the following systems is a lattice.
   a) The set of all subsets of a set S with set-theoretic union and intersection as operations (see definitions 1.1.5 and 1.1.6).
   b) The set of positive integers P together with the operation of greatest common divisor and least common multiple.
   c) The set of positive integers together with the operations of addition and multiplication.
   d) The set of real functions defined in the closed interval $[0,1]$. Functions $f \vee g$ and $f \wedge g$ denote, respectively, functions which assume the greater and smaller of the values of $f(x)$ and $g(x)$ for x in $[0,1]$.
   *e) The set of all the subgroups of a group G, with the operations of set intersection and $\vee$, where $\vee$ is defined as follows: for two subgroups H_1 and H_2, $H_1 \vee H_2$ is the subgroup generated by $H_1 \cup H_2$. (See also exercise 3.2.7 or Note 3.6 for the definition of generators.)

2. Find all lattices with no more than five elements, and draw their Hasse diagrams.

3. Find all sublattices of the lattice given in example 4.3.3.

4. Show that every poset whose Hasse diagram is a planar graph and which has the universal bounds is a lattice.

5. A subset I of a lattice L is called an *ideal* of L if I satisfies the following two conditions: (a) $a,b \in I$ implies that $a \vee b \in I$; (b) for any element $x \in L$ and $a \in I, a \wedge x \in I$. Show that every ideal is a sublattice of L.

6. A function $\varphi:L_1 \to L_2$ is called a homomorphism from lattice $L_1 = [S,\vee,\wedge]$ to lattice $L_2 = [T, \oplus, \cdot]$ if $\varphi(a \wedge b) = \varphi(a) \cdot \varphi(b)$ and $\varphi(a \vee b) = \varphi(a) \oplus \varphi(b)$. Show that if L_2 has a least element O_2, then $\varphi^{-1}(O_2)$ is an ideal in L_1 (see exercise 4.3.5 for definition of ideal).

7. Show that an algebra $[S,\vee,\wedge]$ is a lattice if, for a,b,c in S, the following conditions are satisfied.
   a) $a \vee b = b \vee a, \; a \wedge b = b \wedge a$
   b) $(a \vee b) \vee c = a \vee (b \vee c), \; a \wedge (b \wedge c) = (a \wedge b) \wedge c$
   c) $a \vee a = a, \; a \wedge a = a$
   d) $(a \vee b) \wedge a = a, \; (a \wedge b) \vee a = a$

   (Note that in Section 4.3 we have shown the converse, that is, that (a), (b), (c), and (d) hold for a lattice. You must now show that (a), (b), (c), and (d) imply that S is a poset and that the g.l.b and l.u.b are unique in $[S,\leqslant]$.)

8. A poset $[S,\leqslant]$ is a *join-semilattice* (meet-semilattice) if for arbitrary elements a,b in S, $a \vee b$ $(a \wedge b)$ exists.

  a) Show that a poset is a lattice if and only if it is both a join- and a meet-semilattice
  b) Let A be the set of all continuous strictly upward-convex real-valued functions over the closed interval $[0,1]$. Define a relation $<$ on A such that $f < g$ if and only if $f(x) \leqslant g(x)$, for all $x \in [0,1]$. Show that $[A,<]$ is a meet-semilattice but not a join-semilattice.

4.4 DISTRIBUTIVE LATTICES

Consider three integers, 10, 6, and 15. We know that the least common multiple of 6 and 15 is 30, and that the greatest common divisor of 10 and 30 is 10 itself. This can be expressed concisely by

$$\text{g.c.d } [10, \text{ l.c.m. } (6,15)] = 10.$$

Suppose now that we first determine separately the greatest common divisors of the pairs of integers (10,6) and (10,15), which are easily found to be 2 and 5, respectively. Subsequently, we obtain the least common multiple of these greatest common divisors, which is 10. Formally,

$$\text{l.c.m. } [\text{g.c.d. } (10,6), \text{ g.c.d. } (10,15)] = 10,$$

which coincides with the previous result. Recall now that the given integers can be viewed as members of a lattice, whose partial ordering is the relation of divisibility and whose join and meet are, respectively, l.c.m. and g.c.d. Indeed, 6, 10, and 15 are elements of the lattice whose Hasse diagram is shown in figure 4.3.3. Therefore, in the previous expression l.c.m. and g.c.d. can be replaced, respectively, by the usual symbols $\vee$ and $\wedge$, and the discovered equality can be expressed as $10 \wedge (6 \vee 15) = (10 \wedge 6) \vee (10 \wedge 15)$. In the lattice of figure 4.3.3, we may replace the specific integers selected with the generic integers a, b, and c and ask whether

$$a \wedge (b \vee c) = (a \wedge b) \vee (a \wedge c) \tag{4.4.1}$$

holds; the answer, by inspection of all possible cases (take our word!) is found to be affirmative. We know from corollary 4.1.1 (duality principle) that from (4.4.1) we may obtain another identity by interchanging $\vee$ and $\wedge$; that is,

$$a \vee (b \wedge c) = (a \vee b) \wedge (a \vee c). \tag{4.4.2}$$

Suppose now—just as a game—that we replace $\vee$ with $+$ and $\wedge$ with $\cdot$. Identity (4.4.1) becomes

$$a \cdot (b + c) = a \cdot b + a \cdot c,$$

which is formally identical to the well-known *distributive property* of multiplication over addition in the ordinary algebra of real numbers. Indeed, (4.4.1) is called the distributive property of meet over join. Now, if we perform the same substitution in (4.4.2), we obtain $a + b \cdot c = (a + b) \cdot (a + c)$, a result which will certainly surprise the reader, because it may hold for any three elements of a lattice, but it is generally false if $+$ and $\cdot$ are interpreted as usual in the algebra of real numbers. For example, if a, b, and c are integers, $a + b \cdot c = (a + b) \cdot (a + c)$ holds either if $a = 0$ or, when $a \neq 0$, if $a + b + c = 1$, that is, not for three arbitrary integers.

We may be tempted at this point to conjecture that the pair of dual distributive properties holds in every lattice. If so, consider the lattice whose Hasse diagram is given in figure 4.4.1, and determine separately the elements $d \wedge (c \vee e)$ and $(d \wedge c) \vee (d \wedge e)$. These computations are most easily performed with the Hasse diagram. Indeed, $(c \vee e) = a$, and since $d \leqslant a$, $d \wedge a = d$. Therefore $d \wedge (c \vee e) = d \wedge a = d$. On the other hand, $d \wedge c = g$ and $d \wedge e = e$ (since $e \leqslant d$), whence $(d \wedge c) \vee (d \wedge e) = g \vee e = e$ (because $g \leqslant e$). We conclude that in figure 4.4.1

$$d = d \wedge (c \vee e) \neq (d \wedge c) \vee (d \wedge e) = e;$$

that is, the distributive property does not hold for the triplet c,d,e.

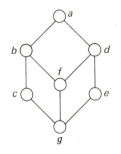

Figure 4.4.1. A nondistributive lattice

We have therefore observed, by direct example, that there are lattices for which the distributive property does not hold. Thus the following definition selects a proper subclass of the family of lattices.

Definition 4.4.1. A lattice L is said to be *distributive* if for any three elements $a,b,c \in L$ the distributive properties (laws)

$$a \wedge (b \vee c) = (a \wedge b) \vee (a \wedge c) \tag{4.4.3}$$

$$a \vee (b \wedge c) = (a \vee b) \wedge (a \vee c) \tag{4.4.4}$$

hold; otherwise it is said to be *nondistributive*.

The preceding discussion has clearly shown that the lattice in figure 4.4.1 is nondistributive. Note now that this lattice, L, contains the sublattice L_1 represented in figure 4.4.2(a). That the represented L_1 is really a sublattice of L is readily verified by referring to definition 4.3.4 (for example, $c \vee d = a$ and $c \wedge d = g$ both in L and in L_1). In other words, if one identifies in a lattice L a sublattice like L_1,

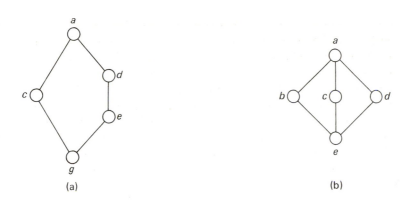

Figure 4.4.2. Nondistributive lattices

that is, the same configuration of five vertices, then L is nondistributive. The same can be said with regard to the five-element lattice L_2 represented in figure 4.4.2(b): The distributive law fails with reference to the three elements b, c, d. Indeed, $(b \wedge c) \vee (b \wedge d) = e \vee e = e$ and $e \neq b = b \vee (c \wedge d)$. It follows that if a lattice L contains a sublattice having the same configuration as L_2, L is nondistributive. But a stronger statement can be made, i.e., not only that the presence in L of sublattices like L_1 and L_2 is a *sufficient* condition for L to be nondistributive, but that it is also *necessary*. A nondistributive lattice L must contain a sublattice like L_1 or L_2. This is a very important condition since distributivity of a lattice can be assessed by visual inspection of the structure of the Hasse diagram (human beings are very powerful "pattern recognizers," and L_1 and L_2 are indeed "patterns"). We shall now formally give this result as a theorem, stated without proof.

Theorem 4.4.1. A lattice L is distributive if and only if it has no sublattices isomorphic with either of the lattices shown in figure 4.4.2.

Finally, it is interesting to examine lattices and distributive lattices from the viewpoint of algebraic structures, as discussed in the preceding chapter. Considering a lattice as a universal algebra (definition 3.8.2), we note that $L = [S, \vee, \wedge]$ is a system $[S, f_1, f_2, f_3, f_4]$, where $f_1 = \vee$ and $f_2 = \wedge$ are binary operations and f_3 and f_4 are the nullary operations corresponding to selecting the universal bounds. Therefore L is *similar* (definition 3.8.3) to a ring with a multiplicative identity. However, we have already noted the rather weak character of the notion of similar algebras (since it induces a very general classification). Upon closer examination, we recognize that a finite lattice contains two commutative monoids $[S, \vee]$ and $[S, \wedge]$, since, for example, $[S, \vee]$ is closed with respect to the associative operation $\vee$ and contains the identity O (likewise, $[S, \wedge]$). If L is also distributive, join distributes over meet, and vice versa. Viewing also a commutative ring R with multiplicative identity as a structure consisting of two commutative monoids, we note that the additive monoid of R is in reality a group (whereas $[S, \vee]$ is not!). By contrast, in R only one distributive law holds, neither operation is idempotent, and absorption does not hold. Thus, although there are various similarities between rings and lattices, there are also specific properties which bring about a clear-cut structural differentiation. However, we surmise that by adding properties to lattices and to rings, we may find that the two classes have a nonempty intersection (that is, there are very special structures which may be viewed both as rings and as lattices). The answer to this conjecture is affirmative, as we shall see in Section 5.6.

- -

EXERCISES 4.4

1. Determine which of the following lattices are distributive.
    a) A chain.
    b) All subsets of a set under union and intersection.
    c) The set of positive integers with the operations of greatest common divisor and least common multiple.
    d) The dual of a distributive lattice.
    e) An arbitrary sublattice of a distributive lattice.
    f) A homomorphic image of a distributive lattice.
    *g) The set of all ideals of a ring with the operation of set-theoretic intersection and the operation $+$, where $I_1 + I_2$ is the ideal generated by the set $I_1 \cup I_2$.

2. Show that whenever a lattice satisfies the identity (4.4.3) it also satisfies the identity (4.4.4), and conversely.

3. Show that the following inequalities hold in any lattice:
    a) $(a \wedge b) \vee (a \wedge c) \leqslant a \wedge (b \vee c)$
    b) $a \vee (b \wedge c) \leqslant (a \vee b) \wedge (a \vee c)$
    c) $(a \wedge b) \vee (b \wedge c) \vee (c \wedge a) \leqslant (a \vee b) \wedge (b \vee c) \wedge (c \vee a)$
    d) $(a \wedge b) \vee (a \wedge c) \leqslant a \wedge (b \vee (a \wedge c))$
    e) If $a \leqslant c$, $a \vee (b \wedge c) \leqslant (a \vee b) \wedge c$.

    (The expressions in (a), (b), and (c) are called *distributive inequalities*, and those in (d) and (e) are called *modular inequalities*.)

4. Add two new elements O' and I' to a distributive lattice L such that $O' \leqslant x \leqslant I'$ for every $x \in L$. Show that the resulting lattice is still distributive.

5. Show that a lattice is distributive if one of the following inequalities is satisfied without restriction.

    $a \wedge (b \vee c) \leqslant (a \wedge b) \vee (a \wedge c)$

    $(a \vee b) \wedge (a \vee c) \leqslant a \vee (b \wedge c)$

*6. A group G is called a *generalized cyclic group* if every finite subset of G generates a cyclic subgroup. Show that the lattice of all subgroups of a generalized cyclic group is distributive. (*Hint*: by exercise 4.4.5, it is sufficient to show that for any subgroups G_1, G_2, and G_3 of G, the following inclusion relation holds

$$G_1 \cap G_2 G_3 \subseteq (G_1 \cap G_2)(G_1 \cap G_3),$$

where $G_1 G_2 \equiv \{g_1 g_2 | g_1 \in G_1 \text{ and } g_2 \in G_2\}$.)

7. A lattice is said to be *modular* if, for any triplet of elements a,b,c satisfying $a \leqslant c$, the identity

$$a \vee (b \wedge c) = (a \vee b) \wedge c$$

holds. Verify that

is a Hasse diagram of a modular lattice.

8. Prove the following statements.

   a) Every distributive lattice is modular but the converse is not true. Show that the Hasse diagram of the smallest modular but not distributive lattice has five vertices.

   b) The dual lattice, every sublattice, and every homomorphic image of a modular lattice is modular.

*9. Show that the set of all normal subgroups of a group is a modular lattice. (*Hint*: Let G_1 and G_2 be arbitrary normal subgroups of G. The two lattice operations are intersection $G_1 \cap G_2$ and product $G_1 G_2$, defined in exercise 4.4.6. One needs to show that $G_1 \cap G_2$ and $G_1 G_2$ are also normal subgroups and that, for normal subgroups G_1, G_2, and G_3 of G with $G_1 \subseteq G_3$, by exercise 4.4.3(e), the following inclusion holds:

$$G_1(G_2 \cap G_3) \supseteq G_1 G_2 \cap G_3.)$$

4.5 ELEMENTS OF REPRESENTATION OF LATTICES

Suppose now that a lattice $L = [S, \vee, \wedge]$ is given and that a, b, c, d are elements of S. Suppose also that we are given the following expression involving the elements a, b, c, d:

$$[(a \vee b) \wedge c] \vee [(a \vee c \vee d) \wedge b].$$

We recognize that this expression merely describes the following operations performed on elements of L. We form the join $(a \vee b)$, which coincides with some element f of L. Then we form the meet of $f \wedge c$, that is, $(a \vee b) \wedge c$, which coincides with $g \in L$. Similarly, we form the join $h = a \vee c \vee d$ ($h \in L$), and subsequently the meet $k = h \wedge b = (a \vee c \vee d) \wedge b$, ($k \in L$). Finally, we form the join $m = g \vee h$, ($m \in L$), which, by repeated substitution, coincides with the expression given above In all the previous steps, the result of the operation, either join or meet, was an element of L by closure of the same operation, and so is the element m *described* by the *expression* containing the symbols of the elements a, b, c, d. In other words, an expression is a device for *representing* elements of L in terms of *other* elements of L, and it contains all the information required for obtaining the "represented" elements from the "representing" ones. This information resides in the parenthesis configuration of the expression and, if necessary, in the precedence rules of the family of expressions (compare example 0.4.3).

The key to the present discussion is the word "representation." Indeed, a fundamental problem in the study of algebraic systems (i.e., not only lattices) is that of expressing the elements of a given system in terms of a subset of the elements of the system. In this connection, very interesting questions are raised as to the structure, the size, the uniqueness, etc., of *subsets* of a system such that *all* the elements of the system can be expressed in terms of members of one such subset. (These subsets may be referred to as *bases* or *generators* of the system under consideration; see definition 3.10.1.)

With this simple background, we now consider a lattice $L = [S, \vee, \wedge]$, which we assume to be given by means of its Hasse diagram. We consider an element a of L and assume that in the diagram there are $k > 1$ edges reaching a from below; i.e., there are k elements $b_1, b_2, \ldots, b_k$ such that, for $j = 1, 2, \ldots, k$, $b_j \leqslant a$, and there is no other element x in S for which $b_j \leqslant x \leqslant a$. This is equivalent to saying that a is the "immediate successor" of $b_1, \ldots, b_k$, where "immediate successor" is the inverse of the relation "immediate predecessor" (definition 4.2.3). Figure 4.5.1 gives the relevant portion of the Hasse diagram of L. We may now select arbitrarily two elements b_i and b_j of the set $\{b_1, \ldots, b_k\}$ and express a as the join of b_i and b_j; that is, $a = b_i \vee b_j$ (we may say that a has been "reduced" to the join of b_i and b_j). The same argument can be applied to b_i and b_j, provided that b_i and b_j are each immediate successors of more than one element of S, and so on. This process of reduction stops any time we reach an element x of S which is the immediate

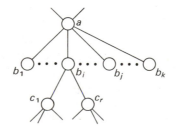

Figure 4.5.1. Illustration of the join-reduction of elements of a lattice

successor of exactly one element of S (in other words, there is only one edge reaching x from below in the diagram of L). The fact that one such element therefore cannot be reduced to the join of other elements motivates the following definition.

Definition 4.5.1. An element a of a lattice L is said to be *join-irreducible* if $x \vee y = a$ implies either $x = a$ or $y = a$.

Note that in each lattice L having the universal lower bound O, the element O itself is join-irreducible. Also, all the elements which are immediate successors of this lower bound (called the *atoms* of L) are join-irreducible.

The preceding discussion seems to indicate that an arbitrary element of a lattice L can be represented as the join of nontrivial (i.e., different from O) join-irreducible elements of L. However, if the lattice is not finite, the process of reduction is, in general, not guaranteed to terminate. For the existence of one such representation for each element of $a \in L$, must we require that L be finite? A slightly weaker condition is sufficient, i.e., that each chain contained in the lattice is finite (the order of a chain is defined, as usual, to be the number of elements in the chain), as we now show. We assume that each chain in L is finite. We know that any finite chain of L has O as its least element and therefore contains at least one atom, which is join-irreducible. Each element $a \in L$, if not join-irreducible itself, belongs to at least two chains of L, and in performing the join-reduction, we proceed one edge downwards on both chains. The procedure is then repeated for each vertex reached from a. Since the chains of L are finite, after a finite number of steps we reach a nontrivial join-irreducible element on each chain; i.e., we express a as a join of join-irreducible elements. Note that the set of the edges traversed in this process forms a rooted binary[3] tree (see definition 2.5.3), with possibly the same vertex occurring more than once. This tree has its root in a—the element to be represented—and each simple path originating from a has finite order (definition 2.5.3) by assumption. Therefore the join-irreducible elements are the terminal vertices of this tree, and they are obviously finite in number (because the tree is finite). Thus we have proved the following theorem.

Theorem 4.5.1. If all chains in a lattice L are finite, then every element a in L can be represented as a join of a finite number of join-irreducible elements of L. (Such lattices are said to be of *finite length*.)

Example 4.5.1. Consider the lattice L_1 of 6 elements illustrated in figure 4.5.2. This lattice contains 5 join-irreducible elements, that is, $\{O,a,b,c,d\}$. We want to represent the element e. We can easily see that $e = b \vee c = c \vee d = b \vee d$ are all equivalent representations and are joins of join-irreducible elements. As a more substantial example, consider the lattice L_2 of 18 elements illustrated in figure

---

3. *Binary* because each join-reducible element is expressed as the join of *two* elements.

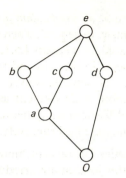

Figure 4.5.2. Diagram of L_1

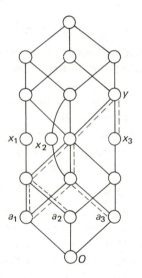

Figure 4.5.3. Diagram of L_2

4.5.3, whose join-irreducible elements are $\{O, a_1, a_2, a_3, x_1, x_2, x_3\}$. We want to obtain a representation of y as a join of join-irreducible elements of L. The join-reduction is clearly indicated by the broken-line poset drawn in figure 4.5.3, which shows the termination of the reduction of the join-irreducible elements. Therefore we obtain

$$y = x_3 \vee a_1 \vee a_2 \vee a_3.$$

This representation of y is undoubtedly correct. However, note that, since $a_2 \leqslant x_3$ and $a_3 \leqslant x_3$, we have $x_3 \vee a_2 \vee a_3 = x_3$, whence

$$y = x_3 \vee a_1;$$

the latter expression is said to be an *irredundant join* in that the join of any subset of $\{a_1, x_3\}$ no longer represents y. In general, when one has obtained an expression of an element a of L as declared by theorem 4.5.1, it is always possible to test for comparability pairs of elements in the join and to remove from the join of any comparable pair its smaller element. However, there is a more fundamental question: Is the representation of an element as an irredundant join of join-irreducible elements unique? The following theorem identifies the class of lattices for which this question is answered affirmatively.

Theorem 4.5.2. In a distributive lattice L of finite length, each element a has a unique representation as the join of an irredundant set of join-irreducible elements.

Proof. The rationale of the proof is to show that, given any two such representations,

$$a = x_1 \vee \cdots \vee x_n = y_1 \vee \cdots \vee y_m,$$

there is a one-to-one correspondence between the terms of the two expressions, thereby leading to the conclusion that they are identical, i.e., that the representation is unique.

The key for this objective is to show that, as the effect of distributivity, for any x_i in $x_1 \vee \cdots \vee x_n$ there is some y_j in $y_1 \vee \cdots \vee y_m$ for which $x_i \leqslant y_j$. Indeed, by definition of join, $x_i \leqslant (x_1 \vee \cdots \vee x_n) = (y_1 \vee \cdots \vee y_k)$. This means that

$$x_i = x_i \wedge (y_1 \vee \cdots \vee y_k),$$

and by *distributivity*,

$$x_i = (x_i \wedge y_1) \vee (x_i \wedge y_2) \vee \cdots \vee (x_i \wedge y_k).$$

In this expression, note that x_i is expressed as a join; but x_i is by hypothesis join-irreducible, whence x_i must coincide with some term $x_i \wedge y_j$, that is $x_i \leqslant y_j$.

Therefore, for each x_i, there is some y_j for which $x_i \leqslant y_j$. But, for *this* y_j, there is some x_k for which $y_j \leqslant x_k$. It follows that $x_i \leqslant y_j \leqslant x_k$, that is, $x_i \leqslant x_k$, but since the expression is irredundant, it must be $x_i = x_k$, that is, $x_i = y_j = x_k$. It follows that the elements of $\{x_1, \ldots, x_n\}$ and of $\{y_1, \ldots, y_m\}$ are pairwise identical; i.e., an irredundant representation is unique.$\|$

This technique of assuming two distinct representations and proving their identity by proving the pairwise identity of their respective terms is a standard scheme to show the uniqueness of representations.

EXERCISES 4.5

1. Consider the distributive lattice consisting of the set $\{1, 2, 3, 5, 6, 10, 15, 30\}$ under the partial ordering of divisibility.
   a) Is this lattice distributive?
   b) Find all of its join-irreducible elements.
   c) Express 15 as a join of join-irreducible elements.

2. Find an example of a lattice containing an infinite number of elements such that all of its chains are finite.

3. a) Define the concept of a meet-irreducible element of a lattice and meet representation of elements of a lattice.
   b) Find all meet-irreducible elements of the lattice given in exercise 4.5.1.
   c) Express the element 3 as an irredundant meet of meet-irreducible elements in the lattice given in exercise 4.5.1.

4. Show that in a distributive lattice L of finite length, each element a has a unique representation as the meet of an irredundant set of meet-irreducible elements.

5. Show that in a lattice of positive integers under the partial ordering of divisibility, the powers of primes are the join-irreducible elements.

*4.6 PARTITION LATTICES

Consider a set $S \equiv \{1,2,3,4\}$. Recall the definition of partition of a set (definition 1.5.7) and the fact (corollary 1.5.6) that there is a one-to-one correspondence between partitions of S and equivalence relations on S. The symbol π is used to denote a partition, and each member of π—a subset of S—is called a *block* of π. For example,

$$\pi = \{\{1,3\}, \{2\}, \{4\}\}$$

is a partition of S. For simplicity of notation, we shall now replace each pair of inner brackets by a bar on top of the elements enclosed. Thus $\pi = \{\overline{1, 3}, \overline{2}, \overline{4}\}$ in the new notation. Members of the same block of π are obviously equivalent under the equivalence relation $\varepsilon(\pi)$ corresponding to π. Consider the set Π of the partitions of (a finite) set S.[4] In Π there is a very natural partial ordering; i.e., for π_1 and π_2 in Π we say

$$\pi_1 \leqslant \pi_2$$

if each block of π_1 is contained (properly or improperly) in a block of π_2. For example, $\{\overline{1}, \overline{2, 4}, \overline{3}\} \leqslant \{\overline{1, 2, 4}, \overline{3}\}$. Therefore Π is a poset with the universal bounds which are readily found to be $\{\overline{1, 2, 3, 4}\} = I$ and $\{\overline{1}, \overline{2}, \overline{3}, \overline{4}\} = O$. Paralleling the development of Section 4.3, we want to see if the g.l.b. and l.u.b. of any two partitions exist in Π in terms of the introduced partial ordering.

As regards the g.l.b. (meet) of π_1 and π_2, we must find a partition π each block of which is contained in some block of π_1 and in some block of π_2 and such that there is no larger block enjoying the same property. Therefore the blocks of π are the intersections of the blocks of π_1 and π_2. For example, we have

$$\text{g.l.b.}[\{\overline{1, 2}, \overline{3, 4}\}, \{\overline{1, 2, 4}, \overline{3}\}] = \{\overline{1, 2}, \overline{3}, \overline{4}\},$$

$$\text{g.l.b.}[\{\overline{1, 2}, \overline{3, 4}\}, \{\overline{1, 3}, \overline{2, 4}\}] = \{\overline{1}, \overline{2}, \overline{3}, \overline{4}\}.$$

---

4. A little thought will show that Π is finite if S is so. In a later section we shall develop techniques for computing the cardinality of Π from that of S.

Obviously, the operation of g.l.b., or meet, has a unique result in Π. This operation has an interesting interpretation in terms of the equivalence relations associated with the partitions. Since equivalence classes in $\varepsilon(\pi_1 \wedge \pi_2)$ are the largest subsets of elements of S which are equivalent both in $\varepsilon(\pi_1)$ and in $\varepsilon(\pi_2)$, the array of $\varepsilon(\pi_1 \wedge \pi_2)$ has a 1 in a given position if and only if both $\varepsilon(\pi_1)$ and $\varepsilon(\pi_2)$ have 1's in that position. In other words, $R_{\varepsilon(\pi_1 \wedge \pi_2)}$ is the intersection of $R_{\varepsilon(\pi_1)}$ and $R_{\varepsilon(\pi_2)}$ (where, of course, these sets $R_{\varepsilon(\pi_1 \wedge \pi_2)}$, $R_{\varepsilon(\pi_1)}$, and $R_{\varepsilon(\pi_2)}$ are considered as sets of pairs of elements of S). For example, letting M_1, M_2, and M denote the arrays of $\varepsilon(\pi_1) = \{\overline{1,2}, \overline{3,4}\}$, $\varepsilon(\pi_2) = \{\overline{1,2,4}, \overline{3}\}$, and $\varepsilon(\pi_1 \wedge \pi_2)$, respectively, we have

$$M_1 = \begin{bmatrix} 1 & 1 & 0 & 0 \\ 1 & 1 & 0 & 0 \\ 0 & 0 & 1 & 1 \\ 0 & 0 & 1 & 1 \end{bmatrix} \quad M_2 = \begin{bmatrix} 1 & 1 & 0 & 1 \\ 1 & 1 & 0 & 1 \\ 0 & 0 & 1 & 0 \\ 1 & 1 & 0 & 1 \end{bmatrix} \quad M = \begin{bmatrix} 1 & 1 & 0 & 0 \\ 1 & 1 & 0 & 0 \\ 0 & 0 & 1 & 0 \\ 0 & 0 & 0 & 1 \end{bmatrix}$$

That is, M identifies the partition $(\overline{1,2}, \overline{3}, \overline{4})$, as expected.

An l.u.b. (join) of π_1 and π_2, if it exists, is a partition π such that each block of π_1 and π_2 is contained in some block of π and such that there is no other partition with smaller blocks enjoying the same property.

The structure of π is better arrived at by considering the equivalence relations associated with the partitions. In fact, if elements s_i and s_j are equivalent in $\varepsilon(\pi_1)$ [or $\varepsilon(\pi_2)$] they must also be equivalent in $\varepsilon(\pi)$; with reference to the relation arrays, each 1 of $\varepsilon(\pi_1)$ [or $\varepsilon(\pi_2)$] is also a 1 of $\varepsilon(\pi)$. Recalling the concepts of Section 1.4, we say that $\varepsilon(\pi)$ covers $\varepsilon(\pi_1)$ and $\varepsilon(\pi_2)$. But $\varepsilon(\pi)$ is an equivalence relation; i.e., it satisfies the transitive property. If we simply superpose the arrays of $\varepsilon(\pi_1)$ and $\varepsilon(\pi_2)$, the relation α described by the resulting array may not be an equivalence relation; however, $\varepsilon(\pi) \geqslant \alpha$, and $\varepsilon(\pi)$ is obtained from α by adding to the array M_α all the entries which render the transitive property satisfied. Repeating the argument developed in Section 4.2 on how to compute "partial ordering" from "immediate predecessor," we conclude that the additional entries needed for transitivity are obtained by repeated composition of the relation α; that is,

$$\varepsilon(\pi) = \alpha^k$$

if k is the smallest positive integer such that $\alpha^k = \alpha^{k+1}$. It is clear that π is unique and is equal to $\pi_1 \vee \pi_2$. Here again, $\varepsilon(\pi_1 \vee \pi_2)$ is said to be the transitive closure of the superposition of $\varepsilon(\pi_1)$ and $\varepsilon(\pi_2)$.

Example 4.6.1. Consider the following partitions of the set $S \equiv \{1,2,3,4,5,6\}$:

$$\pi_1 = \{\overline{1,2}, \overline{3,4}, \overline{5}, \overline{6}\};$$

$$\pi_2 = \{\overline{1}, \overline{2,4}, \overline{3,5}, \overline{6}\}.$$

The arrays M_1 and M_2 of $\varepsilon(\pi_1)$ and $\varepsilon(\pi_2)$, respectively, are

$$M_1 = \begin{array}{c} 1 \\ 2 \\ 3 \\ 4 \\ 5 \\ 6 \end{array}\begin{bmatrix} 1 & 1 & 0 & 0 & 0 & 0 \\ 1 & 1 & 0 & 0 & 0 & 0 \\ 0 & 0 & 1 & 1 & 0 & 0 \\ 0 & 0 & 1 & 1 & 0 & 0 \\ 0 & 0 & 0 & 0 & 1 & 0 \\ 0 & 0 & 0 & 0 & 0 & 1 \end{bmatrix} \qquad M_2 = \begin{bmatrix} 1 & 0 & 0 & 0 & 0 & 0 \\ 0 & 1 & 0 & 1 & 0 & 0 \\ 0 & 0 & 1 & 0 & 1 & 0 \\ 0 & 1 & 0 & 1 & 0 & 0 \\ 0 & 0 & 1 & 0 & 1 & 0 \\ 0 & 0 & 0 & 0 & 0 & 1 \end{bmatrix}$$

The array M, shown below, results from superposing M_1 to M_2. The reader will notice that M does not describe an equivalence relation; however, M^3, obtained with simple manipulations, does describe an equivalence relation and is the array of $\varepsilon(\pi_1 \vee \pi_2)$, where $\pi_1 \vee \pi_2 = \{\overline{1,2,3,4,5},\overline{6}\}$.

$$M = \begin{bmatrix} 1 & 1 & 0 & 0 & 0 & 0 \\ 1 & 1 & 0 & 1 & 0 & 0 \\ 0 & 0 & 1 & 1 & 1 & 0 \\ 0 & 1 & 1 & 1 & 0 & 0 \\ 0 & 0 & 1 & 0 & 1 & 0 \\ 0 & 0 & 0 & 0 & 0 & 1 \end{bmatrix} \qquad M^3 = \begin{bmatrix} 1 & 1 & 1 & 1 & 1 & 0 \\ 1 & 1 & 1 & 1 & 1 & 0 \\ 1 & 1 & 1 & 1 & 1 & 0 \\ 1 & 1 & 1 & 1 & 1 & 0 \\ 1 & 1 & 1 & 1 & 1 & 0 \\ 0 & 0 & 0 & 0 & 0 & 1 \end{bmatrix} = M^4$$

The preceding argument is even more effective in terms of the graphical representation. In figure 4.6.1 edges are depicted as follows: solid lines for $\varepsilon(\pi_1)$, dashed lines for $\varepsilon(\pi_2)$, and dotted lines for the edges added to attain transitivity.

Figure 4.6.1 offers intuitive evidence of the rule for the construction of the partition $\pi_1 \vee \pi_2$. Note that, for two elements s_i and s_j of S to be in the same block of $\pi_1 \vee \pi_2$, there must be a path in the graph of $\varepsilon(\pi_1 \vee \pi_2)$ from s_i to s_j composed of edges either of $\varepsilon(\pi_1)$ or of $\varepsilon(\pi_2)$. For example, in figure 4.6.1, with reference to elements 1 and 5, there is the path

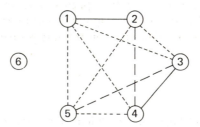

Figure 4.6.1. Illustration of the relation $\varepsilon(\pi_1 \vee \pi_2)$

which is formalized in the following rule: s_i and s_j of S belong to the same block of $(\pi_1 \vee \pi_2)$ if there is a sequence of elements of S, $s_i = t_0, t_1, \ldots, t_k = s_j$ such that either t_i and t_{i+1} belong to the same block of π_1 or they belong to the same block of π_2, for $i = 0, 1, \ldots, k-1$. Note that $t_0, t_1, \ldots, t_k$ belong to the *same* block of $\pi_1 \lor \pi_2$.

The preceding discussion shows that join, as well as meet, of two partitions exists in Π, which is a poset with universal bounds. This result is now stated formally.

Theorem 4.6.1. The family of all partitions of a set forms a lattice with universal bounds.

The diagrams of the partition lattices for sets of three and four elements are shown in figure 4.6.2(a) and (b), respectively.

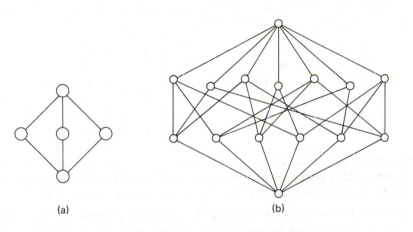

(a)   (b)

Figure 4.6.2. Partition lattices for $S \equiv \{1,2,3\}$ and $S \equiv \{1,2,3,4\}$

Incidentally, note that partition lattices in general are not distributive. A simple example of a nondistributive lattice appears in figure 4.6.2(a).

In Note 4.1 we shall examine very important and interesting sublattices of partition lattices, obtained by additional constraining properties.

EXERCISES 4.6

1. Does theorem 4.6.1 hold if the word "partitions" is replaced by "covers"?

2. Let $L_1 = [S_1, \wedge_1, \vee_1]$ and $L = [S_2, \wedge_2, \vee_2]$ be two lattices. Define $L_1 \times L_2 = [S_1 \times S_2, \wedge, \vee]$ such that

$$(a_1, a_2) \wedge (b_1, b_2) = (a_1 \wedge_1 b_1, a_2 \wedge_2 b_2)$$

and

$$(a_1, a_2) \vee (b_1, b_2) = (a_1 \vee_1 b_1, a_2 \vee_2 b_2).$$

a) Show that $L_1 \times L_2$ is a lattice.

b) If γ_1 and γ_2 are congruence relations on L_1 and L_2, respectively, then define $\gamma_1 \times \gamma_2$ on $L_1 \times L_2$ such that $(a,b) \; \gamma_1 \times \gamma_2 \; (c,d)$ if and only if $a \, \gamma_1 \; c$ and $b \, \gamma_2 \, d$. (An equivalence relation γ on a lattice L is called a *congruence relation* on L if $a_1 \, \gamma \, b_1$ and $a_2 \, \gamma \, b_2$ imply that $(a_1 \wedge a_2) \, \gamma \, (b_1 \wedge b_2)$ and $(a_1 \vee a_2) \, \gamma \, (b_1 \vee b_2)$.)

Show that $\gamma_1 \times \gamma_2$ is a congruence relation on $L_1 \times L_2$, and that, conversely, every congruence relation on $L_1 \times L_2$ is of this form.

4.7 BIBLIOGRAPHY AND HISTORICAL NOTES

The theory of lattices had a rather obscure start in the late nineteenth century, principally in the work of R. Dedekind and E. Schröder. The notion of the lattice was reached both in application to number theory and as a generalization of the algebra of propositions, notably originated by G. Boole (see next chapter). Thus the chronology of development was not the same as is our sequence of exposition (since we have chosen to present the material in the direction of specialization, the opposite of what usually happens in the creative domain). Essentially, however, lattices were not viewed as objects worthy of independent consideration until the fundamental work of G. Birkhoff. Since then, this area of mathematical investigation has exploded.

There exist several valuable texts which we suggest as further readings. Two eminently readable books are

D. E. Rutherford, *Introduction to Lattice Theory*, Hafner, New York, 1965.

G. Grätzer, *Lattice Theory*, Freeman, San Francisco, 1971.

The most classical, complete, and fundamental reference, though advanced and compact, is

G. Birkhoff, *Lattice Theory*, 3rd ed., American Mathematical Society Colloquium Publications, 25, 1967.

The reader may also find quite helpful the chapters on lattices in

G. Birkhoff and T. C. Bartee, *Modern Applied Algebra*, McGraw-Hill, New York, 1970.

Brief expositions of the foundations of lattices are given in many books on boolean algebra and switching theory.

4.8 NOTES AND APPLICATIONS

Note 4.1. Elements of structure theory of finite state machines

We recall from Note 1.1 that a finite state machine $\mathcal{M}$ is a quintuple $[S, I, Z, \delta, \lambda]$, where $S \equiv \{s_1, \ldots, s_n\}$ is a set of states, $I \equiv \{i_1, \ldots, i_m\}$ is a set of inputs, $Z \equiv \{z_1, \ldots, z_r\}$ is a set of outputs, and $\delta : S \times I \to S$ and $\lambda : S \times I \to Z$ are respectively the transition function and the output function of $\mathcal{M}$. We also recall that $\delta : S \times I \rightarrow S$ may be viewed as a collection $\{\delta_1, ..., \delta_m\}$ of functions $S \rightarrow S$; that is, $\delta_k : S \rightarrow S$ is the transition function associated with the input symbol i_k.

We now consider the lattice Π of the partitions in S, discussed in the preceding section. We recall that to each member $\pi \in \Pi$ there corresponds an equivalence relation $\varepsilon(\pi)$ on S (corollary 1.5.6). We now want to identify the $\varepsilon(\pi)$'s which are congruence relations (definition 3.4.1) with respect to each of the functions. Explicitly, we have the following definition.

Definition 4.8.1. A partition Π on S is said to be a *partition with the substitution property* (*SP-partition*) if the equivalence relation $\varepsilon(\pi)$ has the substitution property with respect to each and every one of the functions $\delta_1, \delta_2, ..., \delta_m$ [that is, if for $s_i, s_j \in S$, $s_i \varepsilon(\pi) s_j$ implies $\delta_k(s_i) \varepsilon(\pi) \delta_k(s_j)$, $k = 1, 2, ..., m$].

In other words, π is an SP-partition if, when s_i and s_j belong to the same block of π, for every input i_k the next states $\delta(s_i, i_k)$ and $\delta(s_j, i_k)$ are not in distinct blocks of π.

Recall the important characterization theorem of equivalence relations (theorem 1.5.7), which states that for every equivalence relation ε on S there is at least a function $\alpha : S \rightarrow B$ such that $\varepsilon = \alpha\alpha^{-1}$. This function can be chosen as the one which maps an element of S to its equivalence class under ε. Given an SP-partition π, we have the equivalence relation $\varepsilon(\pi)$ and the function $E : S \rightarrow \pi$; since π is an SP-partition, $s_i \varepsilon(\pi) s_j$ implies $\delta_k(s_i) \varepsilon(\pi) \delta_k(s_j)$, for $k = 1, 2, ..., m$. This means that $E(s_i) = E(s_j)$ implies $E(\delta_k(s_i)) = E(\delta_k(s_j))$; that is, there is a mapping $\delta_k^{(\pi)} : \pi \rightarrow \pi (k = 1, 2, ..., m)$ defined as follows:

$$\delta_k^{(\pi)}(E(s_i)) = E(\delta_k(s_i)). \tag{4.8.1}$$

In other words, if the present state is an arbitrary member s_i of a *fixed* block $E(s_i)$ of π and the present input is i_k, there is a transition to some state of the block $E(\delta_k(s_i))$. We now introduce a *new* finite state machine $\mathcal{M}_\pi$, whose states are the blocks of π and whose transition functions are $\delta_1^{(\pi)}, \delta_2^{(\pi)}, ..., \delta_m^{(\pi)}$, and we consider both $\mathcal{M}$ and $\mathcal{M}_\pi$ as universal algebras (definition 3.8.2): $\mathcal{M} = [S, \delta_1, ..., \delta_m]$ and $\mathcal{M}_\pi = [\pi, \delta_1^{(\pi)}, ..., \delta_m^{(\pi)}]$. We then recognize that $E : S \rightarrow \pi$ is onto and that (4.8.1) coincides, after the appropriate substitutions of symbols, with (3.8.1). That is, E is a homomorphism, and $\mathcal{M}_\pi$ is a *homomorphic image* of $\mathcal{M}$. Note that $\mathcal{M}_\pi$ has, in general, a smaller set of states than $\mathcal{M}$; if we observe $\mathcal{M}_\pi$ instead of $\mathcal{M}$, we have some information about $\mathcal{M}$, but it is incomplete. Indeed, we know that the present state of $\mathcal{M}$ belongs to some block B_i of states (thereby reducing our uncertainty), but we are totally ignorant about which element of B_i is the present state of $\mathcal{M}$. Suppose now that two SP-partitions of S, π and π', be given, and that there is a block B_i of π and a block B_j' of π' whose intersection is a *single* state s. If it happens that the present state of $\mathcal{M}_\pi$ is block B_i and the present state of $\mathcal{M}_{\pi'}$ is block B_j', then the information provided by $\mathcal{M}_\pi$ and $\mathcal{M}_{\pi'}$ together eliminates all uncertainty as to the state of $\mathcal{M}$, which can only be s.

This immediately suggests that if we identify a set $\{\pi_1, \pi_2, ..., \pi_p\}$ of SP-partitions so that the intersection partition $\pi_1 \wedge \pi_2 \wedge \cdots \wedge \pi_p$ consists of blocks of one element (if this happens, $\pi_1 \wedge \pi_2 \wedge \cdots \wedge \pi_p$ is trivially an SP-partition!) then knowledge about the states of $\mathcal{M}_{\pi_1}, \mathcal{M}_{\pi_2}, ..., \mathcal{M}_{\pi_p}$ provides complete information about the state of $\mathcal{M}$. This lends itself to an interesting interpretation. A finite state machine may be viewed as a device which "computes" its present state, given the past state and the past input. Since the collection of machines $\mathcal{M}_{\pi_1}, \ldots, \mathcal{M}_{\pi_p}$ provides complete information about the present state of $\mathcal{M}$, this collection effectively performs the same computation as $\mathcal{M}$. Therefore $\mathcal{M}$ can be replaced by $\mathcal{M}_{\pi_1}, \ldots, \mathcal{M}_{\pi_p}$. Note that each of these machines is independent of any other in the collection, except for the fact that the input symbol is at any time the same for all of them (and for $\mathcal{M}$). It follows that they can be viewed as machines operating in parallel (see figure 4.8.1), and indeed $\mathcal{M}_{\pi_1}, \ldots, \mathcal{M}_{\pi_p}$ are said to specify a *parallel decomposition* of $\mathcal{M}$. This discussion should shed some light on the deep significance of SP-partitions of the state set as formal tools for the structural investigation of finite state machines. Indeed, an elegant structure theory based on SP-partitions has been developed and studied in recent years. This viewpoint, however, will not be pursued further.

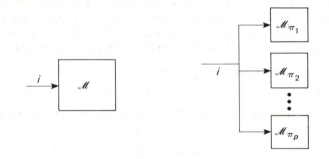

Figure 4.8.1. A finite state machine $\mathcal{M}$ and a parallel decomposition of $\mathcal{M}$

Example 4.8.1. Consider the following finite state machine $\mathcal{M}$ described by means of its transition table (figure 4.8.2), in which s_j is replaced by j (from the reference at the end of this note). We see that the partition $\pi \equiv (\overline{1,2,3}, \overline{4,5,6})$ has the substitution property with respect to δ. Indeed, for the two inputs, we have the transitions shown in figure 4.8.2. Therefore if we consider the machine $\mathcal{M}_\pi$ whose states are $B_1 \equiv \{1,2,3\}$ and $B_2 \equiv \{4,5,6\}$, we obtain the transition table in figure 4.8.3.

	i_0	i_1
1	4	3
2	6	3
3	5	2
4	2	5
5	1	4
6	3	4

i_0: $1 \to 4$, $2 \to 6$, $3 \to 5$, that is, $\{1, 2, 3\} \to \{4, 5, 6\}$

i_0: $4 \to 2$, $5 \to 1$, $6 \to 3$, that is, $\{4, 5, 6\} \to \{1, 2, 3\}$

i_1: $1 \to 3$, $2 \to 3$, $3 \to 2$, that is, $\{1, 2, 3\} \to \{1, 2, 3\}$

i_1: $4 \to 5$, $5 \to 4$, $6 \to 4$, that is, $\{4, 5, 6\} \to \{4, 5, 6\}$

Figure 4.8.2. A transition table of a finite state machine

	i_0	i_1
B_1	B_2	B_1
B_2	B_1	B_2

Figure 4.8.3. Transition table of the image $\mathcal{M}_\pi$ of $\mathcal{M}$

We conclude this section by showing that the set of the SP-partitions of S with respect to a set of functions δ (a subset of the set of all partitions of S) *is a sublattice* of the lattice of the partitions of S. This is stated by the following theorem.

Theorem 4.8.1. The set Σ of the SP-partitions of S with respect to δ is a sublattice of the lattice of the partitions of S.

Proof. We must show that, given two SP-partitions π_1 and π_2 in Σ, both $\pi_1 \vee \pi_2$ and $\pi_1 \wedge \pi_2$ are in Σ. We first consider the meet $\pi_1 \wedge \pi_2$. If s_i and s_j are equivalent under both $\varepsilon(\pi_1)$ and $\varepsilon(\pi_2)$, they are equivalent under $\varepsilon(\pi_1 \wedge \pi_2)$, as shown in Section 4.6. But because of the substitution property, for every input i_k,

$$s_i \, \varepsilon(\pi_1) \, s_j \text{ implies } \delta_k(s_i) \, \varepsilon(\pi_1) \, \delta_k(s_j),$$
$$s_i \, \varepsilon(\pi_2) \, s_j \text{ implies } \delta_k(s_i) \, \varepsilon(\pi_2) \, \delta_k(s_j), \text{ for } k = 1, 2, \ldots, m.$$

But the latter two relationships imply that $\delta_k(s_i)$ and $\delta_k(s_j)$, for every input i_k, are equivalent under $\varepsilon(\pi_1 \wedge \pi_2)$; that is, in summary,

$$s_i \, \varepsilon(\pi_1 \wedge \pi_2) \, s_j \text{ implies } \delta_k(s_i) \, \varepsilon(\pi_1 \wedge \pi_2) \, \delta_k(s_j).$$

This shows that $\pi_1 \wedge \pi_2$ is an SP-partition. We then consider the join $\pi_1 \vee \pi_2$. We know that s_i and s_j belong to the same block of $\pi_1 \vee \pi_2$ if there is a sequence $s_i = t_0, t_1, \ldots, t_k = s_j$ such that either

$$t_i \, \varepsilon(\pi_1) \, t_{i+1} \qquad \text{or} \qquad t_i \, \varepsilon(\pi_2) \, t_{i+1} (i = 0, 1, \ldots, k - 1).$$

However, each block of either π_1 or π_2 is contained within a block of $\pi_1 \quad \pi_2$, and since π_1 and π_2 are SP-partitions, from either

$$\delta_k(t_i) \, \varepsilon(\pi_1) \, \delta_k(t_{i+1}) \qquad \text{or} \qquad \delta_k(t_i) \, \varepsilon(\pi_2) \, \delta_k(t_{i+1})$$

we obtain

$$\delta_k(t_i) \, \varepsilon(\pi_1 \vee \pi_2) \, \delta_k(t_{i+1}).$$

That is, by transitivity

$$s_i \, \varepsilon(\pi_1 \vee \pi_2) \, s_j \text{ implies } \delta_k(s_i) \, \varepsilon(\pi_1 \vee \pi_2) \, \delta_k(s_j).$$

Finally, Σ clearly contains the two trivial partitions as the universal bounds. This concludes the proof of the theorem.$\|$

Reference
J. Hartmanis and R. E. Stearns, *Algebraic Structure Theory of Sequential Machines* Prentice-Hall, Englewood Cliffs, N.J., 1966.

CHAPTER 5

BOOLEAN ALGEBRAS

5.1 THE POWER SET OF A SET (SET ALGEBRA)

A very elementary step in the manipulation of the elements of a set A is the formation of subsets of A. This corresponds to the very fundamental intuition of the relation of set inclusion, and we have already come across this concept on more than one occasion (Sections 1.4, 4.3). Specifically, we noted that set inclusion is a partial ordering relation defined on the set of all the subsets of a given set A. This set, which is a set of sets, has a very interesting structure (as we shall see) and deserves a special denotation.

Definition 5.1.1. The *power set* of a set A, denoted $\mathscr{P}(A)$, is the collection of all the subsets of A.

For example, for $A \equiv \{a_1, a_2, a_3\}$, we have $\mathscr{P}(A) \equiv \{\varnothing, \{a_1\}, \{a_2\}, \{a_3\}, \{a_1, a_2\}, \{a_2, a_3\}, \{a_1, a_3\}, \{a_1, a_2, a_3\}\}$. In other words, for a finite set A, $\mathscr{P}(A)$ is a finite poset having A as its universal upper bound and the empty set $\varnothing$ as its universal lower bound. The partial ordering ("set inclusion," definition 1.1.3) is usually denoted by the special symbol $\subseteq$.

In terms of this partial ordering on $\mathscr{P}(A)$, we consider the problem of determining the join and the meet of two arbitrary elements of $\mathscr{P}(A)$ (definitions 4.3.1 and 4.3.2). For example, with reference to the join, for two subsets A_1 and A_2 of A, $A_1 \vee A_2$ is a subset of A which satisfies the relations $A_1 \subseteq A_1 \vee A_2$ and $A_2 \subseteq A_1 \vee A_2$, and there is no other subset X of A for which $A_1 \subseteq X \subseteq A_1 \vee A_2$, and $A_2 \subseteq X \subseteq A_1 \vee A_2$. In words, every element in A_1 or A_2 is also in $A_1 \vee A_2$, and there is no proper subset of $A_1 \vee A_2$ having the same property; $A_1 \vee A_2$ does not contain any element which does not belong either to A_1 or to A_2. This can be rephrased by saying that $A_1 \vee A_2$ is the set of all and only the elements of A_1 or of A_2 or of both. Clearly, for given A_1 and A_2, the join $A_1 \vee A_2$ is uniquely determined and coincides with a concept introduced quite early in our presentation, that is, the concept of union of two sets (definition 1.1.4). We recognize now that the join of two subsets of A is the binary operation of *set union* of the two subsets. The symbol usually adopted to denote set union is $\cup$ (sometimes read "cup"), and in this context it replaces the more general $\vee$.

Exactly the same argument can be repeated for the (dual) concept of meet of A_1 and A_2 in $\mathscr{P}(A)$. We find that $A_1 \wedge A_2$ coincides with the intersection of A_1 and A_2 (definition 1.1.5). The symbol usually adopted for the binary operation of *set intersection* is $\cap$, and in this context it replaces the more general $\wedge$. Note that $\cap$ (sometimes read "cap") is not so mnemonic as $\cup$, but it is easily remembered as the reverse of the mnemonic $\cup$.

Because of the uniqueness of union and intersection in $\mathscr{P}(A)$, the immediate conclusion of the preceding discussion is that the poset $[\mathscr{P}(A), \subseteq]$ is a lattice $[\mathscr{P}(A), \cup, \cap]$ (definition 4.3.3). The next natural question is whether this lattice possesses additional properties. As noted in Section 4.4, an important property to be tested is distributivity (definition 4.4.1). In this regard, we have the following theorem.

Theorem 5.1.1. The lattice $[\mathscr{P}(A), \cup, \cap]$ is distributive.

Proof. For $A_1, A_2, A_3 \in \mathscr{P}(A)$, we need to prove only one of the distributive laws, say,

$$A_1 \cap (A_2 \cup A_3) = (A_1 \cap A_2) \cup (A_1 \cap A_3);$$

the other follows by duality. To prove this identity, we show that each element of the set on the left side also belongs to the set on the right side, and vice versa.

Assume that $a \in A_1 \cap (A_2 \cup A_3)$. Then, by definition of intersection, $a \in A_1$ and $a \in A_2 \cup A_3$. Suppose that $a \in A_2$ (this entails no loss of generality, because if a does not belong to A_2, it must belong to A_3, and the same argument holds with A_3 replacing A_2). Then a belongs to both A_1 and A_2; that is, $a \in A_1 \cap A_2$. It follows that a belongs to the set on the right side, which is defined as $\{x \mid x \in A_1 \cap A_2 \text{ or } x \in A_1 \cap A_3\}$.

To prove the converse, assume, without loss of generality, that $a \in A_1 \cap A_2$. By definition of intersection, $a \in A_1$ and $a \in A_2$. But if a belongs to A_2, it also belongs to $A_2 \cup A_3$, by definition of union. From $a \in A_1$ and $a \in A_2 \cup A_3$, it follows that a belongs to $A_1 \cap (A_2 \cup A_3)$, which completes the proof of the identity.$\|$

Suppose now that $A \equiv \{a_1, a_2, a_3, a_4, a_5\}$, and in $\mathscr{P}(A)$ consider an arbitrary set, say, $A_1 \equiv \{a_1, a_3, a_4\}$. Then we recognize that there exists a unique set in $\mathscr{P}(A)$ which contains all and only the elements of A which are not in A_1. This set is obviously $\{a_2, a_5\}$, and it is very natural to call it the "complement" of A_1 in A. This is now formalized for easy reference.

Definition 5.1.2. The *set-theoretic complement* of a subset A_1 of a set A is the set of all and only the elements of A which are not in A_1. It is denoted by $\overline{A}_1$.

The determination of $\overline{A}_1$ from A_1 may be viewed either as a function $f: \mathscr{P}(A) \to \mathscr{P}(A)$ or, equivalently, as a unary operation on $\mathscr{P}(A)$. (See definition 3.8.1 and the following discussion.)

Note not only that the complement, as a function, is one-to-one (distinct sets of $\mathscr{P}(A)$ have distinct complements) and onto (in $\mathscr{P}(A)$ every set is the complement of a set), but also that the complement of $\overline{A}_1$ in $\mathscr{P}(A)$ is A_1 itself. In other words, A_1 and $\overline{A}_1$ in $\mathscr{P}(A)$ are complements of each other.

There are two additional interesting properties of the complement, which we now consider. First, for $A_1 \in \mathscr{P}(A)$, the sets A_1 and $\overline{A}_1$ have no element in common, since $\overline{A}_1$ contains *only* the elements of A which are not in A_1. Formally we can say that their intersection is the empty set $\varnothing$ [note that $\varnothing$ is the universal lower bound of $\mathscr{P}(A)$]; that is,

$$A_1 \cap \overline{A}_1 = \varnothing \text{ for } A_1 \in \mathscr{P}(A).$$

Second, since $\overline{A}_1$ contains *all* the elements of A which are not in A_1, the union of A_1 and $\overline{A}_1$ contains all the elements of A; that is, it coincides with A [note that A is the universal upper bound of $\mathscr{P}(A)$]. Formally, this is expressed by

$$A_1 \cup \overline{A}_1 = A \text{ for } A_1 \in \mathscr{P}(A).$$

In summary, the elements of the power set $\mathscr{P}(A)$ of A can be subdivided into pairs of complementary subsets, whose intersection is void and whose union is the entire set A. Obviously, A and $\varnothing$ form a complementary pair.

Example 5.1.1. For $A \equiv \{a_1, a_2, a_3\}$ there are eight subsets of A, as we noted before, and they form $\mathscr{P}(A)$. The Hasse diagram of $\mathscr{P}(A)$ is shown in figure 5.1.1. We have also shown, linked with broken lines, complementary pairs of subsets of A.

Observing figure 5.1.1 carefully, the reader will notice that the Hasse diagram of $\mathscr{P}(A)$ for a set A of three elements may be viewed as the (axonometric) projec-

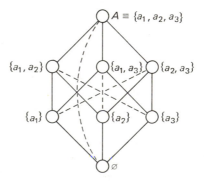

Figure 5.1.1. The Hasse diagram of $\mathscr{P}(A)$ for $A \equiv \{a_1, a_2, a_3\}$

tion of a three-dimensional cube, defined, for example, by the following ortho-
gonal edges: $(\varnothing,\{a_1\})$, $(\varnothing,\{a_2\})$, $(\varnothing,\{a_3\})$. This intriguing connection between
the Hasse diagram of $\mathscr{P}(A)$ and a well-known geometric configuration spon-
taneously suggests the question as to the general structure of the Hasse diagram
of $\mathscr{P}(A)$ for a set A of arbitrary finite cardinality. Fortunately, the solution of this
problem is remarkably simple. First, we define an interesting geometric structure.

Definition 5.1.3. The *unitary n-cube* (briefly, *n-cube*) in the cartesian *n*-dimensional
space of coordinates $x_1, x_2, \ldots, x_n$ is the set V of the vertices $(x_n, x_{n-1}, \ldots, x_1)$,
where the generic $x_j (j = 1, 2, \ldots, n)$ is either 0 or 1; in the usual notation,
$V \equiv \{v \mid v = (x_n, \ldots, x_1), x_j \in \{0, 1\}\}$.

We are able to conceive of an *n*-cube as a geometric figure only for $n \leqslant 3$.
Indeed, a 0-cube is a vertex, a 1-cube is a segment, a 2-cube is a square, and a
3-cube is a proper cube. For $n \geqslant 4$, our intuition fails with respect to the entire
geometric figure; however, for any *n*, a somewhat limited (yet totally adequate)
representation of an *n*-cube is still possible on a sheet of paper. Specifically, we
represent the *n*-cube as an undirected graph, whose vertices are the vertices of the
n-cube and whose edges connect vertices differing in only one coordinate (referred
to as *adjacent* vertices). For example, $(1, 1, 0, 1)$ and $(1, 0, 0, 1)$ are adjacent because
they differ only in the coordinate x_3 in the set of vertices $V \equiv \{v \mid v = (x_4, x_3, x_2, x_1)\}$.
Diagrams of *n*-cubes for $n = 1, 2, 3$ are shown in figure 5.1.2.

Diagrams of *n*-cubes for arbitrary *n* can be generated recursively. Assume that
we know how to construct the diagram of the $(n-1)$-cube. The diagram of the
n-cube, whose vertices are $v = (x_n, x_{n-1}, \ldots, x_1)$, can be constructed as follows:
We partition the vertices into two subsets, $V_0 \equiv \{v \mid x_n = 0\}$ and $V_1 \equiv \{v \mid x_n = 1\}$.
Both V_0 and V_1 are clearly $(n-1)$-cubes since the variables $x_1, x_2, \ldots, x_{n-1}$ can
assume the two values 0 and 1. Now the generic vertex of V_0, say, $(0, \xi_{n-1}, \ldots, \xi_1)$,
is adjacent to one and only one vertex of V_1, that is, $(1, \xi_{n-1}, \ldots, \xi_1)$. This completely reveals the structure of the *n*-cube. It consists of two $(n-1)$-cubes, of

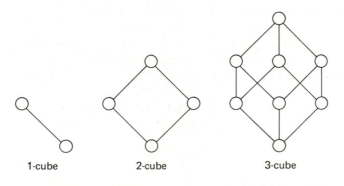

1-cube   2-cube   3-cube

Figure 5.1.2. Diagrams of *n*-cubes for $n = 1, 2, 3$

which one is the translation of the other, with edges between corresponding vertices of the two $(n-1)$-cubes (see figure 5.1.3, where, to exemplify, we have chosen $n=4$).

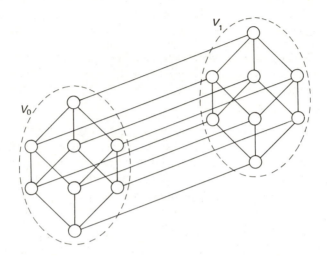

Figure 5.1.3. The structure of a 4-cube

We now return to power sets and prove the following theorem.

Theorem 5.1.2. The Hasse diagram of $\mathscr{P}(A)$, for $\#(A) = n$, is the n-cube.

Proof. Let $A \equiv \{a_1, a_2, \ldots, a_n\}$, and consider the n-cube in the space of coordinates $x_{,1}x_2, \ldots, x_n$. We assign an n-tuple $(x_1, x_2, \ldots, x_n)$ to a member A_h of $\mathscr{P}(A)$ in the following manner: $x_j = 1$ if $a_j \in A_h$, and $x_j = 0$ otherwise $(j = 1, 2, \ldots, n)$. With this rule, each member of $\mathscr{P}(A)$ will correspond to a unique vertex of the n-cube, and vice versa (i.e., we have a one-to-one correspondence between subsets of A and vertices of the n-cube). Furthermore, a subset A_h is an immediate successor of the subset A_k in the poset $\mathscr{P}(A)$ if and only if A_h contains exactly one more element than A_k. But this entails the requirement that the vertices associated with A_h and A_k differ in exactly one coordinate; that is, they are adjacent. Conversely, if two vertices are adjacent in the n-cube, the corresponding subsets are in a successor-predecessor relation in $\mathscr{P}(A)$. Since edges in the n-cube exist only between adjacent nodes, there is an edge in the n-cube if and only if there is an edge in the Hasse diagram of $\mathscr{P}(A)$. This completes the proof.$\|$

Theorem 5.1.2 is illustrated in figure 5.1.4, in which the set $A \equiv \{a_1, a_2, a_3\}$ of example 5.1.1 is again used.

Corollary 5.1.3. The cardinality of $\mathscr{P}(A)$, for $\#(A) = n$, is 2^n.

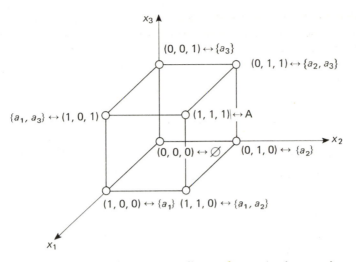

Figure 5.1.4. The cube corresponding to the set $A \equiv \{a_1, a_2, a_3\}$

Proof. By the preceding theorem, $\mathscr{P}(A)$ has as many members as there are vertices in the n-cube. We claim that the n-cube has 2^n vertices and prove it with an induction argument. Clearly the 1-cube has 2^1 vertices. Assume that the $(n-1)$-cube has 2^{n-1} vertices. Since the n-cube consists of two parallel $(n-1)$-cubes, it has twice as many vertices as the $(n-1)$-cube, that is, 2^n.||

The statement of corollary 5.1.3 gives intuitive substance to the choice of the phrase "power set" to denote $\mathscr{P}(A)$.

A very interesting case of power sets occurs when the elements $a_1, a_2, \ldots, a_n$ are connected domains of points (for example, of the plane) and constitute a partition of a connected domain A. The area of the intersection of any two a_i, a_j is null, and the area of the union of all the a_i's coincides with the area of A. If one regards the points as elements (in the same fashion, the a_i's of the preceding discussion), then the set of the elements is not only infinite but even uncountable (see Section 1.1), and the preceding analysis of power sets does not apply. However, if one regards the domains $a_1, a_2, \ldots$ as elements which cannot be further subdivided, as long as the number of these domains is finite, the preceding considerations apply in full. For example, in figure 5.1.5 we exhibit pictorially a

Figure 5.1.5. A Venn diagram

connected domain A (the square), partitioned into four connected subdomains a_1, a_2, a_3, a_4, which are not further subdivided. We turn again to the Venn diagram, introduced in Section 1.1, to intuitively illustrate the notions of union and intersection. We note here that complementation also acquires a very intuitive appeal, as shown in figure 5.1.6. Moreover, consider the "indivisible"

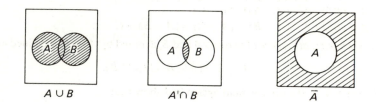

$$A \cup B \qquad\qquad A' \cap B \qquad\qquad \bar{A}$$

Figure 5.1.6. Set-theoretic operations illustrated by Venn diagrams

subdomains of A: a_1, a_2, a_3, and a_4. As elements of $\mathscr{P}(A)$, a_1, a_2, a_3, and a_4 appear in the Hasse diagram of $\mathscr{P}(A)$ as immediate successors of the lower bound $\varnothing$. But in Section 4.5, we called the immediate successors of the lower bound the *atoms* of a lattice, a Greek-derived word meaning "indivisible."

This concludes temporarily our discussion of the power set of a finite set. Our main objective was to show that $\mathscr{P}(A)$ is a distributive lattice and that a very special unary operation, *complementation*, is naturally introduced in $\mathscr{P}(A)$. We shall return to power sets in subsequent sections.

——

EXERCISES 5.1

1. Construct the power set of the set $S \equiv \{1, 2, 3, 4\}$, and illustrate it by means of its Hasse diagram.

2. Draw the Venn diagrams of the following sets.

$$\overline{A \cap B}, \qquad \bar{A} \cup \bar{B}, \qquad (A \cup B) \cap \bar{A}, \qquad B \cap \bar{A}$$

3. Illustrate each of the following conditions with a Venn diagram.

$$A \nsubseteq \bar{B}, \qquad B \nsubseteq \bar{A}, \qquad A \neq B.$$

4. Let A and B be two arbitrary sets. Prove the following identities.
   a) $(A \cup B) \cap \bar{A} = B \cap \bar{A}$
   b) $(A \cap B) \cup A = A$
   c) $\overline{A \cap B} = \bar{A} \cup \bar{B}$
   d) $\overline{A \cup B} = \bar{A} \cap \bar{B}$
   e) $A \cup (\bar{A} \cap B) = A \cup B$
   f) $A \cap (\bar{A} \cup B) = A \cap B$
   g) $(A \cap B) \cup (\bar{A} \cap \bar{B}) = (\bar{A} \cap B) \cup (A \cap \bar{B})$

5. If A and B are sets, the *relative complement* of B in A, denoted by $A - B$, is defined as the set $\{a | a \in A$ but $a \notin B\}$. Show that
   a) $A - B \subseteq A$
   b) $A - B = A$ if and only if A and B are disjoint
   c) $A - B$ and B are disjoint and $(A - B) \cup B = A \cup B$
   d) $A \cap B = A - (A - B)$
   e) $A - \bar{B} = \varnothing$ if and only if A and B are disjoint
   f) $A - (B \cap C) = (A - B) \cup (A - C)$
   g) $A - (B \cup C) = (A - B) \cap (A - C) = (A - B) - C$

6. The *symmetric difference* of two sets A and B, denoted by $A \oplus B$, is defined as follows:

$$A \oplus B = (\bar{A} \cap B) \cup (A \cap \bar{B}).$$

   Illustrate $A \oplus B$ with a Venn diagram and show that
   a) $A \oplus B = B \oplus A$
   b) $(A \oplus B) \oplus B = A$
   c) $(A \oplus B) \oplus C = A \oplus (B \oplus C)$
   d) $A \oplus A = \varnothing$
   e) $A \oplus \varnothing = A$
   f) If $A \oplus B = C$, then $B = A \oplus C$
   g) $A \oplus B \oplus (A \cap B) = A \cup B$
   h) $A \cap (B \oplus C) = (A \cap B) \oplus (A \cap C)$

7. Let A, B, C be subsets of a set U. Show that
   a) If $A \cup B = \varnothing$, then $A = \varnothing$ and $B = \varnothing$
   b) If $A \cup B = B$ and $A \cap B = \varnothing$, then $A = \varnothing$
   c) If $A \cap B = A$ and $B \cup C = C$, then $A \cap \bar{C} = \varnothing$
   d) If $A \cap \bar{B} = \varnothing$ and $A \cap B = \varnothing$, then $A = \varnothing$
   e) $A = \varnothing$ if and only if $(A \cap \bar{B}) \cup (\bar{A} \cap B) = B$ for any $B \subseteq U$
   f) $A \subseteq B$ if and only if $\bar{B} \subseteq \bar{A}$

8. a) Verify that the Venn diagram in figure 5.1.7 describes all possible memberships to any subset of four sets A, B, C, and D.

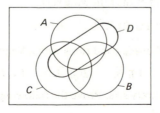

Figure 5.1.7.

   b) Draw the modified diagrams in the following special cases.
      i) any element in $B \cap C$ is in either A or D
      ii) $C \subseteq (\bar{B} \cap D) \cup (B \cap \bar{D})$

5.2 BOOLEAN ALGEBRAS AS LATTICES

The notion of complement, introduced in the preceding section in a very intuitive way which motivated the choice of the term, will now be considered in a somewhat more general way. Specifically, lattices $[S, \vee, \wedge]$ will be considered instead of the more restrictive class of power sets. As usual, I and O will denote the universal upper and lower bounds.

Cthe notion of complement is characterized by two properties which were shown to be possessed by the set-theoretic complement in Section 5.1, as stated by the following definition.

Definition 5.2.1. In a lattice $[S, \vee, \wedge]$ with universal bounds, a *complement* of an element $a \in S$ is any element b such that

$$a \wedge b = O \qquad \text{and} \qquad a \vee b = I. \tag{5.2.1}$$

Definition 5.2.1 states that it is meaningful to talk about complements only when lattices have the universal bounds. Such lattices are called *bounded*. Moreover, by definition of upper bound, for every $a \in S$, $I \wedge a = a$. It follows that the *only* x which satisfies $I \wedge x = O$ is $x = O$ itself. Since $I \vee O = I$, we conclude that O is a complement of I. By a dual argument, we can equally easily show that I is a complement of O; that is, in a bounded lattice the universal bounds O and I are the complements of each other (a fact already observed in power sets).

Note that definition 5.2.1 specifies *a* complement, not *the* complement, thereby allowing for more than one. Indeed, it is possible for a lattice element to have more than one complement satisfying conditions (5.2.1). In figure 5.2.1 we see that $a \vee b = I$ and $a \wedge b = O$; that is, b qualifies as a complement of a. But the same holds for c; that is, both b and c are complements of a. In general, "complement" is a relation on the set S (not necessarily a function).

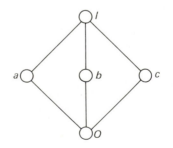

Figure 5.2.1. An example of a complemented lattice

In figure 5.2.1, we find that each element of the lattice has at least one complement; in fact, the relation "complement" on the lattice under consideration is described by the following table, each row of which has a nonzero entry.

$$
\begin{array}{c c}
 & I\ a\ b\ c\ O \\
\begin{array}{c} I \\ a \\ b \\ c \\ O \end{array} &
\left[\begin{array}{ccccc}
0 & 0 & 0 & 0 & 1 \\
0 & 0 & 1 & 1 & 0 \\
0 & 1 & 0 & 1 & 0 \\
0 & 1 & 1 & 0 & 0 \\
1 & 0 & 0 & 0 & 0
\end{array}\right]
\end{array}
$$

Definition 5.2.2. A bounded lattice $[S,\vee,\wedge]$ is said to be *complemented* if for each element $a \in S$ there exists a complement $b \in S$.

Of course, there are lattices which are not complemented. For example, in the lattice of the partitions of the integer 4 (example 4.3.1), whose Hasse diagram is repeated for convenience in figure 5.2.2, the elements a, b, and c have no

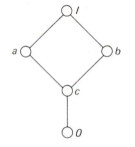

Figure 5.2.2. Lattice of the partitions of the integer 4

complement. Considering, for example, the element a, we note that the simple equation $a \vee x = I$ has solutions $x = b$, I, and the other simple equation $a \wedge x = O$ has solution $x = O$; that is, there is no lattice element which satisfies them simultaneously (hence a has no complement in S).

It is now interesting to ask under which conditions a complement of an element, if it exists, is unique. The reader may already have a clue as to the answer, since the lattice of figure 5.2.1, in which the complement is not unique, is a typical case of *nondistributive* lattices (figure 4.4.2(b)). The reader may check the correctness of his expectation in the statement of the following theorem.

Theorem 5.2.1. In a bounded distributive lattice $[S,\vee,\wedge]$ complements are unique when they exist.

Proof. Assume that an element $a \in S$ has two complements b_1 and b_2 in $[S,\vee,\wedge]$. Then, by definition 5.2.1 of complements,

$$a \vee b_1 = I,\ a \wedge b_1 = O;\ a \vee b_2 = I,\ a \wedge b_2 = O.$$

We now use the substitution principle in replacing a generic element b of S by an expression of b. Specifically, since $O \leqslant b_1$, by definition of lower bound (definition 4.2.5), we have

$$b_1 = b_1 \vee O = b_1 \vee (a \wedge b_2),$$

where we have used the identity $O = a \wedge b_2$, which we have assumed to be true. The distributive law is crucial in establishing that

$$b_1 \vee (a \wedge b_2) = (b_1 \vee a) \wedge (b_1 \vee b_2).$$

Since $b_1 \vee a = I$ by hypothesis, we have in summary

$$b_1 = b_1 \vee O = b_1 \vee (a \wedge b_2) = (b_1 \vee a) \wedge (b_1 \vee b_2) = I \wedge (b_1 \vee b_2) = b_1 \vee b_2.$$

Similarly, exchanging b_1 and b_2, we have

$$b_2 = b_2 \vee O = b_2 \vee (a \wedge b_1) = (b_2 \vee a) \wedge (b_2 \vee b_1) = I \wedge (b_2 \vee b_1) = b_2 \vee b_1.$$

Therefore, since two elements having the same expression are identical, we have

$$b_1 = b_1 \vee b_2 = b_2 \vee b_1 = b_2,$$

which clearly shows that the two hypothesized complements in fact coincide; that is, the complement of an element of a distributive lattice is unique when it exists.‖

Henceforth we shall use the notation $\bar{a}$ to denote the *unique* complement of an element a.

Example 5.2.1. Consider the lattice L of the divisors of the integer 60 whose partial ordering is divisibility (already considered in example 4.3.3). Its Hasse diagram is repeated in figure 5.2.3. This lattice is known to be distributive. Note that the integers 3, 4, 5, 12, 15, and 20 have the complement in L (now unique because of distributivity!) and so do, trivially, 1 and 60; but 2, 6, 10, and 30 do not. For example, $12 = \bar{5}$, since l.c.m $(12,5) = 60$ and g.c.d. $(12,5) = 1$; but l.c.m. $(6,x) = 60$, considered as an equation in x, has the solution $x = 20$, 60,

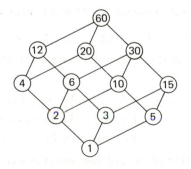

Figure 5.2.3. The lattice L of the divisors of 60

whereas g.c.d. $(6, x) = 1$ has the solution $x = 5$, which proves the nonexistence of the complement of 6 in L.

Theorem 5.2.1 can be paraphrased by saying that in a distributive lattice $[S, \vee, \wedge]$ the complement is a function, or a *unary operation* (definition 3.8.1), defined on a subset of S. However, distributivity is such a remarkably powerful attribute that the properties illustrated in the two following theorems are also brought about by its introduction.

Theorem 5.2.2. In a bounded distributive lattice $[S, \vee, \wedge]$, if $a \in S$ has the complement $\bar{a}$, then $\bar{a}$ also has the complement, and

$$(\overline{\bar{a}}) = a.$$

(This property is referred to as *involution*; complement is an *involutory* function on a subset of S.)

Proof. By definition of complement, $a \vee \bar{a} = I$ and $a \wedge \bar{a} = O$. To determine (if it exists) the complement of $\bar{a}$, we must find an element $x \in S$ which simultaneously satisfies

$$\bar{a} \vee x = I \qquad \text{and} \qquad \bar{a} \wedge x = O.$$

But a is one such element, and since complements are unique in $[S, \vee, \wedge]$ by theorem 5.2.1, a is the complement of $\bar{a}$. ‖

Theorem 5.2.3. In a bounded distributive lattice $[S, \vee, \wedge]$, if for a and b in S, the complements $\bar{a}$ and $\bar{b}$ exist, then also $\overline{(a \vee b)}$ and $\overline{(a \wedge b)}$ exist, and

$$\overline{a \vee b} = \bar{a} \wedge \bar{b},$$
$$\overline{a \wedge b} = \bar{a} \vee \bar{b}.$$

(This property is referred to as the *dualization* law, or De Morgan's Law.)

Proof. Since the two identities are dual, we need to prove only one and invoke the duality principle for the other. We choose to prove that $\overline{a \vee b}$ exists and that

$$\overline{a \vee b} = \bar{a} \wedge \bar{b}.$$

Indeed, the complement of the element $(a \vee b)$ of S is an element x of S for which

$$(a \vee b) \vee x = I \qquad \text{and} \qquad (a \vee b) \wedge x = O.$$

We construct $\bar{a} \wedge \bar{b}$ (which exists, because $\bar{a}$ and $\bar{b}$ exist) and claim that it satisfies the requirements for x. Indeed,

$$\begin{aligned}
(a \vee b) \vee (\bar{a} \wedge \bar{b}) &= [(a \vee b) \vee \bar{a}] \wedge [(a \vee b) \vee \bar{b}] \\
&= [(a \vee \bar{a}) \vee b] \wedge [a \vee (b \vee \bar{b})] \\
&= [(I \vee b) \wedge (a \vee I)] = I \wedge I = I,
\end{aligned}$$

where the first equality follows from distributivity, the second from associativity and commutativity, the third from complementarity, the fourth from the property of the bound I, and the fifth from idempotency. With a parallel sequence of steps we have

$$(a \vee b) \wedge (\overline{a} \wedge \overline{b}) = [a \wedge (\overline{a} \wedge \overline{b})] \vee [b \wedge (\overline{a} \wedge \overline{b})]$$
$$= [(a \wedge \overline{a}) \wedge \overline{b}] \vee [(b \wedge \overline{b}) \wedge \overline{a}]$$
$$= (O \wedge \overline{b}) \vee (O \wedge \overline{a}) = O \vee O = O,$$

which proves our claim. Therefore not only do the complements of $(a \vee b)$ and $(a \wedge b)$ exist, but the dualization laws hold.‖

Example 5.2.2. Referring again to the lattice L of the divisors of 60 (example 5.2.1, figure 5.2.3), we note that $\overline{4} = 15$ and $\overline{3} = 20$; the join of 3 and 4, that is, their l.c.m., is 12, and the complement of 12 is $\overline{12} = 5$. Indeed, the dualization law holds:

$$\overline{\text{l.c.m.}(3,4)} = \overline{12} = 5 = \text{g.c.d.}(20,15) = \text{g.c.d.}(\overline{3},\overline{4}).$$

Thus far we have considered two interesting classes of bounded lattices, distributive lattices and complemented lattices, and we have convinced ourselves that their respective characterizing properties are independent (i.e., there are distributive lattices which are not complemented, and vice versa). When the two characterizing properties coexist in a lattice, we obtain one of the most fundamental algebraic structures.

Definition 5.2.3. A *boolean algebra* or *boolean lattice* is a bounded, distributive, and complemented lattice. The usual symbol for a boolean algebra is $\mathscr{B} = [S, \vee, \wedge, {}^{-}, O, I]$, where S is a set of elements, $\vee$ and $\wedge$ are binary operations, ${}^{-}$ is a unary operation, and O and I are the universal bounds.

Example 5.2.3. We have previously encountered a boolean algebra without so naming it: the power set $\mathscr{P}(A)$ of a set A. To realize that $\mathscr{P}(A)$ is indeed a boolean algebra is quite simple. $\mathscr{P}(A)$ is a bounded lattice, having the empty set $\varnothing$ as its lower bound and the set A as its upper bound. In $\mathscr{P}(A)$ the join is set union $\cup$, the meet is set intersection $\cap$. The lattice $[\mathscr{P}(A), \cup, \cap]$ is distributive, by theorem 5.1.1. Finally, the set-theoretic complement (definition 5.1.2) is the complement in $\mathscr{P}(A)$; since each member of $\mathscr{P}(A)$ has the complement, $\mathscr{P}(A)$ is a complemented lattice. We conclude that $\mathscr{P}(A)$ is indeed a boolean algebra $[\mathscr{P}(A), \cup, \cap, {}^{-}, \varnothing, A]$, with the following correspondence of symbols.

Boolean algebras		Power sets $\mathscr{P}(A)$
$\vee$	$\leftrightarrow$	$\cup$
$\wedge$	$\leftrightarrow$	$\cap$
${}^{-}$	$\leftrightarrow$	${}^{-}$
$\leqslant$	$\leftrightarrow$	$\subseteq$
I	$\leftrightarrow$	A
O	$\leftrightarrow$	$\varnothing$

Boolean algebras will constitute the subject of the rest of this chapter, both as a calculus (i.e., a set of manipulative rules) and as formal models of systems of various nature. The applications of boolean algebra are very numerous and fascinating.

_____

EXERCISES 5.2

1. Determine all pairs of complementary elements in the lattice L of the divisors of the integer 75, whose partial ordering is divisibility.

2. Show that any lattice-homomorphism φ of a boolean algebra A onto another boolean algebra B preserves complementation.

3. Give an example of a bounded distributive lattice in which some elements do not have complements.

4. Show that in a bounded distributive lattice the elements which have complements form a sublattice.

5. Let F denote the set of all binary relations on a set S. Define an ordering $\leqslant$ on F such that

$$\theta \leqslant \rho \text{ if and only if } x \, \theta \, y \text{ implies } x \, \rho \, y, \text{ for all } x, y \in S.$$

Show that F is a boolean algebra with respect to the ordering $\leqslant$. (See exercise 4.1.6.)

6. Let $\mathscr{B}$ be the set of the divisors of the integer 30. Define $a \vee b = \text{l.c.m.}(a,b)$, $a \wedge b = \text{g.c.d.}(a,b)$ and $\bar{a} = 30/a$.

   a) Show that $\mathscr{B}$ with these operations is a boolean algebra.
   b) What is the underlying partial ordering?
   c) Can the same statement be made for the set of the divisors of the integer 45?

7. Let $\mathscr{B}$ be a boolean algebra. For $a, b \in \mathscr{B}$, define

$$a \oplus b = (a \wedge \bar{b}) \vee (\bar{a} \wedge b).$$

Show that the following identities hold.

   a) $a \oplus b = b \oplus a$
   b) $a \oplus (b \oplus c) = (a \oplus b) \oplus c$
   c) $a \wedge (b \oplus c) = (a \wedge b) \oplus (a \wedge c)$
   d) $a \oplus a = 0, \, a \oplus 0 = a$
      $a \oplus \bar{a} = 1, \, a \oplus 1 = \bar{a}$
      $x \oplus a = b$ if and only if $x = a \oplus b$

(See exercise 5.1.6.)

*8. Show that if a bounded lattice L is uniquely complemented and the dualization laws (theorem 5.2.3) hold in it, then L is a boolean algebra.

9. Let R be the set of real numbers in the closed interval $[0,1]$. For $a, b \in I$, define $a \vee b = \max(a,b)$, $a \wedge b = \min(a,b)$ and $\bar{a} = 1 - a$. Is R with these operations a boolean algebra?

_____

5.3 THE REPRESENTATION OF BOOLEAN ALGEBRAS; CANONICAL EXPRESSIONS

In Section 4.5 we introduced the problem of representation as one of the fundamental questions in the analysis of algebraic structures. As we mentioned, the objective is the analysis of elements of the system in terms of a subset of these elements or, equivalently, the *description* of all the elements by means of *expressions* which involve only a subset of them. With regard to distributive lattices of finite length, we reached the remarkable conclusion, summarized as theorem 4.5.2, that "each element has a unique representation as the join of an irredundant set of join-irreducible elements." The reader is referred to definition 4.5.1 to refresh the pertinent notions.

Henceforth we will restrict ourselves to considering only boolean algebras of finite length (in the lattice sense), since boolean algebras of infinite length involve notions which are beyond the scope of this book. We now determine the join-irreducible elements of a boolean algebra (of finite length).

Lemma 5.3.1. In a boolean algebra $x \neq O$ is join-irreducible if and only if x is an atom.

Proof. Certainly the atoms of $\mathscr{B}$ are join-irreducible. We must show that there is no other join-irreducible element in $\mathscr{B}$. We prove this by contradiction. Let us assume that there is a join-irreducible element a of $\mathscr{B}$ which is not an atom. This means that the only immediate predecessor of a is an element $b \neq O$. Now consider an element x for which $a \nleqslant x$ (see figure 5.3.1). It follows, by the consistency principle (4.3.5), that $(a \wedge x) \neq a$; that is, $(a \wedge x)$ is an element belonging to a down-directed chain issuing from a in the Hasse diagram of $\mathscr{B}$. But each such chain passes through the element b, since a is join-irreducible; it follows immediately that $(a \wedge x) = (b \wedge x)$.

Consider now the element $\bar{b}$. Clearly $a \nleqslant \bar{b}$. Indeed, if $a \leqslant \bar{b}$ were true, then we would have $b \leqslant a \leqslant \bar{b}$; that is, by the consistency principle, $b \wedge \bar{b} = b$, which violates the definition of complement. Similarly, if $\bar{b} \leqslant a$, we would also have $\bar{b} \leqslant b \leqslant a$, which is impossible for the same reason. Using the preceding result, we have $(a \wedge \bar{b}) = (b \wedge \bar{b})$; but $(b \wedge \bar{b}) = O$ by definition of complement, whence $a \wedge \bar{b} = O$. We now consider the element $(a \vee \bar{b})$. Since $a = a \vee b$, we have $a \vee \bar{b} = (a \vee b) \vee \bar{b} = a \vee (b \vee \bar{b}) = a \vee I = I$. In summary $a \wedge \bar{b} = O$, and $a \vee \bar{b} = I$; that is, a satisfies the requirements for being the complement of $\bar{b}$. But we know that b is the unique complement of $\bar{b}$ (since $\mathscr{B}$ is distributive, theorems 5.2.1 and 5.2.2) and that a and b are distinct, whence a contradiction has been reached. Note that this contradiction would not have been reached if $b = O$. This proves that the atoms are all and the only join-irreducible elements of a boolean algebra $\mathscr{B}$.$\|$

---

1. Recall that the *atoms* are the immediate successors of the lower bound, which is their only immediate predecessor (whence they are join-irreducible).

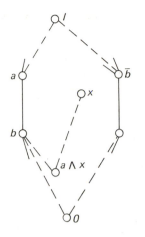

Figure 5.3.1. Illustration of the proof of lemma 5.3.1

The preceding result, which makes crucial use of the property that each element has a unique complement in boolean algebras, is the key to the solution of the representation problem. In fact, the atoms of $\mathscr{B}$ form a set

$$M \equiv \{m_1, m_2, \ldots, m_n\},$$

no two elements of which are comparable, since the O element is their only predecessor [i.e., for distinct m_i and m_j, $(m_i \wedge m_j) = O$]. Therefore the join of any subset of atoms of $\mathscr{B}$ is irredundant, and any such join represents a unique element of $\mathscr{B}$, by virtue of theorem 4.5.2, which holds in distributive lattices. It follows that there is a one-to-one correspondence between elements of $\mathscr{B}$ and subsets of the set M of the atoms of $\mathscr{B}$, that is, elements of the power set $\mathscr{P}(M)$ of M! We denote this correspondence with the mapping $\Phi : \mathscr{B} \rightarrow [\mathscr{P}(M), \cup, \cap]$, and $\Phi(a)$ denotes the subset of M whose join represents $a \in \mathscr{B}$. This mapping does much more than simply map elements of the lattice $\mathscr{B}$ to elements of the lattice $[\mathscr{P}(M), \cup, \cap]$. It also preserves the operations in the two lattices; that is, for a and b in $\mathscr{B}$,

$$\Phi(a \vee b) = \Phi(a) \cup \Phi(b),$$
$$\Phi(a \wedge b) = \Phi(a) \cap \Phi(b),$$
$$\Phi(\bar{a}) = \overline{\Phi(a)}.$$

The first of these three identities can be easily proved by showing that the set on the left side is contained in the set on the right side, and vice versa; i.e., the two sets are identical. Assume that $m_j \in \Phi(a \vee b)$. We claim that either $m_j \in \Phi(a)$ or $m_j \in \Phi(b)$ or m_j belongs to both. Suppose the contrary. Then, expressing a as a (unique) join of atoms $a = (m_{i_1} \vee m_{i_2} \vee \cdots \vee m_{i_r})$ with $i_k \neq j$ $(k = 1, 2, \ldots, r)$, we have

$$m_j \wedge a = m_j \wedge (m_{i_1} \vee \cdots \vee m_{i_r}) = (m_j \wedge m_{i_1}) \vee \cdots \vee (m_j \wedge m_{i_r}) = O,$$

since $(m_j \wedge m_{i_r}) = O$ for $i_r \neq j$. Similarly, we obtain $(m_j \wedge b) = O$, whence

$$m_j \wedge (a \vee b) = (m_j \wedge a) \vee (m_j \wedge b) = O \vee O = O.$$

However, expressing $(a \vee b)$ as a join of atoms, we have, by hypothesis, an expression containing m_j, whence

$$m_j \wedge (a \vee b) = m_j,$$

a contradiction. Therefore if, say $m_j \in \Phi(a)$, then also $m_j \in \Phi(a) \cup \Phi(b)$, which proves the first part of our assertion,

$$\Phi(a \vee b) \subseteq \Phi(a) \cup \Phi(b).$$

Conversely, assume that $m_j \in \Phi(a) \cup \Phi(b)$. Then m_j must be in at least one of the sets $\Phi(a)$ and $\Phi(b)$; say, $m_j \in \Phi(a)$. By definition, then, m_j appears in the join of atoms representing a and, being join-irreducible, it also appears in the expression of $a \vee b$; that is, $m_j \in \Phi(a \vee b)$. This shows that

$$\Phi(a \vee b) \supseteq \Phi(a) \cup \Phi(b)$$

and completes the proof of the first identity. The other two identities are proved by using analogous arguments. We reach the conclusion that the mapping Φ from $\mathscr{B}$ to $[\mathscr{P}(M), \cup, \cap]$ is invertible and preserves the operations of join, meet, and complementation, as defined in the two lattices; that is, Φ is a lattice isomorphism. This is summarized by the following classical theorem (*Stone representation theorem*).

Theorem 5.3.2. A boolean algebra $\mathscr{B}$ (of finite length) is isomorphic to the power set of its atoms.

This result opens the road to various considerations. In Section 5.1 we introduced and discussed the structure of the power set $\mathscr{P}(A)$ of a set A. In Section 5.2 we showed that $\mathscr{P}(A)$ was an instance of boolean algebras. Theorem 5.3.2 states the much stronger result that every boolean algebra (of finite length) can be represented (or thought of) as a power set. Therefore, boolean algebras (of finite length) are in no way more general than power sets. The previously stated representation theorem can also be generalized to arbitrary boolean algebras. However, we will not pursue it here.

Combining theorem 5.3.2 with theorem 5.1.2 and corollary 5.1.3, we have the following corollary.

Corollary 5.3.3. The Hasse diagram of a boolean algebra $\mathscr{B}$ with n atoms is the n-cube and $\mathscr{B}$ contains 2^n elements.

The corollary says that the structure of a boolean algebra is entirely specified when we know the number n of its atoms. Therefore we shall hereafter use the symbol $\mathscr{B}_n$ to denote one such algebra.

In connection with boolean lattices $\mathscr{B}_n$, the join and the meet are frequently renamed *disjunction* and *conjunction*, respectively. It is now convenient to introduce a notational change, since we shall later be doing a considerable amount of formula manipulation. The reader may have noticed the awkwardness of operating with the symbols $\vee$ and $\wedge$. In Section 4.4 we pointed out the analogies between join and meet in lattices and sum and product in the algebras of integers or of real numbers (rings and fields). For this reason we choose to replace hereafter the symbol $\vee$ with $+$ and the symbol $\wedge$ either with $\cdot$ or with simple juxtaposition of the operand symbols; we shall maintain unaltered the symbol for complementation. Finally, we mention that atoms of $\mathscr{B}_n$ are customarily called *minterms*. The reason for this word may be the fact that in the power set $\mathscr{P}(M)$, the atoms themselves are the *terms* of *minimal* nonzero cardinality.

If an element a of $\mathscr{B}_n$ is expressed as a disjunction

$$a = m_{j_1} + m_{j_2} + \cdots + m_{j_r} \, (r \leqslant n),$$

then obviously $m_{j_i} \leqslant a \; (i = 1, 2, \ldots, r)$; moreover, there is no minterm m_k not appearing in the expression above (i.e., $k \neq j_i$) for which $m_k \leqslant a$. Indeed, assume that $m_k \leqslant a$. Then we have

$$m_k = m_k a = m_k (m_{j_1} + \cdots + m_{j_r}) = m_k m_{j_1} + \cdots + m_k m_{j_r} = O.$$

The first equality follows from the consistency principle, the second from the substitution principle, the third from distributivity, the fourth from the fact that $m_i m_j = O$ for two distinct minterms m_i and m_j; the entire line is obviously a contradiction. This result, combined with theorem 4.5.2 and lemma 5.3.1, leads to the following theorem.

Theorem 5.3.4. Each element a of a boolean algebra $\mathscr{B}_n$ has a unique representation as the disjunction of the minterms m_j of $\mathscr{B}_n$ for which $m_j \leqslant a$ (this representation is referred to as the *canonical disjunctive representation*).

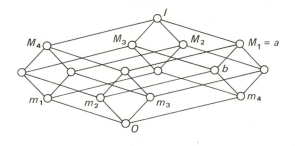

Figure 5.3.2. The Hasse diagram of $\mathscr{B}_4$

Example 5.3.1. Consider the boolean algebra $\mathscr{B}_4$ with four atoms (and $2^4 = 16$ elements) represented in figure 5.3.2. The elements m_1, m_2, m_3, and m_4 are the minterms (atoms) of $\mathscr{B}_4$. Consider, for example, the element a. Since m_2, m_3, and m_4 are $\leqslant a$ in the partial ordering "$\leqslant$" of $\mathscr{B}_4$, and m_1 is incomparable with a, we have the canonical disjunctive representation

$$a = m_2 + m_3 + m_4.$$

We now turn our attention to the complements of the minterms of $\mathscr{B}_n$. Let M_j denote the complement of minterm m_j; that is, $\overline{m}_j = M_j$. Note that, by definition of a join-irreducible element, for the minterm m_j, $x + y = m_j$ implies that either $x = m_j$ or $y = m_j$. If we apply De Morgan's dualization theorem we have

$$\overline{x} \cdot \overline{y} = \overline{m}_j = M_j$$

implies that either

$$\overline{x} = M_j \qquad \text{or} \qquad \overline{y} = M_j.$$

In other words, the complement of a join-irreducible element is a meet-irreducible element (and conversely). By involution (theorem 5.2.2) $\overline{M}_j = m_j$, and by the uniqueness of the complement, there is no other meet-irreducible element m in $\mathscr{B}_n$ for which $\overline{m} = m_j$ (if there were, $\overline{m}_j = M_j$ and $\overline{m}_j = m$, with $m \neq M_j$, which is impossible). We summarize this as a lemma.

Lemma 5.3.5. In a boolean algebra $\mathscr{B}_n$ the complements of the minterms are the meet-irreducible elements of $\mathscr{B}_n$.

In the set-theoretic sense, i.e., in the lattice $[\mathscr{P}(M), \cup, \cap, ^-, M, \varnothing\,]$, the set $\Phi(M_j)$, being the complement of the set $\Phi(m_j)$, contains all the $m_i \neq m_j$. The cardinality of $\Phi(M_j)$ is $(n-1)$, the *maximum* for a proper subset of $M \equiv \{m_1, \ldots, m_n\}$; for this reason the complement of a minterm is called a *maxterm*. In the boolean algebra $\mathscr{B}_n$, isomorphic to $[\mathscr{P}(M), \cup, \cap, ^-, M, \varnothing\,]$, M_j is simply the disjunction of all the m_i's distinct from m_j.

Suppose now that we invert the partial ordering relation $\leqslant$ in $\mathscr{B}_n$, that is, consider the structure whose Hasse diagram is that of $\mathscr{B}_n$ turned upside-down. The structure is still a boolean algebra $\mathscr{B}_n$, since distributivity, boundedness, and complementedness are preserved through inversion of $\leqslant$; O and I exchange their roles, and so do $+$ and $\cdot$ (and the join-irreducible minterms become the meet-irreducible maxterms). Therefore we can immediately rephrase theorem 5.3.4 in the following way.

Theorem 5.3.6. (Dual of theorem 5.3.4) Each element a of a boolean algebra $\mathscr{B}_n$ has a unique representation as the conjunction of the maxterms M_j of $\mathscr{B}_n$ for which $M_j \geqslant a$ (this representation is referred to as the *canonical conjunctive representation*).

Example 5.3.2. In $\mathscr{B}_4$ (figure 5.3.2) the element b is comparable exactly with the minterms m_2 and m_4 and the maxterms M_1 and M_3. Therefore, the two canonical representations of b are

$$b = m_2 + m_4 = M_1 \cdot M_3.$$

How are the two canonical representations related? In other words, is there a simple way to obtain one from the other for a given element of $\mathscr{B}_n$? The answer is affirmative. First of all, we show that each minterm m_j is comparable with all the maxterms except its complement; indeed $m_j M_i \leqslant m_j$ and $M_i \leqslant m_j + M_i$. Since m_j is an atom in the partial ordering $\leqslant$, either $m_j M_i = O$ or $m_j M_i = m_j$. Similarly, by duality, either $m_j + M_i = I$ or $m_j + M_i = M_i$. If m_j and M_i are incomparable, then $m_j M_i = O$ and $m_j + M_i = I$; that is, M_i satisfies the conditions of the complement of m_j, and, by the definition of M_j $(M_j = \overline{m}_j)$ and because of the uniqueness of the complement, $M_i = M_j$; moreover, M_j is the only maxterm incomparable with m_j. It follows that, for $i \neq j$, we have $m_j M_i = m_j$; this is equivalent, by the consistency principle (4.3.5) to $m_j \leqslant M_i$. Therefore, by theorem 5.3.6,

$$m_j = \prod_{i \neq j} M_i, \tag{5.3.1}$$

where the symbol $\prod_{i \neq j} M_i$ denotes "conjunction of M_i over all indexes $i \neq j$."

Theorem 5.3.7. For $a \in \mathscr{B}_n$ the two canonical dual representations of a,

$$m_{i_1} + \cdots + m_{i_s} = M_{j_1} \cdots M_{j_r},$$

are such that $(\{i_1, i_2, \ldots, i_s\}, \{j_1, j_2, \ldots, j_r\})$ is a dichotomy of the set of integers $\{1, 2, \ldots, n\}$.

Proof. The proof is by induction on r. Without loss of generality, we may assume that $\{i_1, i_2, \ldots, i_s\} = \{1, 2, \ldots, s\}$. Now, for a single m_1 the claim is true, as shown by (5.3.1) with $j = 1$. Assume now that for some $r > 1$, we have

$$m_1 + m_2 + \cdots + m_{r-1} = M_r M_{r+1} \cdots M_n.$$

Thus, using identity (5.3.1) again, we have

$$m_r = \prod_{i \neq r} M_i,$$

whence

$$m_1 + \cdots + m_{r-1} + m_r = (m_1 + \cdots + m_{r-1}) + m_r = M_r M_{r+1} \cdots M_n + \prod_{i \neq r} M_i.$$

When we apply the distributive property, the last expression becomes

$$M_{r+1} \cdots M_n(M_r + M_1 M_2 \cdots M_{r-1}).$$

We claim that

$$M_r + M_1 \cdots M_{r-1} = I.$$

Indeed, if we again apply the distributive property,

$$M_r + M_1 \cdots M_{r-1} = (M_r + M_1) \cdots (M_r + M_{r-1}) = I \cdots I = I,$$

since for $i \neq j$, $M_i + M_j = I$. Hence

$$m_1 + m_2 + \cdots + m_r = M_{r+1} \cdots M_n.$$

This completes the proof.$\|$

It is convenient to summarize the results of this section. First of all, a boolean algebra is as general a structure as a power set. Second, each element of a boolean algebra has two unique canonical representations, as a disjunction of minterms or as a conjunction of maxterms. These representations have far-reaching consequences in the development of a calculus of boolean algebras.

_____

EXERCISES 5.3

1. Let k be an integer greater than 1, and $\mathscr{B}$ be the set of divisors of k. If $a, b \in \mathscr{B}$, then define $a \vee b = \text{l.c.m.}(a, b)$, $a \wedge b = \text{g.c.d.}(a, b)$ and $\bar{a} = k/a$. Show that $\mathscr{B}$ is a boolean algebra with respect to these operations if and only if k is not divisible by a square greater than 1. (This is a generalization of exercise 5.2.6.) What are the atoms in this algebra?

2. How many subalgebras are there of the boolean algebra of the power set of a set with four elements?

3. Let $\mathscr{B}$ be a boolean algebra with 2^n elements. Show that the number of subalgebras of $\mathscr{B}$ is equal to the number of partitions of a set with n elements (*Hint*: Recall that $\mathscr{B}$ has n atoms).

_____

5.4 THE CALCULUS OF BOOLEAN ALGEBRAS

Boolean algebras $\mathscr{B}_n$ have been introduced and analyzed in the preceding sections as formal structures, that is, as sets of elements possessing certain properties. Among these properties, we saw that certain operations (disjunction, conjunction, and complementation) were defined on $\mathscr{B}_n$. These operations are used to manipulate elements of $\mathscr{B}_n$ to obtain other elements of $\mathscr{B}_n$. Therefore the properties of the defined operations constitute a set of rules of calculation which are applicable to *form* and *transform* expressions involving elements of $\mathscr{B}_n$. This set of rules is referred to as the *calculus of boolean algebras*. We remark, at this point, that the word "algebra" is commonly used in two connotations, either as a formal structure or as the set of rules applicable to that formal structure. For the sake of precision, we reserve the word "algebra" for the former notion and adopt for the latter the seemingly more appropriate phrase "calculus of boolean algebras."

It is now appropriate to summarize synoptically the formal properties of boolean algebras. They are encompassed by the adjectives "bounded, distributive, and complemented" of definition 5.2.3. For simplicity of exposition, we shall use the symbols "0" and "1" to denote the lower and upper bounds of $\mathscr{B}_n$, respectively.

First of all, a boolean algebra $\mathscr{B}_n$ is a lattice; thus properties (4.3.1) through (4.3.4), common to all lattices, hold for a, b, and c in $\mathscr{B}_n$.

$a + a = a$	$a \cdot a = a$	(idempotency)	(5.4.1)
$a + b = b + a$	$ab = ba$	(commutativity)	(5.4.2)
$(a + b) + c = a + (b + c)$	$(ab)c = a(bc)$	(associativity)	(5.4.3)
$a + ab = a$	$a(a + b) = a$	(absorption)	(5.4.4)

Second, since $\mathscr{B}_n$ is a distributive lattice, we have (definition 4.4.1)

$a(b + c) = ab + ac$	$a + bc = (a + b)(a + c)$	(distributivity)	(5.4.5)

Third, $\mathscr{B}_n$ is a bounded lattice; i.e., it contains 0 and 1, and

$a + 0 = a$	$a \cdot 1 = a$	(universal bounds)	(5.4.6)
$a0 = 0$	$a + 1 = 1$	(universal bounds)	(5.4.7)

Finally, since $\mathscr{B}_n$ is a complemented lattice, we have

$\bar{1} = 0$			(5.4.8)
$a + \bar{a} = 1$	$a\bar{a} = 0$	(complementarity)	(5.4.9)
$\overline{(\bar{a})} = a$		(involution)	(5.4.10)
$\overline{a + b} = \bar{a} \cdot \bar{b}$	$\overline{ab} = \bar{a} + \bar{b}$	(dualization)	(5.4.11)

This set of eleven identities (5.4.1) through (5.4.11)—wherever they involve the operations $+$ and $\cdot$, they are offered in dual pairs—represent a summary of the formal properties of boolean algebras as inherited, so to speak, from the parent structures of bounded lattices, distributive lattices, and complemented lattices. These same properties can now be grouped so as to lead to a reformulation of the definition of boolean algebras which is more suitable for considering their calculus.

Definition 5.4.1. (Alternative of definition 5.2.3) A boolean algebra $\mathscr{B}_n$ is a set S of at least two elements (the bounds) with the following properties.

1. $\mathscr{B}_n$ has two binary operations, $+$ and $\cdot$, which are commutative, associative, idempotent, and mutually distributive.
2. $\mathscr{B}_n$ has a unary operation, $^-$, which satisfies the properties of complementarity, involution, and dualization.

3. $\mathscr{B}_n$ contains two mutually complementary universal bounds 0 and 1, such that for $a \in S$, $a + 0 = a \cdot 1 = a$, $a \cdot 0 = 0$, $a + 1 = 1$.

4. S is a partially ordered set.

It can be shown that the identities (5.4.1) through (5.4.11), which characterize boolean algebras, do not form an independent set of axioms (see Section 0.2). The extraction of a subset of the given identities, to be taken as axioms, so that the others are deducible as theorems from them is an interesting exercise. However, since it is largely a side topic in our current presentation, it is discussed in Note 5.3 at the end of this chapter. The reduced set of axioms is presented in the following definition.

Definition 5.4.2. (Alternative of definitions 5.2.3 and 5.4.1) A boolean algebra $\mathscr{B}_n$ is a set S of at least two elements with the following properties.

1. $\mathscr{B}_n$ has two binary operations, $+$ and $\cdot$, which are commutative and mutually distributive; i.e., for a and b in S,

$$a + b = b + a, \qquad ab = ba \qquad (5.4.2)$$

$$a(b + c) = ab + ac, \qquad a + bc = (a + b)(a + c) \qquad (5.4.5)$$

2. $\mathscr{B}_n$ has a unary operation, which obeys complementarity; i.e., for a in S,

$$a + \bar{a} = 1, \qquad a\bar{a} = 0 \qquad (5.4.9)$$

3. $\mathscr{B}_n$ contains two mutually complementary bounds, 0 and 1, such that, for a in S

$$a + 0 = a, \qquad a \cdot 1 = a \qquad (5.4.6)$$

$$\bar{1} = 0 \qquad (5.4.8)$$

Note 5.3 is essentially devoted to justifying the nontrivial equivalence of definitions 5.4.2 and 5.4.1, that is, that boolean algebras are characterized by the axioms (5.4.2), (5.4.5), (5.4.6), (5.4.8), and (5.4.9), which constitute the celebrated Huntington's postulates.

The similarity between disjunction and conjunction in $\mathscr{B}_n$, on one side, and ordinary addition and multiplication of numbers, on the other, has already been noted. To exploit the convenient vocabulary developed in ordinary algebra, since no ambiguity will arise, we shall use the following pairs of words and phrases as synonyms.

disjunction $\leftrightarrow$ sum

to form the disjunction $\leftrightarrow$ to add

conjunction $\leftrightarrow$ product

to form the conjunction $\leftrightarrow$ to multiply

The notion of expression is not new in this book (Sections 0.4, 2.5, 4.5, 5.3), and certainly all readers were exposed to it much earlier. We now give a formal definition (patterned after the definition in example 0.4.3) of the class of expressions which represent our current interest, that is, boolean expressions.

Definition 5.4.3. Let $X \equiv \{x_1, x_2, x_3, \ldots\}$ be a set of symbols called *literals*. *Boolean expressions* $E(x_1, x_2, \ldots)$ in the literals x_1, x_2, ... are defined inductively as follows:

1. Literals and the symbol O are boolean expressions.

2. Disjunction, conjunction, and complementation of boolean expressions are boolean expressions.

3. No other boolean expressions exist that cannot be generated by a finite number of applications of rules (1) and (2).

A boolean expression containing literals $x_1, \ldots, x_n$ is said to be "an expression *in the n literals* $x_1, \ldots, x_n$" and is denoted by $E(x_1, \ldots, x_n)$ or simply E. We are able to write a boolean expression as a string of symbols comprising—beside literals and $+$, $\cdot$, $^-$—nested symmetric pairs of parentheses, ( ), enclosing the expressions operated upon according to part (2) of the preceding definition. For example, if E_1 and E_2 are expressions, then the disjunction of E_1 and E_2—itself an expression—is written as $(E_1 + E_2)$. Therefore, for some literals x_1, x_2, x_3, and x_4,

$$((x_1 + x_2 + ((x_3 + x_4)((\overline{x_1\,x_2}))))(x_2 + (x_2\,x_4)))$$

is an expression written according to the given rule.

The writing of a boolean expression is governed by the following *precedence rule* (see Section 0.4): complementation—conjunction—disjunction. Recall that in ordinary algebraic expressions we have the precedence rule: multiplication—addition. The use of the same symbols in boolean algebras ($+$ for addition and disjunction, $\cdot$ for multiplication and conjunction) takes advantage of a well-established habit. Thus the expression

$$((x_1 + x_2 + ((x_3 + x_4)((\overline{x_1\,x_2}))))(x_2 + (x_2\,x_4)))$$

may be rewritten as

$$[x_1 + x_2 + (x_3 + x_4)\,\overline{x_1\,x_2}](x_2 + x_2\,x_4),$$

which is certainly easier to interpret.

Concerning the notion of literal, a literal x of an expression E is an undefined symbol until some element a of a boolean algebra $\mathscr{B}_n$ is substituted for every occurrence of x in E. When elements of $\mathscr{B}_n$ are substituted for all the distinct literals of an expression E, we say that the expression has been evaluated over $\mathscr{B}_n$. Since expressions involve only the application of the three operations of $\mathscr{B}_n$, which are closed, each valuated expression is an element of $\mathscr{B}_n$.

Definition 5.4.3 is reminiscent of the synthesis of oriented rooted trees (Section 2.5). Indeed, each expression is described by one such tree, whose terminal vertices are labeled with literals and whose intermediate vertices are labeled with operation symbols. Moreover, since the binary operations of a boolean algebra are commutative and associative, the rooted tree is not oriented. Example 2.5.8, though concerning an algebraic expression (that is, not a boolean expression), entirely illustrates the point.

A basic task is the evaluation of an expression E over a given boolean algebra $\mathscr{B}_n$. This task can be accomplished in two distinct ways.

Method 1. First substitute the specified elements of $\mathscr{B}_n$ for literals of E, and then use the structure of $\mathscr{B}_n$ to calculate the result. Express this structure by a set of identities or equivalently by means of the Hasse diagram $\mathscr{B}_n$.

Method 2. Manipulate the expression E using exclusively the general identities (5.4.1) through (5.4.11) until an expression is obtained which is considered simpler to evaluate than the original expression. Apply method (1) to the expression obtained.

Before discussing the two methods, we illustrate them with specific examples.

Example 5.4.1. The expression

$$E(x_1, x_2, x_3) = \bar{x}_1 + (\bar{x}_3 + x_1 + \overline{x_2 x_3})(x_3 + \bar{x}_1 x_2) \qquad (5.4.12)$$

is to be evaluated over the boolean algebra $\mathscr{B}_3$, illustrated in figure 5.4.1 by its Hasse diagram, for the following substitution:

$$x_1 = a_2, \qquad x_2 = a_5, \qquad x_3 = a_6.$$

We have

$$E(a_2, a_5, a_6) = \bar{a}_2 + (\bar{a}_6 + a_2 + \overline{a_5 a_6})(a_6 + \bar{a}_2 a_5) \qquad (5.4.13)$$

Note that in $\mathscr{B}_3$, $\bar{a}_2 = a_5$, $\bar{a}_1 = a_6$, $\bar{a}_3 = a_4$, and $\bar{a}_0 = a_7$. Whence $E(a_2, a_5, a_6) = a_5 + (a_1 + a_2 + \bar{a}_3)(a_6 + a_5 a_5) = a_5 + (a_1 + a_2 + a_4)(a_6 + a_5)$. Since, in $\mathscr{B}_3$, $a_1 + a_2 + a_4 = a_4$ and $a_6 + a_5 = a_7$, $E(a_2, a_5, \bar{a}_6) = a_5 + a_4 a_7 = a_5 + a_4 = a_7$.

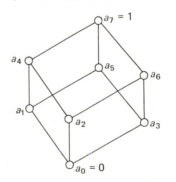

Figure 5.4.1. The Hasse diagram of $\mathscr{B}_3$

The same expression $E(x_1, x_2, x_3)$ may also be manipulated by use of the general identities. Beside each intermediate step we indicate in brackets which identities are being used.

$$E(x_1, x_2, x_3) = \bar{x}_1 + (\bar{x}_3 + x_1 + \overline{x_2 x_3})(x_3 + \bar{x}_1 x_2)$$
$$= \bar{x}_1 + (\bar{x}_3 + x_1 + \bar{x}_2 + \bar{x}_3)(x_3 + \bar{x}_1 x_2)$$
$$[5.4.11]: \overline{x_2 x_3} = \bar{x}_2 + \bar{x}_3$$
$$= \bar{x}_1 + (\bar{x}_3 + x_1 + \bar{x}_2)(x_3 + \bar{x}_1 x_2)$$
$$[5.4.1]: \bar{x}_3 + \bar{x}_3 = \bar{x}_3$$
$$= \bar{x}_1 + (\bar{x}_3 x_3 + \bar{x}_1 x_2 \bar{x}_3 + x_1 x_3 + x_1 \bar{x}_1 x_2 + \bar{x}_2 x_3 + \bar{x}_1 x_2 \bar{x}_2)$$
$$[5.4.5]$$
$$= \bar{x}_1 + \bar{x}_1 x_2 \bar{x}_3 + x_1 x_3 + \bar{x}_2 x_3$$
$$[5.4.9]: x_1 \bar{x}_1 = 0, \ x_2 \bar{x}_2 = 0, \ x_3 \bar{x}_3 = 0$$

At this point we apply method (1) to the resulting expression and obtain for the specified substitution

$$E(a_2, a_5, a_6) = \bar{a}_2 + \bar{a}_2 a_5 \bar{a}_6 + a_2 a_6 + \bar{a}_5 a_6$$
$$= a_2 + 0 \cdot a_1 + a_2 + a_2 a_6 \qquad (\bar{a}_2 a_5 = 0, \ a_2 a_6 = a_2)$$
$$= \bar{a}_2 + a_2 \qquad (a_1 0 = 0, \ a_2 + a_2 a_6 = a_2)$$
$$= a_7,$$

which coincides with the result obtained previously.

The difference between the two methods is not confined only to the mechanics of the various steps; rather, it is of a much more fundamental nature. In fact, expression (5.4.13) is itself an element of $\mathscr{B}_n$, and once the substitution is made, all the specific identities of $\mathscr{B}_n$ are utilized. Instead, when manipulating $E(x_1, x_2, x_3)$, we treat each literal as an element of some boolean algebra $\mathscr{B}$, and literals with different subscripts are *entirely* unrelated. In other words, although $\bar{x}_i$ is the complement of x_i, no identity relates x_i to x_j for $i \neq j$. What is the structure of the algebra $\mathscr{B}$, mentioned before? This crucial point is worth some additional study.

Suppose that we have an expression $E(x_1, x_2, \ldots, x_n)$ in n literals $x_1, x_2, \ldots, x_n$. By systematically applying some of the identities (5.4.1) through (5.4.11), we can always transform it into another expression $E'(x_1, x_2, \ldots, x_n)$, whose form is the sum of products of literals, complemented or uncomplemented. An expression of this type is said to be in *disjunctive normal form*. In example 5.4.1, $E(x_1, x_2, x_3)$ is transformed into $\bar{x}_1 + \bar{x}_1 x_2 \bar{x}_3 + x_1 x_3 + \bar{x}_2 x_3$. The steps required to put an expression in disjunctive normal form follow.

1. Complements are placed on the literals by application of the *dualization* laws (5.4.11; De Morgan's laws). For example,

$$\overline{(x_1 + x_2) \overline{x_3 x_4}} = \overline{(x_1 + x_2)} + \overline{\overline{x_3 x_4}} = \bar{x}_1 \bar{x}_2 + x_3 x_4.$$

2. The distributive property is applied and repeated literals are eliminated. For example, $(x_1 + x_2 x_3)x_2 = x_1 x_2 + x_2 x_3 x_2 = x_1 x_2 + x_2 x_3$. Referring to example 5.4.1, we convince ourselves that, by exactly applying this procedure, we obtain an expression in disjunctive normal form.

Assume now that we have an expression in normal form, $E(x_1, x_2, x_3, x_4)$, and let $x_1 \bar{x}_2 x_3$ be a product term of it. We note, by identity (5.4.6), that $x_1 \bar{x}_2 x_3 = x_1 \bar{x}_2 x_3 \cdot 1$. Moreover, by identity (5.4.9), $x_4 + \bar{x}_4 = 1$, whence, by the substitution principle and the distributive property,

$$x_1 \bar{x}_2 x_3 = x_1 \bar{x}_2 x_3 \cdot 1 = x_1 \bar{x}_2 x_3 (x_4 + \bar{x}_4)$$
$$= x_1 \bar{x}_2 x_3 x_4 + x_1 \bar{x}_2 x_3 \bar{x}_4.$$

In other words, by applying valid identities, we have "transformed" a product of three literals, $x_1 \bar{x}_2 x_3$, into the disjunction of two terms, each of four literals, $x_1 \bar{x}_2 x_3 x_4 + x_1 \bar{x}_2 x_3 \bar{x}_4$. Note that each of these terms has as many distinct literals (that is, x_1, x_2, x_3, and x_4) as, by hypothesis, appear in the expression $E(x_1, x_2, x_3, x_4)$. Such terms are characterized in the following definition.

Definition 5.4.4. A *fundamental product* in n literals $x_1, x_2, \ldots, x_n$ is a conjunction in which each literal x_j appears exactly once, either complemented or uncomplemented, for $j = 1, 2, \ldots, n$.

For example, $x_1 \bar{x}_2 x_3$ and $\bar{x}_1 \bar{x}_2 x_3$ are fundamental products in three literals, x_1, x_2, and x_3. It is now clear that any normal form expression $E(x_1, \ldots, x_n)$ can be transformed into a sum of fundamental products in n literals as follows: Each term of $E(x_1, \ldots, x_n)$ is multiplied by an expression $(x_i + \bar{x}_i)$ for each literal x_i not appearing in that term. Thus an arbitrary expression can be transformed into the disjunction of fundamental products, by first obtaining an expression in normal form, then replacing each product term by a disjunction of fundamental products, and finally eliminating (by virtue of idempotency) repeated fundamental products.

Example 5.4.2. The expression

$$E(x_1, x_2, x_3) = \bar{x}_2 + (\bar{x}_3 + x_1 + \overline{x_2 x_3})(x_3 + \bar{x}_1 x_2)$$

is at first transformed (cf. example 5.4.1) into the normal expression

$$\bar{x}_2 + x_1 x_3 + \bar{x}_2 x_3 + \bar{x}_1 x_2 \bar{x}_3.$$

Subsequently we transform this normal expression into

$$\bar{x}_2(x_1 + \bar{x}_1)(x_3 + \bar{x}_3) + x_1 x_3(x_2 + \bar{x}_2) + \bar{x}_2 x_3(x_1 + \bar{x}_1) + \bar{x}_1 x_2 \bar{x}_3,$$

and by the distributive property we obtain

$$x_1 \bar{x}_2 x_3 + x_1 \bar{x}_2 \bar{x}_3 + \bar{x}_1 \bar{x}_2 x_3 + \bar{x}_1 \bar{x}_2 \bar{x}_3 + x_1 x_2 x_3 + x_1 \bar{x}_2 x_3$$
$$+ x_1 \bar{x}_2 x_3 + \bar{x}_1 \bar{x}_2 x_3 + \bar{x}_1 x_2 \bar{x}_3.$$

Note that $x_1 \bar{x}_2 x_3$ is repeated three times and $\bar{x}_1 \bar{x}_2 x_3$ twice; thus, eliminating the repeated products, we have the expression

$$x_1 \bar{x}_2 x_3 + x_1 \bar{x}_2 \bar{x}_3 + \bar{x}_1 \bar{x}_2 x_3 + \bar{x}_1 \bar{x}_2 \bar{x}_3 + x_1 x_2 x_3 + \bar{x}_1 x_2 \bar{x}_3.$$

What is the significance of a fundamental product in n literals $x_1, x_2, \ldots, x_n$? First of all, note that for two distinct such products p_i and p_j, we always have

$$p_i p_j = 0.$$

Indeed, since they are distinct, there is at least one literal x_k which appears complemented in one product and uncomplemented in the other, whence the product $p_i p_j$ contains a pair of factors $x_k \bar{x}_k$ and is 0 by (5.4.9) and (5.4.7). Second, assume that a fundamental product p can be expressed as

$$p_i = E_1 + E_2, \tag{5.4.14}$$

where E_1 and E_2 are two expressions in $x_1, \ldots, x_n$ such that, when transformed into sums of fundamental products, they are not the disjunctions of identical sets of fundamental products. It follows that the expression $E_1 + E_2$ contains more than one fundamental product; therefore $E_1 + E_2$ contains at least a p_k such that $p_k \neq p_i$. Now, if we multiply each side of (5.4.14) by p_k, we obtain

$$0 = p_k p_i = (E_1 + E_2) p_k = p_k \neq 0,$$

a contradiction. Then it follows that the fundamental products are join-irreducible; i.e., by lemma 5.3.1, they are the atoms of a boolean algebra. We then have the following lemma.

Lemma 5.4.1. There are 2^n fundamental products in n literals.

Proof. By induction. For one literal x there are 2^1 such products, x and $\bar{x}$. Assume that there are 2^{n-1} fundamental products in $(n-1)$ literals $x_1, \ldots, x_{n-1}$. To each of them we can append either x_n or $\bar{x}_n$, thereby generating all the fundamental products in n literals. Their number is clearly $2 \cdot 2^{n-1} = 2^n$. ‖

Therefore the fundamental products in n literals are the minterms of a boolean algebra $\mathscr{B}_{2^n}$ containing 2^{2^n} elements. The algebra $\mathscr{B}_{2^n}$ is called the *free boolean algebra* with n generators $x_1, \ldots, x_n$, because all its elements are obtainable by freely applying the algebra operations on expressions in $x_1, \ldots, x_n$. We note that, when manipulating an expression $E(x_1, \ldots, x_n)$, we are really using the identities valid on the free boolean algebra $\mathscr{B}_{2^n}$.

We recognize that the sum of fundamental products is a *canonical disjunctive expression* (theorem 5.3.4). Previously we have shown the existence of a canonical disjunctive expression into which any expression can be transformed. Since by theorem 5.3.4 a canonical expression represents a unique element of $\mathscr{B}_{2^n}$, we have the following important result.

Theorem 5.4.2. Any expression $E(x_1, \ldots, x_n)$ can be transformed into a unique disjunctive canonical expression (sum of minterms).

Another consequence of the preceding discussion is that an expression $E(x_1, \ldots, x_n)$ represents a unique element of $\mathscr{B}_{2^n}$. Therefore we can establish the following equivalence relation among expressions in n literals.

Definition 5.4.5. Two expressions in n literals are *equivalent* if they represent the same element of $\mathscr{B}_{2^n}$.

We then have the following straightforward consequence.

Corollary 5.4.3. Two expressions in n literals are equivalent if and only if they have the same canonical expression.

This corollary not only gives us an operational method for establishing the equivalence of expressions, with no necessity of employing ingenious manipulative tricks; it also tells us that, when manipulating an expression by means of identities (5.4.1) through (5.4.11), we are really generating new expressions equivalent to the original one. These expressions, indeed, do represent the *same* element of $\mathscr{B}_{2^n}$ but describe *different* procedures for its calculation.

This concludes our quick analysis of the calculus of boolean algebras.

Note. We wish to draw attention to a very important point in the manipulation of boolean expressions. The similarity between the rules of the calculus of boolean algebras and those of ordinary arithmetic is indeed very useful because the familiarity acquired in the latter transfers to the former. However, because this similarity may also be a trap, it must be viewed with extreme caution—the same caution one should use when, for example, trying without thorough knowledge to speak a foreign language very similar to one's own. A very important difference between the two calculi is the absence of a "cancellative property" (see Section 3.5) in boolean algebra; for example,

$$ab = cb \quad \text{does not imply} \quad a = c$$

although

$$a = c \quad \text{implies} \quad ab = cb.$$

The same obviously holds true for the dual statement.

EXERCISES 5.4

1. Write the following boolean expressions in disjunctive normal form.
   a) $(x + \bar{y}z)(\bar{z} + y)$
   b) $(x + y)(x + z)\overline{(\bar{x} + y)}$
   c) $(x + y(z + x))$
   d) $\overline{(\bar{x} + \bar{y})} + xu$
   e) $\bar{y}\bar{w} + (x + z)(\bar{y}\bar{z} + xw)$
   f) $[xy + (\bar{y} + z)\bar{x}](y + \bar{z})$

2. A boolean expression is said to be in *conjunctive normal form* if it is a product of maxterms. Give a systematic procedure to transform a boolean expression into its conjunctive normal form (use theorem 5.3.7). Write the boolean expressions of exercise 5.4.1 in conjunctive normal form.

3. Consider the expression

$$E(x_1, x_2, x_3) = \overline{x_1 + x_2(x_3 + \overline{x}_1)}$$

and the boolean algebra $\mathscr{B}_3$ illustrated in figure 5.4.2 by means of its Hasse diagram.

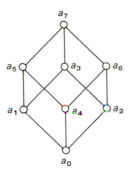

Figure 5.4.2.

a) Evaluate $E(x_1, x_2, x_3)$ in $\mathscr{B}_3$ with the following assignments

   i) $x_1 = a_2$, $x_2 = a_6$, $x_3 = a_3$
   ii) $x_1 = a_1$, $x_2 = a_6$, $x_3 = a_5$

b) Repeat part (a) for the expression $E^*(x_1, x_2, x_3)$ obtained by transforming $E(x_1, x_2, x_3)$ into disjunctive canonical form.

4. Can $\mathscr{B}_3$ be viewed as a free boolean algebra? Justify your answer.

5. A Byrne algebra is an algebra $[B, \wedge, ^-, 0]$ where B is a set, $\wedge$ is a binary operation on B, $^-$ is a unary operation on B, and 0 is an element of B. For any elements x, y, and z in B, we have the following.

   a) $x \wedge y = y \wedge x$
   b) $x \wedge (y \wedge z) = (x \wedge y) \wedge z$
   c) $x \wedge x = x$
   d) $x \wedge \bar{y} = 0$ if and only if $x \wedge y = x$
   e) $0 \neq \bar{0}$

   Show that for any boolean algebra $[B, +, \cdot, ^-, 0, 1]$, the structure $[B, \cdot, ^-, 0]$ is a Byrne algebra.

*6. If $[B, \wedge, ^-, 0]$ is a given Byrne algebra (see exercise 5.4.5), then define $1 = \bar{0}$ and $x + y = (\overline{\bar{x} \wedge \bar{y}})$. Then show that the structure $[B, +, \wedge, ^-, 0, 1]$ is a boolean algebra. (Thus exercises 5 and 6 give another axiom system for boolean algebras.)

5.5 BOOLEAN FUNCTIONS

In the preceding section we asserted that an important task is the evaluation of boolean expressions. More specifically, we have an expression $E(x_1, x_2, \ldots, x_n)$ which represents an element of the free boolean algebra $\mathscr{B}_{2^n}$ generated by $x_1, x_2, \ldots, x_n$, and this expression is to be evaluated over another boolean algebra $\mathscr{B}_s$ (frequently referred to as the *ground algebra*). As before, this procedure entails assigning each of the literals x_i to an element a_i of $\mathscr{B}_s$ and then calculating $E(a_1, a_2, \ldots, a_n)$, which is an element of $\mathscr{B}_s$. For every assignment $x_1 \to a_1$, $x_2 \to a_2$, $\ldots$, $x_n \to a_n$, the boolean expression E determines a mapping of the ordered n-tuple $(a_1, a_2, \ldots, a_n)$ to an element of $\mathscr{B}_s$. In other words, denoting by A_s the set of elements of $\mathscr{B}_s$ and by A_s^n the n times cartesian product of A_s, we may view the expression $E(x_1, x_2, \ldots, x_n)$ as describing a function $f_E : A_s^n \to A_s$ if, for every $(a_1, a_2, \ldots, a_n) \in A_s^n$, $E(a_1, \ldots, a_n) = f_E(a_1, \ldots, a_n)$. This way of looking at a boolean expression is very important. It would be natural to call the function described by E a boolean function, but some remarks are in order before we reach a formal definition.

First of all, we note that, given $E(x_1, \ldots, x_n)$ and $\mathscr{B}_s$, we have a function $f_E : A_s^n \to A_s$, but it is not clear whether, given a function $\varphi : A_s^n \to A_s$, there exists an expression $E(x_1, \ldots, x_n)$ which describes φ. To clarify this problem, we may use the following argument. For a fixed $\mathscr{B}_s$, each expression $E(x_1, \ldots, x_n)$ clearly describes a *unique* function $f_E : A_s^n \to A_s$. We recall that E belongs to $\mathscr{B}_{2^n}$, the free boolean algebra with n generators; we know by lemma 5.4.1 that $\mathscr{B}_{2^n}$ has 2^n atoms (the fundamental products of n literals), and by corollary 5.3.3, that a boolean algebra with 2^n atoms has 2^{2^n} elements. Therefore, recalling also definition 5.4.5 and corollary 5.4.3, we conclude that there are 2^{2^n} distinct equivalence classes of expressions in the literals $x_1, x_2, \ldots, x_n$. Next, we inquire how many are the distinct functions $\varphi : A_x^n \to A_s$. We recall again corollary 5.3.3, which states that a boolean algebra $\mathscr{B}_s$ with s atoms contains 2^s elements. It can be shown with arguments to be developed in the next chapter (Section 6.2), that we can form $(2^s)^n$ distinct ordered n-tuples $(a_1, a_2, \ldots, a_n)$, with $a_i \in A_s$, for $i = 1, 2, \ldots, n$. Furthermore, a function $\varphi : (A_s)^n \to A_s$ is completely specified when to each n-tuple we assign an element of A_s; if we imagine forming a list of the 2^{sn} n-tuples, each member of this list can be assigned a value in 2^s ways; thus the first member can be assigned a value in 2^s ways, the first two in $2^s \cdot 2^s = 2^{2s}$ ways, and, in general, the entire list can be assigned values in

$$\underbrace{2^s \times 2^s \times \cdots \times 2^s}_{2^{sn} \text{ factors}} = 2^{s \cdot 2^{sn}}$$

ways. This is precisely the number of functions $\varphi : A_s^n \to A_s$. We recognize that, for $s > 1$

$$2^{s 2^{sn}} > 2^{2^n} \ (s > 1), \tag{5.5.1}$$

which shows that not every function φ can be described by a boolean expression. To better grasp this interesting fact, consider the following simple example.

Example 5.5.1. Let $s = 2$ and $n = 1$. Set A_2 contains 4 elements, $\{0, 1, a, b\}$, and the n-tuples are the single elements of A_2. Consider the following function $\varphi : A_2 \to A_2$:

$$\begin{pmatrix} 0 & 1 & a & b \\ b & b & 0 & a \end{pmatrix}.$$

We claim that there is no boolean expression $E(x)$ in the single variable x which describes φ. In fact, there are only four nonequivalent expressions in one variable x $(0, x, \bar{x}, \text{ and } 1)$ which correspond, respectively, to the following functions $A_2 \to A_2$.

$$\begin{pmatrix} 0 & 1 & a & b \\ 0 & 0 & 0 & 0 \end{pmatrix} \quad \begin{pmatrix} 0 & 1 & a & b \\ 0 & 1 & a & b \end{pmatrix} \quad \begin{pmatrix} 0 & 1 & a & b \\ 1 & 0 & b & a \end{pmatrix} \quad \begin{pmatrix} 0 & 1 & a & b \\ 1 & 1 & 1 & 1 \end{pmatrix}$$

None of these functions coincides with the given φ.

We are now ready to formally define boolean functions.

Definition 5.5.1. For a given boolean algebra $\mathscr{B}_s = [A_s, +, \cdot, ^-, 0, 1]$, a boolean expression $E(x_1, x_2, \ldots, x_n)$ defines a *boolean function in n variables over* $\mathscr{B}_s$, $f : A_s^n \to A_s$ according to the rule:

$$f(a_1, a_2, \ldots, a_n) = E(a_1, a_2, \ldots, a_n) \text{ for every } (a_1, \ldots, a_n) \in (A_s)^n.$$

We say that E *describes* the function f.

In other words, a boolean function $f : A_s^n \to A_s$ is one for which there exists a boolean expression describing it.

Returning to inequality (5.5.1) we note that it becomes an equality for $s = 1$. This appears not to preclude the fact that each function $\varphi : A_1^n \to A_1$ is describable by a boolean expression $E(x_1, \ldots, x_n)$. Remarkably, this is true, as we shall now prove. Hereafter, we call a constant $a \in \{0, 1\}$ a "binary" constant. Also, we ignore the equivalence among expressions by restricting ourselves to expressions in disjunctive canonical form (theorem 5.4.2). Since each such expression describes a function, to establish the one-to-one correspondence between canonical expressions $E(x_1, x_2, \ldots, x_n)$ and functions $\varphi : A_1^n \to A_1$, we must show only that each function is described by a unique canonical expression. We use finite induction on the integer n. First, for $n = 1$ there are four functions $\varphi : \{0, 1\} \to \{0, 1\}$, which are represented as follows:

a	$\varphi(a)$		a	$\varphi(a)$		a	$\varphi(a)$		a	$\varphi(a)$
0	0		0	0		0	1		0	1
1	0		1	1		1	0		1	1

These functions are respectively described by the canonical expression $0, x, \bar{x}, x + \bar{x} = 1$. We also note that each of these expressions contains the minterm x or $\bar{x}$, depending on whether, in the corresponding function, $\varphi(1) = 1$ or $\varphi(0) = 1$. This can be rephrased as follows: If, for $a \in \{0, 1\}$, we define the notation x^a as

$$x^a = \begin{cases} x & \text{if } a = 1, \\ \bar{x} & \text{if } a = 0, \end{cases} \tag{5.5.2}$$

then for every $a \in \{0, 1\}$ for which $\varphi(a) = 1$, the expression describing φ contains a minterm x^a.

Next, assume that any function $f: \{0, 1\}^{n-1} \to \{0, 1\}$ is a boolean function and that the canonical expression $E_f(x_1, \ldots, x_{n-1})$ describing it contains a minterm $x_1^{a_1} \cdots x_{n-1}^{a_{n-1}}$ if and only if $f(a_1, \ldots, a_{n-1}) = 1$. If we are given a function $\varphi: \{0, 1\}^n \to \{0, 1\}$, we represent it by the two-column array shown below.

a_1	a_2	$\cdots$	a_{n-1}	a_n	φ
0	0	$\cdots$	0	0	$\varphi(00 \cdots 00)$
1	0	$\cdots$	0	0	$\varphi(10 \cdots 00)$
		$\vdots$			$\vdots$
1	1	$\cdots$	1	0	$\varphi(11 \cdots 10)$
0	0	$\cdots$	0	1	$\varphi(00 \cdots 01)$
1	0	$\cdots$	0	1	$\varphi(10 \cdots 01)$
		$\vdots$			$\vdots$
1	1	$\cdots$	1	1	$\varphi(11 \cdots 11)$

This array contains 2^n rows [as many rows as there are binary n-tuples $(a_1, a_2, \ldots, a_n)$]. The top 2^{n-1} n-tuples have $a_n = 0$, and the bottom 2^{n-1} n-tuples have $a_n = 1$. We consider the two halves of the array as describing, respectively, the functions

$$f_0(a_1, \ldots, a_{n-1}) = \varphi(a_1, \ldots, a_{n-1}, 0),$$
$$f_1(a_1, \ldots, a_{n-1}) = \varphi(a_1, \ldots, a_{n-1}, 1).$$

By the inductive hypothesis, f_0 and f_1 are boolean functions; i.e., they are described by the canonical expressions $E_{f_0}(x_1, \ldots, x_{n-1})$ and $E_{f_1}(x_1, \ldots, x_{n-1})$. We now claim that

$$E(x_1, \ldots, x_n) = \bar{x}_n E_{f_0}(x_1, \ldots, x_{n-1}) + x_n E_{f_1}(x_1, \ldots, x_{n-1}) \tag{5.5.3}$$

describes the function φ. In fact, for a binary n-tuple $(a_1, \ldots, a_n)$ we have

$$E(a_1, \ldots, a_{n-1}, a_n) = \bar{a}_n E_{f_0}(a_1, \ldots, a_{n-1}) + a_n E_{f_1}(a_1, \ldots, a_{n-1})$$
$$= \bar{a}_n f_0(a_1, \ldots, a_{n-1}) + a_n f_1(a_1, \ldots, a_{n-1})$$
$$= \bar{a}_n \varphi(a_1, \ldots, a_{n-1}, 0) + a_n \varphi(a_1, \ldots, a_{n-1}, 1).$$

If $a_n = 1$, we have $E(a_1, \ldots, a_{n-1}, 1) = 0 \cdot \varphi(a_1, \ldots, a_{n-1}, 0) + 1 \cdot \varphi(a_1, \ldots, a_{n-1}, 1) =$

$\varphi(a_1,\ldots,a_{n-1},1)$; similarly, we show that $E(a_1,\ldots,a_{n-1},0) = \varphi(a_1,\ldots,a_{n-1},0)$, whence we conclude that $E(a_1,\ldots,a_n) = \varphi(a_1,\ldots,a_n)$; that is, the boolean expression E describes φ. Moreover, suppose that $\varphi(a_1,\ldots,a_{n-1},1) = 1$; this means that $f_1(a_1,\ldots,a_{n-1}) = 1$, and, by the inductive hypothesis, E_{f_1} contains the minterm $x_1^{a_1} \cdots x_{n-1}^{a_{n-1}}$; this in turn implies by (5.5.3), that $E(x_1,\ldots,x_n)$ contains the minterm $x_1^{a_1} \cdots x_{n-1}^{a_{n-1}} x_n$. The argument is similar if $\varphi(a_1,\ldots,a_{n-1},0) = 1$; the converse to both cases is straightforward. We have proved the following theorem.

Theorem 5.5.1. Every function $\varphi:\{0,1\}^n \to \{0,1\}$ is a boolean function, and conversely, every canonical boolean expression $E(x_1,\ldots,x_n)$ describes one such function. The expression E_φ describes φ if and only if, for every binary n-tuple $(a_1,\ldots,a_n)$ for which $\varphi(a_1,\ldots,a_n) = 1$, E_φ contains the minterm $x_1^{a_1} \cdots x_n^{a_n}$.

Example 5.5.2. Consider the function $\varphi:\{0,1\}^3 \to \{0,1\}$, represented below in tabular form. (The tabular representation of a function $\varphi:\{0,1\}^n \to \{0,1\}$ is universally known in the literature as the *truth table* of φ. Section 5.7 will justify such terminology.)

$a_1\,a_2\,a_3$	$\varphi(a_1\,a_2\,a_3)$	Corresponding minterm
0 0 0	0	
1 0 0	1	$x_1\,\overline{x}_2\,\overline{x}_3$
0 1 0	1	$\overline{x}_1\,x_2\,\overline{x}_3$
1 1 0	0	
0 0 1	1	$\overline{x}_1\,\overline{x}_2\,x_3$
1 0 1	0	
0 1 1	1	$\overline{x}_1\,x_2\,x_3$
1 1 1	0	

The canonical expression describing φ is the disjunction of the minterms corresponding to the triplets (a_1,a_2,a_3) for which $\varphi(a_1,a_2,a_3) = 1$; that is,

$$E(x_1, x_2, x_3) = x_1\,\overline{x}_2\,\overline{x}_3 + \overline{x}_1\,x_2\,\overline{x}_3 + \overline{x}_1\,\overline{x}_2\,x_3 + \overline{x}_1\,x_2\,x_3.$$

On the set of the boolean functions over a boolean algebra $\mathscr{B}_s$, we can also define two binary operations, $*$ and $\circ$, and a unary operation, $'$, in the following manner. Define $\Phi_s^{(n)}$ as the set of boolean functions in n variables over $\mathscr{B}_s$, and for any $\varphi \in \Phi_s^{(n)}$, let $E(\varphi)$ be the canonical expression describing φ (that is, E is an invertible mapping from $\Phi_s^{(n)}$ to the free boolean algebra $\mathscr{B}_{2^n}$). For φ_1 and φ_2 in in $\Phi_s^{(n)}$ we define $*$, $\circ$, and $'$ by

$$E(\varphi_1 * \varphi_2) = E(\varphi_1) + E(\varphi_2),$$
$$E(\varphi_1 \circ \varphi_2) = E(\varphi_1)\,E(\varphi_2),$$
$$E(\varphi_1') = \overline{E(\varphi_1)}.$$

Recalling definition 3.8.4, we recognize that $E: \Phi_s^{(n)} \to \mathscr{B}_{2^n}$ is an isomorphism between $\mathscr{B}_{2^n}$ and $\Phi_s^{(n)}$. Thus $\Phi_s^{(n)}$ is a boolean algebra. Moreover, since by theorem 5.5.1 there is a one-to-one correspondence between members of $\mathscr{B}_{2^n}$ and of $\Phi_1^{(n)}$, and since $\Phi_1^{(n)}$—the set of the boolean functions in n variables over $\mathscr{B}_1$—coincides with the set of *all* the functions $\varphi: \{0, 1\}^n \to \{0, 1\}$, we reach the following important conclusion.

Corollary 5.5.2. The boolean algebra of the functions $\varphi: \{0, 1\}^n \to \{0, 1\}$ (the boolean functions in n variables over $\mathscr{B}_1$) is isomorphic to the free boolean algebra with n generators; i.e., it is a representation of the latter.

For reasons that will become apparent in Section 5.8, a function $\varphi: \{0, 1\}^n \to \{0, 1\}$ is termed a *switching function* of n arguments, and the boolean algebra of switching functions is called *switching algebra*. Note that, by identifying binary n-tuples with vertices of the n-cube (definition 5.1.3), a switching function determines a partition of the n-cube vertices into two sets: those which are valued 0 and those which are valued 1. Because of the isomorphism between $\Phi_s^{(n)}$ and $\Phi_s^{(n)}$, for any s, and between $\Phi_1^{(n)}$ and $\mathscr{B}_{2^n}$ (corollary 5.5.2), the phrases "switching function" and "boolean function" are used interchangeably, and it is also common practice to call boolean functions $f(x_1, \ldots, x_n)$ the elements of $\mathscr{B}_{2^n}$. In other words, $f(x_1, \ldots, x_n)$ becomes the identifier of all the expressions equivalent to the canonical expression describing $f(x_1, \ldots, x_n)$.

- -

EXERCISES 5.5

1. Show that every boolean function of n variables $f(x_1, \ldots, x_n)$ may be written in the form

$$f(x_1, \ldots, x_n) = f(1, x_2, \ldots, x_n) \, x_1 + f(0, x_2, \ldots, x_n) \, \bar{x}_1.$$

2. Show that every boolean function of n variables $f(x_1, \ldots, x_n)$ may be written in the *disjunctive normal form*

$$f(x_1, \ldots, x_n) = \bigvee_{a_1, \ldots, a_n} f(a_1, a_2, \ldots, a_n) \, x_1^{a_1} x_2^{a_2} \ldots x_n^{a_n},$$

   where a_i is defined as in (5.5.2). (*Hint*: use theorem 5.5.1.)

3. Show that every boolean function of n variables $f(x_1, \ldots, x_n)$ may be written in the *conjunctive normal form*

$$f(x_1, \ldots, x_n) = \prod_{a_1, \ldots, a_n} [f(\bar{a}_1, \ldots, \bar{a}_n) + x_1^{a_1} + \ldots + x_n^{a_n}].$$

   (Refer to exercise 5.5.2.)

4. Consider the boolean algebra $\mathscr{B}_2$ illustrated below. State which of the following functions $\varphi_i: \mathscr{B}_2^2 \to \mathscr{B}_2$ ($i = 1, 2, 3$) is representable by a boolean expression $E(x_1, x_2)$ and exhibit it.

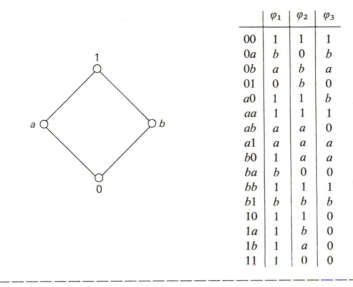

	φ_1	φ_2	φ_3
00	1	1	1
0a	b	0	b
0b	a	b	a
01	0	b	0
a0	1	1	b
aa	1	1	1
ab	a	a	0
a1	a	a	a
b0	1	a	a
ba	b	0	0
bb	1	1	1
b1	b	b	b
10	1	1	0
1a	1	b	0
1b	1	a	0
11	1	0	0

*5.6 THE BOOLEAN RING

In Section 4.4 we noted the similarity between distributive lattices and rings. We also noted that, together with common properties, there were some important distinguishing properties of the two structures. We also raised the conjecture that, by adding an appropriate set of properties to either structure, we might have found a nonempty *intersection* of the two classes. The objective of this section is precisely to provide an affirmative answer to this conjecture. We shall show that the same structure can be interpreted either as a lattice or as a ring, in that the lattice operations are definable in terms of the ring operations, and vice versa.

Theorem 5.6.1. Every boolean algebra $\mathscr{B}_n = [S, +, \cdot, \bar{\ }, 0, 1]$ is a ring $R_n = [S, \oplus, \circ]$ with respect to the ring operations defined by the following rules:

$$a \oplus b = a\bar{b} + \bar{a}b \tag{5.6.1}$$

$$a \circ b = ab \tag{5.6.2}$$

Proof. We must show that R_n is indeed a ring. Verification of the closure for $\oplus$ and $\circ$ in R_n is immediate. The operation $\oplus$ has the following properties:

1. *Commutativity.* Since $+$ and $\cdot$ are commutative, we have

$$a \oplus b = a\bar{b} + \bar{a}b = \bar{a}b + a\bar{b} = b\bar{a} + \bar{b}a = b \oplus a.$$

2. *Associativity.* We have

$$(a \oplus b) \oplus c = (a\bar{b} + \bar{a}b)\bar{c} + \overline{(a\bar{b} + \bar{a}b)}\, c = (a\bar{b} + \bar{a}b)\bar{c} + (ab + \overline{a}\overline{b})\, c$$
$$= a\bar{b}\bar{c} + \bar{a}b\bar{c} + abc + \bar{a}\bar{b}c = a(\bar{b}\bar{c} + bc) + \bar{a}(b\bar{c} + \bar{b}c)$$
$$= a\overline{(b\bar{c} + \bar{b}c)} + \bar{a}(b\bar{c} + \bar{b}c)$$
$$= a \oplus (b \oplus c).$$

3. *Existence of identity.* Since $a \oplus 0 = a\bar{0} + \bar{a}0 = a \cdot 1 + 0 = a$, 0 is the additive identity of R_n.

4. *Existence of inverse.* Since for each $a \in S$, we have $a \oplus a = a\bar{a} + \bar{a}a = 0 + 0 = 0$, each element is its own additive inverse.

Since the operation $\cdot$ is associative by definition, we need only to prove the distributive property. We have

$$a \circ (b \oplus c) = a \cdot (b\bar{c} + \bar{b}c) = ab\bar{c} + a\bar{b}c.$$

Since $a\bar{a}b + a\bar{a}c = 0$, we also have

$$a \circ (b \oplus c) = ab\bar{c} + a\bar{b}c + a\bar{a}b + a\bar{a}c = ab\bar{c} + ab\bar{a} + ac\bar{b} + ac\bar{a}$$
$$= ab(\bar{a} + \bar{c}) + ac(\bar{a} + \bar{b}) = ab\overline{ac} + ac\overline{ab}$$
$$= (a \circ b) \oplus (a \circ c),$$

which proves the distributive property. We conclude that R_n is indeed a ring.$\|$

The preceding theorem shows that every bounded, distributive, complemented lattice may be interpreted as a ring, although a distributive lattice in general does not enjoy this property. We now ask an analogous question with respect to rings: Which additional properties should be introduced into a ring so that it is interpretable as a lattice? Upon examination of the lattice properties (4.3.1)–(4.3.4), it appears that idempotency, commutativity, dual distributivity, and absorption should be introduced. Surprisingly, however, we have a much weaker requirement. To approach the main result we need the following definition.

Definition 5.6.1. A ring $R = [S, \oplus, \circ]$ such that for every element $a \in S$, $a \circ a = a$ is said to be a *boolean ring*.

The postulated idempotency in the multiplicative structure of R brings about a remarkable wealth of properties, as we shall see.

Lemma 5.6.2. In a boolean ring R, every element is its own additive inverse.

Proof. From idempotency and distributivity, we have $a \oplus a = (a \oplus a) \circ (a \oplus a) = a \circ a \oplus a \circ a \oplus a \circ a \oplus a \circ a = (a \oplus a) \oplus (a \oplus a)$; but by the cancellation law in the additive structure of R, $(a \oplus a) = (a \oplus a) \oplus (a \oplus a)$ implies $0 = a \oplus a$, which proves the lemma.$\|$

Lemma 5.6.3. A boolean ring R is commutative.

Proof. We recall that a commutative ring is one whose multiplicative structure is commutative. For a and b in a boolean ring $R = [S, \oplus, \circ]$ we have

$$a \oplus b = (a \oplus b) \circ (a \oplus b) = a \circ a \oplus a \circ b \oplus b \circ a \oplus b \circ b$$
$$= a \oplus b \oplus (a \circ b \oplus b \circ a),$$

where the first equality follows from multiplicative idempotency, the second from distributivity, and the third from idempotency, commutativity, and associativity of addition. But by the cancellative property, $a \oplus b = a \oplus b \oplus (a \circ b \oplus b \circ a)$ implies $a \circ b \oplus b \circ a = 0$. We recall that, by lemma 5.6.2, $a \circ b \oplus a \circ b = 0$; whence by the transitive property of equality

$$a \circ b \oplus b \circ a = 0 = a \circ b \oplus a \circ b.$$

Applying again the cancellative rule, we have $b \circ a = a \circ b$. ‖

Theorem 5.6.4. A finite boolean ring $R = [S, \oplus, \circ]$ is a bounded distributive lattice $L = [S, +, \cdot]$ with respect to the lattice operations defined by the following rules:

$$a + b = a \oplus b \oplus a \circ b \qquad \text{(disjunction)} \qquad\qquad (5.6.3)$$

$$a \cdot b = a \circ b \qquad\qquad \text{(conjunction)} \qquad\qquad (5.6.4)$$

Proof. Verification of closure of both $+$ and $\cdot$ in L is immediate. The operations of L also have the following properties.

1. *Commutativity.* Indeed, $\cdot$ is commutative by lemma 5.6.3 and the definition of $\cdot$ (5.6.4); with respect to $+$, we have

$$a + b = a \oplus b \oplus a \circ b = b \oplus a \oplus b \circ a = b + a,$$

where the first and last equalities follow from (5.6.3), and the second one follows from the commutativity of $\oplus$ and $\circ$ in R.

2. *Idempotency.* Conjunction is idempotent because $a \cdot a = a \circ a = a$. Disjunction is also idempotent because

$$a + a = a \oplus a \oplus a \circ a = a \oplus a \oplus a = a \oplus (a \oplus a) = a \oplus 0 = a.$$

3. *Distributivity.* We have

$$a \cdot (b + c) = a \circ (b \oplus c \oplus b \circ c) = a \circ b \oplus a \circ c \oplus a \circ b \circ c;$$

but since $a \circ a = a$ and R is commutative, we have

$$a \circ b \circ c = a \circ a \circ b \circ c = a \circ b \circ a \circ c,$$

whence

$$a \cdot (b + c) = a \circ b \oplus a \circ c \oplus (a \circ b) \circ (a \circ c) = ab + ac.$$

By similar arguments one proves the dual distributive property in L.

4. *Associativity.* The operation $\cdot$ is associative because $\circ$ is associative. With reference to $+$, we have

$$(a + b) + c = (a \oplus b \oplus a \circ b) \oplus c \oplus (a \oplus b \oplus a \circ b) \circ c$$
$$= a \oplus b \oplus a \circ b \oplus c \oplus a \circ c \oplus b \circ c \oplus a \circ b \circ c.$$

We form the following associations of terms in the expression above,

$$\{1st\}, \{2nd, 4th, 6th\}, \{3rd, 5th, 7th\},$$

and obtain

$$(a + b) + c = a \oplus (b \oplus c \oplus b \circ c) \oplus a \circ (b \oplus c \oplus b \circ c) = a + (b + c).$$

5. *Absorption.* Consider the expressions

$$a + ab = a \oplus a \circ b \oplus a \circ a \circ b = a \oplus a \circ b \oplus a \circ b = a,$$

where the first equality follows from (5.6.3), the second from idempotency of $\circ$, the third from lemma 5.6.2 ($a \circ b + a \circ b = 0$). Similarly, one proves that $a(a + b) = a$.

Finally we must show the validity of the principle of consistency (4.3.5) in L; that is, $a \oplus b \oplus a \circ b = a$ is equivalent to $a \circ b = b$. Indeed, if $a \oplus b \oplus a \circ b = a$, by cancellation we have $b \oplus a \circ b = 0$, which in turn means $a \circ b = b$.

Thus we have shown that in L disjunction and conjunction are idempotent, commutative, associative, mutually distributive, and absorptive, and that the consistency law holds true. Since S is finite, L is a bounded lattice.‖

Merely as a consequence of the property $a \circ a = a$, a ring R is interpretable as a distributive lattice. In reality, the property $a \circ a = a$ is so powerful that it even induces the structure of a boolean algebra. We need to show that complementation can be defined in L in terms of the ring operations. By definition of complement, for $a \in S$ we must have

$$a + \bar{a} = 1, a \cdot a = 0$$

for some unknown $\bar{a}$ in S. Now, from (5.6.3) and (5.6.4) and the identities defining $\bar{a}$, we obtain

$$1 = a + \bar{a} = a \oplus \bar{a} \oplus a \circ \bar{a} = a \oplus \bar{a} \oplus a \cdot \bar{a} = a \oplus \bar{a},$$

whence

$$a \oplus 1 = a \oplus a \oplus \bar{a} = \bar{a},$$

which defines the complement in terms of the ring operations and the constant 1. We summarize the results as a corollary.

Corollary 5.6.5. A finite boolean ring $R = [S, \oplus, \circ]$ is a finite boolean algebra $\mathscr{B} = [S, +, \cdot, ^-, 0, 1]$ with respect to the boolean operations defined as follows:

$$a + b = a \oplus b \oplus a \circ b,$$
$$a \cdot b = a \circ b,$$
$$\bar{a} = 1 \oplus a.$$

5.7 AN ALGEBRA OF LOGIC

To illustrate certain operations of logic, let us consider something familiar to most of us: regulated central heating. Although the situation may be rather unlikely, suppose that a guest from a tropical island visits us on a chilly day and expresses surprise that the furnace turns on without any apparent action of ours. A typical explanation may be "The room temperature is less than that at which the thermostat is set, the furnace pilot flame is on, and of course, there is electric power." (At this point, however, one should perhaps add "and everything else is properly functioning," but let us assume that we live, at least for a short time, in a perfect world, where malfunction is unknown.)

If we examine the explanatory statement, we recognize three simpler statements: "The room temperature is less than that at which the thermostat is set," "The furnace pilot flame is on," and "There is electric power." We may say that we have analyzed a given statement into three new statements. It we try to repeat the same type of analysis on any of these three statements, we do not succeed in producing any decomposition into simpler statements. We may therefore call them *simple* statements, and for economy of writing, we denote them with some symbolic designation, say, the letters c, p, and e, respectively (there is some attempt at mnemonic aid in the chosen letters: c for "cold," p for "pilot," e for "electricity").

We see that we have formed a more complex statement by combining the simple statements c, p, and e. It is therefore natural to call the former a *compound* statement. How did we combine the simple statements? We used the grammatical connective "and". Thus we may symbolically represent the compound statement as (c AND p AND e). An equivalent viewpoint considers the simple statements as *events* and obtains the compound event (c AND p AND e) from the simple events c, p, and e. This viewpoint is quite fruitful since there is a *different* event, "The furnace is on" (designated here by the letter f), which in our situation occurs exactly when (c AND p AND e) occurs and thus may be considered equivalent to the latter; that is, $f = c$ AND p AND e. (We must keep in mind, however, that the two sides of the given "equality" are not the *same* event, but only *co-occurring* events.)

In a similar view, the statement "The pilot light is off" is obviously the negation of the statement "The pilot light is on" and may be thought of as obtained by *negating* the latter; we denote it as NOT(p). Suppose now that the room temperature is lower than that called for by the thermostat, but that the furnace is not running; asked why, we may answer, "The power is off or the furnace pilot is off." Here again we are combining statements, NOT(e) and NOT(p), to obtain a compound statement through the grammatical connective "or"; we may denote it as [NOT(e) OR NOT(p)]. Note, however, that the English conjunction OR is ambiguous, since it is not immediately clear—unless some additional contextual elements resolve the ambiguity—whether OR means "one or the other or both" (INCLUSIVE OR) or "one or the other but not both" (EXCLUSIVE OR). (This ambiguity is accidental, though present in a number of natural languages.) When we use OR in what follows, we shall mean INCLUSIVE OR.

Thus, given statements a and b (each declaratory, like those mentioned before), we can "connect" them to form (a AND b) and (a OR b). For this reason, AND and OR are commonly referred to as connectives, and by extension, the word "connective" is also used for NOT (improperly, however, because NOT does not connect two or more statements). Note that the statements being combined may themselves be compound statements.

If we now consider a finite set S of statements and regard AND and OR as binary operations on S and NOT as a unary operation on S, a most remarkable finding is that the system generated by S under the operations of AND, OR, and NOT is a boolean algebra! Specifically, we may refer to the set of axioms obtained in Section 5.4 (definition 5.4.2) and regard AND and OR as the binary operations of *conjunction* and *disjunction*, respectively, and NOT as the unary operation of *complementation*. We make the following observations.

1. We regard, for example, the statement "The pilot flame is on and there is power" equivalent to the statement "There is power and the pilot flame is on"; that is, conjunction is commutative (check 5.4.2). An analogous remark can be made for disjunction.

2. We regard, for example, the statement "It is cold and either there is no power or the pilot flame is off" equivalent to the statement "It is cold and there is no power or it is cold and the pilot flame is off"; that is, conjunction distributes over disjunction (check 5.4.5). (Equally intuitive plausibility can be given to the dual distributive law (5.4.6).)

3. We consider a generic statement, say, "It is cold," and its negation, "It is not cold," and form two new statements.

    a) "It is cold and it is not cold."
    b) "It is cold or it is not cold."

The first is a plain absurdity since it declares simultaneously a statement and its negation. Such a statement is referred to as an *inconsistency* and is denoted here by the symbol F. The second statement is the ultimate in triviality and is referred to as a *tautology*[2] (denoted here by the symbol T). Note, then, that for a generic statement $a \in S$ we always have

$$a \text{ AND } (\text{NOT } a) = F,$$
$$a \text{ OR } (\text{NOT } a) = T.$$

---

2. From the Greek ταὐτόν: "the same" and λέγειν: "to say", i.e., "to say the same," "to say the obvious."

That is, F and T play the respective roles of the lower and upper universal bounds of a boolean algebra; therefore, we assume that S contains them (check (5.4.9) and (5.4.8)).

4. For a generic statement a, it is intuitively quite acceptable that saying "statement a or the inconsistency" is equivalent to saying "statement a" alone, and that saying "statement a and the tautology" is also equivalent to saying "statement a" alone. Symbolically,

$$a \text{ OR } F = a, \qquad a \text{ AND } T = a.$$

In terms of events, T is interpreted as the "certain" event and F as the "impossible" event. Therefore (a OR F)—the occurrence either of a or of the impossible event—is simply the occurrence of a, and (a AND T)—the simultaneous occurrence of a and of the certain event—is also simply the occurrence of a (check (5.4.6)).

Verification of the acceptability of axioms (5.4.2), (5.4.5), (5.4.6), (5.4.8), and (5.4.9) substantiates our claim that an algebra of statements (or an algebra of events) is a boolean algebra. The corresponding calculus (Section 5.4) is commonly referred to as *sentential* or *propositional calculus*.

It must be pointed out that the elements of the boolean algebra $\mathscr{B} = [S, \text{ OR}, \text{ AND, NOT}, F, T]$ are objects called "propositions" or "statements," and that absolutely no hypothesis is made concerning their meaning or their relatedness. In other words, if we have a set $\Sigma \equiv \{\sigma_1, \ldots, \sigma_p\}$ of simple statements (i.e., statements that we cannot or will not further analyze), $\mathscr{B}$ is nothing but the free boolean algebra generated by $\sigma_1, \ldots, \sigma_p$ (see Section 5.3). It follows that each element of $\mathscr{B}$ is represented by some boolean expression over the arguments $\sigma_1, \ldots, \sigma_p$ and is a *boolean function* of those arguments.

Thus far, propositional calculus appears to be a formal tool for generating new propositions from given propositions, or for proving the equivalence of two differently expressed statements. These, by themselves, are not trivial tasks, especially in connection with the complicated wording of sentences with multiple negations.

Example 5.7.1. "It is not true that either I do not have a dog or it is not true that I do not have a cat" is a statement whose interpretation may give one a little headache (there is widespread feeling that some contractual clauses fall in this category). However, we can formalize it by first introducing parentheses and other enclosure symbols, as follows: (It is not true that) {either (I do *not* have a dog) or [(It is not true that) (I do *not* have a cat)]}, whence we obtain compactly

$$\text{NOT}\{\text{NOT(I have a dog) OR NOT[NOT(I have a cat)]}\}. \qquad (5.7.1)$$

In order to use more expeditiously the rules of the calculus of boolean algebra, it is convenient to simplify the notation by replacing AND and OR with the more familiar $\cdot$ and $+$, respectively. With the same intent, we replace NOT by a bar above the statement to which NOT applies. (Several slightly different notations are common in symbolic logic, the prevailing ones perhaps being ∧ for AND, ∨ for OR, ⌐ for NOT.) Then we define the symbols d and c in the following way.

d: I have a dog.

c: I have a cat.

Thus statement (5.7.1) can be rewritten as

$$\overline{(\overline{d} + \overline{c})},$$

and the following manipulations are easily justified

$$\overline{(\overline{d} + \overline{c})} = \overline{(\overline{d} + c)} = d\overline{c}.$$

The right side is then interpreted verbally as "I have a dog but I do not have a cat," which is a very reasonable and easily understood statement.

However, a more fundamental task of propositional calculus arises from the notion of *truth values*. That the attributes *true* (T) or *false* (F) naturally apply to propositions is not surprising; in fact, in the preceding example it seemed obvious that the phrase "It is not true that" was synonymous with the negation of its dependent proposition. It must be observed, however, that there are propositions which, because of our present ignorance, are neither true nor false. An example is "Life exists outside of the solar system." In the following discussion we shall restrict ourselves to propositions which are either true or false. If T is regarded as the negation (complement) of F, then $\mathscr{B}_1 = [\{T, F\}, +, \cdot, ^-, T, F]$ is a boolean algebra of only two elements (that is, $\mathscr{B}_1$ is a smallest boolean algebra), and the truth evaluation of propositions of $\mathscr{B}$ may be viewed as a function $\theta : S \rightarrow \{T, F\}$. Even more intuitively acceptable is the notion of computing truth values as the evaluation of $\mathscr{B}$ over $\{T, F\}$ (see Section 5.5). In other words, given a proposition which is an expression $f(\sigma_1, \ldots, \sigma_p)$ in the propositions $\sigma_1, \ldots, \sigma_p$, the evaluation consists of computing the truth value of f from the truth values of $\sigma_1, \sigma_2, \ldots, \sigma_p$.

Example 5.7.2. A room has two swinging doors, D_1 and D_2. We denote by I_i the proposition "D_i is open toward the inside of the room," and by O_i the proposition "D_i is open toward the outside of the room." Thus the propositional expression (compound proposition)

$$(I_1 + O_1) \cdot (I_2 + O_2)$$

is equivalent to the statement "There is a clear path through the room," which we denote by P. Therefore $P = (I_1 + O_1)(I_2 + O_2)$. Now, assume that D_1 is closed, and that D_2 is open toward the inside. It follows that I_1, O_1, and O_2 are assigned to F in $\mathscr{B}_1$, whereas I_2 is assigned to T, and the truth value of P is

$$(F + F) \cdot (T + F) = F \cdot T = F.$$

That is, there is no clear path through the room.

Thus far we have considered a propositional calculus based on the connectives AND, OR, and NOT, and we also know that this set of operations is perfectly adequate for the description (Section 5.3) of each compound statement. Although there is no essential need for them, we may introduce other connectives which are quite useful for the formalization of logical discourse. We shall principally concentrate on two of them, the so-called *conditional* and *biconditional*.

Definition 5.7.1. Given two elements a and b in $\mathscr{B}$ the *conditional* of the ordered pair (a, b), denoted by $(a \rightarrow b)$, is the element $(\bar{a} + b)$; a is termed the *antecedent* and b the *consequent*.

For the present, the conditional $(a \rightarrow b)$ must be viewed exclusively as an element of the algebra $\mathscr{B}$, and the three words "conditional," "antecedent," and "consequent," as well as the symbol "$\rightarrow$" should not evoke any extra meaning to be attached to the function just defined. In other words, if a represents the statement "Ice is hot" and b represents the statement "Grass is green," $(\bar{a} + b)$ represents the compound statement "Either ice is not hot or grass is green or both." Note that, in normal circumstances, a is false and b is true, from which we can compute that $(a \rightarrow b)$ is true; indeed, we evaluate $(a \rightarrow b)$ in $\mathscr{B}_1$, having assigned a and b elements of $\mathscr{B}_1$.

Suppose now that a is defined as "$2 + 2 = 4$" and b is a yet-undisclosed proposition (let us say, a proposition written on paper enclosed in a sealed envelope marked b). Suppose that we have been informed by a reliable source that $(a \rightarrow b)$ is true; that is, $\bar{a} + b = T$. Since a is true, the left side of the equality $\bar{a} + b = T$ becomes $\bar{T} + b = F + b = b$; that is, we obtain the result $b = T$. In other words, we have concluded that b must be some true proposition—for example, "5 is a prime number," "There are twelve months in the year," etc. It is important to note that a *deduction* has been made; that is, from

$$a = T, \qquad \bar{a} + b = T$$

we have deduced $b = T$. The scope of the deduction, however, is *confined to the truth values* of the proposition and *does not extend to their meanings*. However, when a and b are related, it is usual (and perfectly comfortable) to assert that proposition b can be logically deduced from proposition a, although, we stress, the deduction is always limited to truth values. The latter situation justifies the terms "conditional," "antecedent," and "consequent" and the symbol "$\rightarrow$". Moreover, it is customary to rewrite the equality

$$a \rightarrow b = T \qquad \text{as} \qquad a \Rightarrow b.$$

The notation $a \Rightarrow b$ is referred to as an *implication* and is read "*a* implies *b*" although this terminology is misleading except when *a* and *b* are related.

An important remark is now in order. The expression $a \Rightarrow b$, which can also be read "$(\bar{a} + b)$ is true," is a statement. However, although a, b, $(a \rightarrow b)$, and $a \Rightarrow b$ are all statements, they do not belong to the same algebraic structure. Indeed, whereas a, b, and $(a \rightarrow b)$ are members of the boolean algebra $\mathscr{B}$—the free boolean algebra over a set S of undefined statements—$(a \Rightarrow b)$ is a statement *about* a statement of $\mathscr{B}$ (precisely, about the truth value of a statement of $\mathscr{B}$). Therefore $a \Rightarrow b$ is a member of a distinct algebra of statements, and it will be briefly denoted as a *metastatement*.

Thus far we have not mentioned that the notation $(a \rightarrow b)$ is commonly read "if *a*, then *b*," whose semantics is perfectly plausible only when *a* and *b* are related. We feel that this verbal description should be avoided since its linguistic inadequacy has been the origin of an enormous literature on the "paradoxes" of the conditional. We now illustrate one such amusing paradox.

Example 5.7.3. Let *a* stand for "I have blond hair" and *b* for "Tomorrow it will rain" (two clearly unrelated statements, which can be either true or false). Then, by simple manipulations, we have

$$(\bar{a} + a) + (\bar{b} + b) = (\bar{a} + b) + (\bar{b} + a). \tag{5.7.2}$$

Obviously, the truth values of the left and right sides of any identity are identical. Since both $(\bar{a} + a)$ and $(\bar{b} + b)$ are tautologies, we have

$$(\bar{a} + a) + (\bar{b} + b) = T + T = T;$$

that is, the left side of (5.7.2) is a tautology. Therefore the right side must be one as well,

$$(\bar{a} + b) + (\bar{b} + a) = T.$$

This equality says that either $(\bar{a} + b) = T$ or $(\bar{b} + a) = T$, which is rewritten symbolically as

$$(a \Rightarrow b) + (b \Rightarrow a).$$

A careless translation of this symbolic metastatement into natural language would be "Either the fact that I have blond hair implies that tomorrow it will rain or the fact that tomorrow it will rain implies that I have blond hair," whose plain nonsense is due to the inappropriate use of the word "implies."

The notion of implication, however, is extremely useful in formalizing the process of reasoning, that is, in the deduction of propositions called *consequences* from other propositions called *premises*, which are assumed to be true. A deduction, which would derive the truth of consequence *r* from the truths of premises $p_1, p_2, \ldots, p_n$, is *valid* when and only when

$$(p_1 p_2 \cdots p_n \rightarrow r) = T,$$

or equivalently,

$$p_1 p_2 \cdots p_n \Rightarrow r.$$

Example 5.7.4. Let a, b, and c be (related) propositions of $\mathscr{B}$; assume that $n = 2$, and let $p_1 = (a \to b)$ and $p_2 = (b \to c)$. We then construct the new function (in $\mathscr{B}$)

$$f = [(a \to b)(b \to c)] \to (a \to c).$$

We claim that $f = T$, that is, that f is a tautology. Indeed, f can be rewritten as

$$
\begin{aligned}
f &= \overline{(a \to b)(b \to c)} + (a \to c) \\
&= \overline{(\bar{a} + b)(\bar{b} + c)} + (\bar{a} + c) = a\bar{b} + b\bar{c} + \bar{a} + c \\
&= (\bar{a} + a\bar{b}) + (c + \bar{c}b) = \bar{a} + \bar{b} + c + b = \bar{a} + c + T = T,
\end{aligned}
$$

since $b + \bar{b} = T$. Therefore

$$(a \to b)(b \to c) \to (a \to c) = T;$$

that is, $(a \to b)(b \to c) \Rightarrow (a \to c)$, where $(a \to c)$ is the consequence. Now, if $a \Rightarrow b$ (that is, $a \to b = T$) and $b \Rightarrow c$ (that is, $b \to c = T$)—both premises are true—we obtain $(T \to (a \to c)) = T$; that is, $(a \to c) = T$, which is equivalent to $(a \Rightarrow c)$. This formal deductive process is referred to as a *syllogism* and was formalized by Aristotle. It is read "If a implies b and b implies c, then a implies c." Let us now consider a classical example of the syllogism. We first define the symbols a, b, and c.

    a: This being is Socrates.

    b: This being is a man.

    c: This being is mortal.

Then we know, as premises, that $a \Rightarrow b$ and $b \Rightarrow c$, and we deduce $a \Rightarrow c$, which is read "If Socrates is a man and man is mortal, then Socrates is mortal."

    We now turn our attention to the connective *biconditional*.

Definition 5.7.2. Given two elements a and b in $\mathscr{B}$, the *biconditional* of the (unordered) pair $\langle a, b \rangle$, denoted by $(a \leftrightarrow b)$, is the element $(\bar{a}\bar{b} + ab)$.

    Note that the expression $(a \leftrightarrow b)$ is equivalent to the expression $(a \to b) \cdot (b \to a)$. Indeed, converting the latter to the usual connectives, we have

$$(a \to b)(b \to a) = (\bar{a} + b)(\bar{b} + a) = \bar{a}\bar{b} + \bar{a}a + b\bar{b} + ba = \bar{a}\bar{b} + ab,$$

which coincides with the definition of biconditional. This also motivates the choice of the word "biconditional" (that is, a *bi*directional *conditional*).

    The function biconditional is used to define the concept of equivalence between propositions. Here again, equivalence refers to the truth values only and not to the meanings of the propositions (fortunately, the word "equivalence" seems to be semantically less misleading than the word "implication").

Definition 5.7.3. Two propositions a and b in B are said to be *equivalent* if $(a \leftrightarrow b) = T$. This is also denoted by $a \Leftrightarrow b$ (a metastatement).

Example 5.7.5. We proved some theorems in the preceding chapters by a device referred to as "cyclic implication." See, for example, theorem 2.5.3, where we had the following statements.

a: G is a tree.
b: G contains no cycle and has $(n-1)$ edges.
c: G is connected and has $(n-1)$ edges.

Our objective was to show the equivalence of any pair of the statements, and we accomplished it by showing that the three statements cyclically imply each other. We now formalize that otherwise intuitive claim. Cyclic implication means $a \Rightarrow b$, $b \Rightarrow c$, $c \Rightarrow a$, and this sequence is equivalent to the equality

$$f_1 = (a \to b)(b \to c)(c \to a) = T.$$

(Indeed, $f_1 = T$ if and only if each term of the conjunction is also equal to T.) Similarly, pairwise equivalence means $a \Leftrightarrow b$, $b \Leftrightarrow c$, $a \Leftrightarrow c$, which, by the same argument, is equivalent to

$$f_2 = (a \leftrightarrow b)(b \leftrightarrow c)(a \leftrightarrow c) = T.$$

We must now show that $f_1 = f_2$. Indeed, we have

$$f_1 = (\bar{a} + b)(\bar{b} + c)(\bar{c} + a) = \bar{a}\bar{b}\bar{c} + abc$$

and

$$f_2 = (\bar{a}b + ab)(\bar{b}\bar{c} + bc)(\bar{a}\bar{c} + ac) = \bar{a}\bar{b}\bar{c} + abc;$$

that is, f_1 and f_2, being equal to the same expression, are the same function.

Example 5.7.6. Repeated application of the syllogism is the essence of so-called *direct proofs*, as we saw in Chapter 0. We now formalize, in terms of the propositional calculus just developed, the validity of *contrapositive proofs*; that is, "P implies Q if and only if $\bar{Q}$ implies $\bar{P}$." In symbols we have

$$(P \Rightarrow Q) \Leftrightarrow (\bar{Q} \Rightarrow \bar{P}),$$

and we want to show the validity of this expression.

Using definitions 5.7.1 and 5.7.2, we have

$$[(\bar{P} + Q = T) \Leftrightarrow (Q + \bar{P} = T)] = [(\overline{\bar{P} + Q = T})(\overline{Q + \bar{P} = T}) + \\ + (\bar{P} + Q = T)(Q + \bar{P} = T)]$$

By idempotency, the right side becomes

$$(\overline{\bar{P} + Q = T}) + (\bar{P} + Q = T).$$

i.e., a tautology, which proves the claim.

We close this section with an additional application of propositional calculus.

Example 5.7.7. Suppose that a large computing system consists of n processing units $U_1, U_2, \ldots, U_n$. Each of these units is sufficiently complex to perform tests on other units of the set and to determine their correct operation. More specifically, U_i tests U_j by applying inputs to U_j and by comparing the outputs from U_j with their correct form, stored in the memory of U_i. The result of the test is expressed by a signal a_{ij}, which is conventionally 1 if the test passes and 0 if the test fails. Clearly, a_{ij} is reliable only when U_i is itself fault-free. We want to know whether the inspection of the set of signals a_{ij} permits the identification of the faulty units in the set $U \equiv \{U_1, \ldots, U_n\}$. We reformulate the problem in the following terms. The symbol u_i is defined as

$$u_i : \text{unit } U_i \text{ is fault-free.}$$

If $a_{ij} = 1$, the unit U_i declares that U_j is fault-free, which is equivalent to saying, "If U_i is fault-free, then U_j is fault-free." Symbolically,

$$a_{ij} = 1 \qquad \text{corresponds to} \qquad u_i \Rightarrow u_j.$$

Similarly, if $a_{ij} = 0$, we have

$$a_{ij} = 0 \qquad \text{corresponds to} \qquad u_i \Rightarrow \bar{u}_j.$$

Suppose now that in the following three-unit system the tests have given the results shown in figure 5.7.1.

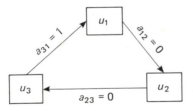

Figure 5.7.1.

We express the test results as

$$u_1 \Rightarrow \bar{u}_2, \, u_2 \Rightarrow \bar{u}_3, \, u_3 \Rightarrow u_1$$

or, equivalently,

$$\bar{u}_1 + \bar{u}_2 = T, \, \bar{u}_2 + \bar{u}_3 = T, \, \bar{u}_3 + u_1 = T.$$

It follows that the conjunction of these statements is also true; that is,

$$(\bar{u}_1 + \bar{u}_2)(\bar{u}_2 + \bar{u}_3)(\bar{u}_3 + u_1) = T.$$

Distributing the product, we have

$$\bar{u}_2 \bar{u}_3 + u_1 \bar{u}_2 + \bar{u}_1 \bar{u}_3 = T.$$

If we now put the expression on the left in canonical disjunctive form, we obtain

$$u_1 \bar{u}_2 \bar{u}_3 + u_1 \bar{u}_2 u_3 + \bar{u}_1 \bar{u}_2 \bar{u}_3 + \bar{u}_1 u_2 \bar{u}_3 = T. \qquad (5.7.3)$$

Each of these fundamental products of the literals u_1, u_2, and u_3 (definition 5.4.4) identifies a unique state of the system; for example, $u_1 \bar{u}_2 \bar{u}_3$ is read "u_1 is fault-free, but both u_2 and u_3 are faulty." Identity (5.7.3) tells us that any of the four states listed is compatible with the given test results; that is, we are not able to pronounce a diagnosis. However, if we have some reason to prefer the hypothesis that a state with a least number of faults is more likely, we would prefer the interpretation $u_1 \bar{u}_2 u_3$ ("u_2 is the only faulty unit") over the other three possible interpretations.

This concludes our rapid presentation of propositional calculus. We make absolutely no pretense either of depth or of extension, and the reader whose interest has been stimulated should proceed to more specialized texts on symbolic logic (see Section 5.9). Our primary objective was to illustrate one of the most important applications of boolean algebra. Indeed, boolean algebra was formulated by George Boole as a formal model for the systematic treatment of logic, and an important part of logic is precisely the manipulation of propositions.

EXERCISES 5.7

1. Represent the following statements as propositional expressions.
   a) It is not true that he will seek another term as governor.
   b) Grass will die if the drought continues.
   c) He will die today unless he receives medical treatment right away.
   d) Mr. Smith will be elected president if and only if he wins all the primaries.
   e) It is raining and the sun is also shining.
   f) A function is continuous if it is differentiable.
   g) If the government does not raise taxes or cut its spending, it will not be able to curb inflation.

2. A *propositional form* is defined as a statement containing variables such that their replacement by constants yields a proposition. For example, the following algebraic equations are propositional forms (here $+$ denotes addition of real numbers). Which of them can be made tautologies (i.e., true propositions) by restricting the domain of the variable x to certain nonempty intervals?

   a) $x = 7.5$ b) $x + x = 2x$
   c) $x = 1050x$ d) $x \leqslant x^2 - 1$
   e) $x < 10 + x$ f) $x + 10 = 4x$

3. Under what domain of the variable x are the following two propositional forms equivalent?

   a) $\dfrac{x}{x^2 + 1} = 1$
   b) $x^2 - x + 1 = 0$

4. Let P and Q be two given propositions. A convenient way to describe the truth value, T or F, of compound statements is by means of the so-called *truth table*, which lists, for each configuration of truth values of the component statements, the truth value of the compound statement. For example, the truth tables of $\bar{P}$, PQ, and $P + Q$ are given below.

P	$\bar{P}$
T	F
F	T

P	Q	PQ
T	T	T
T	F	F
F	T	F
F	F	F

P	Q	$P + Q$
T	T	T
T	F	T
F	T	T
F	F	F

Write the truth table for each of the following propositions.

a) $\overline{P + Q}$
b) $PQ + R$
c) $PQ + \bar{P}Q + P\bar{Q} + \overline{PQ}$
d) $(PQ + R)(\bar{P} + QR)$
e) $(P\bar{Q}) \to (RP)$
f) $(PQ + R)(P + R) + (\overline{PQ + R})$
g) $(P \leftrightarrow \bar{B}) + (B \to A)$

5. How many are the distinct compound statements of two statements? List all of them.

6. Determine which of the following are tautologies, which are inconsistencies, and which are neither.

a) $PQ + \bar{P}Q + P\bar{Q} + \overline{PQ}$
b) $[(P \to Q)(P \to Q)] \to Q$
c) $(P + Q)(P + \bar{Q})(\bar{P} + Q)(\bar{P} + \bar{Q})$
d) $(P \to Q) + P$
e) $B(A + B)$
f) $(P \to Q) + (Q \to P)$
g) $((P \to Q) \to Q) \to Q$

7. Prove the following equivalences:

a) $PQ + \bar{P}\bar{Q} + PR \Leftrightarrow PQ + \bar{P}\bar{Q} + \bar{Q}R$
b) $PS + QS + \bar{Q}R \Leftrightarrow PS + RS + \bar{Q}R$
c) $P(Q + R) \Leftrightarrow PQ + PR$
d) $P + QR \Leftrightarrow (P + Q)(P + R)$
e) $P + \bar{P}Q \Leftrightarrow P + Q$
f) $(P + Q)(\bar{P} + R)(Q + R) \Leftrightarrow (P + Q)(\bar{P} + R)$
g) $PQ + \bar{P}R + QR \Leftrightarrow PQ + \bar{P}R$

h) $P + \prod_{i=1}^{n} Q_i \Leftrightarrow \prod_{i=1}^{n} (P + Q_i)$

i) $P\left(\sum_{i=1}^{n} Q_i \right) \Leftrightarrow \sum_{i=1}^{n} PQ_i$

j) $\overline{\prod_{i=1}^{n} P_i} \Leftrightarrow \sum_{i=1}^{n} \overline{P}_i$

k) $\overline{\sum_{i=1}^{n} P_i} \Leftrightarrow \prod_{i=1}^{n} \overline{P}_i$

*8. Prove the following equivalences.

     a) $P\overline{Q} + Q\overline{P} \Leftrightarrow (P+Q)\overline{PQ}$
     b) $P\overline{Q} + Q\overline{R} + R\overline{P} \Leftrightarrow (P+Q+R)\overline{PQR}$

Can you generalize (b) to a larger number of propositions?

*9. An *adequate set of connectives* is one for which any propositional expression can be written using only connectives of the set. Thus {NOT ($^-$), AND ($\cdot$), OR (+)} is an adequate set of connectives. Show that each of the following is also an adequate set of connectives.

     a) $\{^-, +\}, \{^-, \cdot\}, \{^-, \rightarrow\}$
     b) Is $\{\rightarrow\}$ an adequate set?

5.8 THE ALGEBRA OF SWITCHING

In this section, we present an extremely important application of propositional calculus, hence of boolean algebras: the algebra of switching. Although for pedagogical reasons we have placed this topic at the end of this chapter, the present widespread interest in boolean algebras draws its motivation from the fact that binary switching circuits are the basis of practically all of today's digital computers.

As introduction, we consider a rather familiar example. The oil pressure warning light in a car is normally off except when the engine is running and the oil pressure is dangerously low. It is desirable, however, for the driver to be able to ascertain that the warning light is working properly. He can do so, whatever the oil pressure, by turning on the light when he turns on the ignition. We assume that the engine is running whenever the alternator is on. We now define the following events.

     w: Oil warning light is on.

     o: Oil pressure is below minimum.

     k: Ignition key is turned.

     a: Alternator is on.

We conclude that event w is equivalent either to k or to oa; that is,

$$w = k + oa. \tag{5.8.1}$$

Suppose now that we want to realize a physical arrangement of parts which lights the warning lamp according to the specifications. This corresponds to *detecting* the occurrence of the compound event described by the right side of (5.8.1). We must therefore begin by detecting the simple events it results from. We can do so by means of appropriate sensors, which will indicate the occurrence or non-occurrence of the event they are designed to detect by the position (open or closed) of a switch. In other words, the pressure sensor converts the event "The oil pressure is below minimum" into the equivalent event "The oil pressure switch O is closed." That event may, in turn, be replaced by "There is electrical *continuity* between the *two terminals* of switch O."

Analogous correspondences are established between the pairs of events a,k and switches A,K. It is easy to verify that the compound event $(k + oa)$ occurs

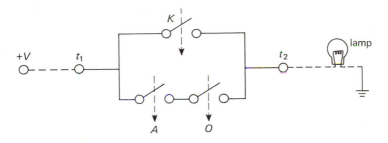

Figure 5.8.1. A simple contact network

exactly when there is continuity between the two terminals t_1 and t_2 of the network illustrated in figure 5.8.1 (where the arrows indicate the direction of motion of the switch when the corresponding event occurs). Thus the lamp must be lit exactly when there is continuity between t_1 and t_2. This is readily established by connecting t_1 to a voltage source of value V and measuring the voltage at t_2. It turns out that this voltage may be used directly to light the lamp.

If we examine critically what we have done, we recognize that we have first transformed statements about simple events into statements about the condition of two-position, two-terminal switches. Subsequently we have constructed a two-terminal network consisting of two-terminal switches and we have interpreted the continuity of the former as the occurrence of the compound event. This shows that the network *evaluates* a boolean expression $w = k + oa$ over the boolean algebra {closed, open}, which is a specialization of $\mathcal{B}_1 = \{T, F\}$, defined in the preceding section.

This very simple example is readily generalized in two main directions.

The first generalization corresponds to assuming that not only "normally open" but also "normally closed" switches may appear in the network. A "normally open" switch is one which is closed when the corresponding event occurs, whereas a normally closed switch is one which is closed when the corresponding event *does not* occur. Thus, in terms of continuity between the switch terminals, normally open and closed switches correspond to complemented or uncomplemented literals in the boolean expressions. Moreover, we may assume that the same event controls a set of several switches (referred to as contacts). This multiple control may be physically implemented by relays, that is, devices consisting of a coil and a set of contacts (in general, a mixture of normally open and normally closed contacts). The contact positions are changed when the coil is energized, that is, when it is traversed by a specified current (see figure 5.8.2).

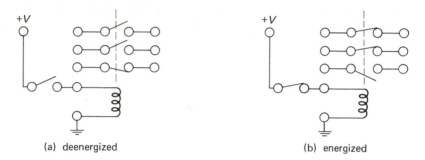

(a) deenergized   (b) energized

Figure 5.8.2. Scheme of a relay

Note that a contact network, such as the one in figure 5.8.1, is equivalently interpreted either in terms of continuity between its two terminals t_1 and t_2, or in terms of t_2 being at voltage V, given that t_1 is at voltage V. If t_2 is connected to voltage $V' \neq V$ through a convenient resistance, t_2 is either at voltage V or at voltage V'. Thus the ordered pair of *logical voltages* (V, V') may be taken to correspond to the ordered pair (T, F). With this interpretation, we are not restricted to contact networks, and we may consider appropriate interconnections of elements realizing, for example, the connectives AND, OR, and NOT. Without entering in details which are beyond our present scope, we note, by way of example, that the diode circuit in figure 5.8.3, effectively computes the AND of

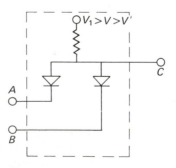

Figure 5.8.3. A circuit realizing the AND operation (AND gate)

two signals. In fact, if $V > V'$ and if the diodes are perfect, the (output) voltage V_C at C is the smallest of V_A and V_B; that is, $V_C = V$ if and only if $V_A = V_B = V$ (which reflects the definition of AND).

Extending these considerations, we may consider networks, called *switching networks*, having n input terminals and one output terminal such that, when the input terminals are at voltages V or V', the output terminal is also at V or V'. Such networks, therefore, realize the mappings of an ordered n-tuple in $\{V', V\}^n$ to $\{V', V\}$. If we now denote conventionally by $\{0, 1\}$ the pair of logical voltages $\{V', V\}$, the mappings just mentioned coincide with the functions $\varphi:(\{0, 1\})^n \to \{0, 1\}$, that is, the set of switching functions introduced at the end of Section 5.5. Thus a switching network computes or, as we say, "realizes" a switching function. The analysis and the synthesis of a switching network which realizes a specific boolean function, as well as the study of the problems pertaining to the satisfaction of some criteria of simplicity in this synthesis, are the subject of a very important field, called switching theory and logical design (see Note 5.1).

EXERCISES 5.8

1. A committee of three persons makes decisions by majority vote. Each committee member can register a "yes" vote by pressing a button. Design a contact network which will light a lamp when and only when there is a majority of "yes" votes.

2. A light in a room is controlled by three wall switches and can be turned on or off from any one of them. Design a contact network for such a light.

3. Consider the elementary networks (gates) in figure 5.8.4, whose inputs are 0,1-valued variables. The gates are defined as follows: (1) AND—the output is 1 if and only if all the inputs are 1; (2) OR—the output is 1 if at least one of the inputs is 1; (3) NOT—the output is 1 if the input is 0, and vice versa.

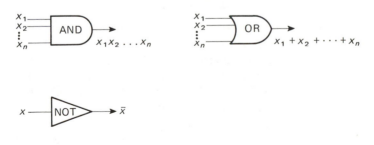

Figure 5.8.4.

A *switching network* is defined as an interconnection of these gates so that any number of gate inputs may be connected to any gate output, and no two outputs may be connected together. Show that the switching networks in figure 5.8.5 realize the same function $\varphi: \{0,1\}^2 \to \{0,1\}$. Give φ in tabular form (truth table). [Note that network (b) is planar!]

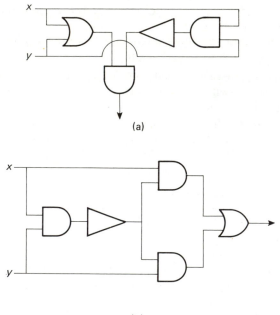

(a)

(b)

Figure 5.8.5.

5.9 BIBLIOGRAPHY AND HISTORICAL NOTES

The study of boolean algebras began around the middle of the past century. George Boole's work must be considered the first formal system for the description of logical thinking, although attempts in this direction had been made throughout the centuries, originating with the Greek philosopher Aristotle. Boolean algebra is the foundation of symbolic logic, which is now a solidly developed area (Russell, Whitehead, Carnap, etc.). Since the time of Boole, most of the studies related to boolean algebra have been concerned with its axiomatic structure and with the nature of the systems which are obtained by removing some axioms (Brouwer, Birkhoff, Heyting, Tarski, etc.).

Recently, because of the great relevance of boolean algebra to switching and computers and because of the rapid growth of the computer population, boolean algebra has become a very "popular" topic.

The literature on, or related to, boolean algebras is extremely extensive.

F. E. Hohn, *Applied Boolean Algebra*, Macmillan, New York, 1966
is a clear and pleasant treatment of boolean algebra applied to logic and switching.
Excellent among the mathematical treatments of boolean algebras is

P. Halmos, *Lectures on Boolean Algebra*, Van Nostrand, Princeton, 1963.

As additional readings, with no intention to slight many other valuable references,
we suggest

A. Tarski, *Introduction to Logic and to the Methodology of Deductive Sciences*,
Oxford University Press, New York, 1939

and the first half of

Z. Kohavi, *Switching and Finite Automata Theory*, McGraw-Hill, New York,
1970.

5.10 NOTES AND APPLICATIONS

Note 5.1. The Design of Switching Networks

We have stated more than once that the design of switching networks—the so-
called *logical design*—is a fully developed discipline in the engineering of digital
systems. We now illustrate some of the interesting problems in this area.

Abstractly, a switching network computes, or realizes, a switching function.
However, since a given switching function can be realized by an infinite number of
networks, it is desirable in selecting a network to use some reasonable criterion of
simplicity.

A traditional criterion of simplicity has been based on the cost of the electronic
components used. Here we shall assume, for the sake of simplicity, the use of only
logical blocks (*gates*) which realize the AND, OR, and NOT connectives; for each
of these, it is reasonable to assume that the cost is proportional to the number of
inputs they have. Thus a network which realizes the expression

$$E = x_1 x_2 + \bar{x}_3 x_4 \bar{x}_1,$$

as shown in figure 5.10.1, has a total cost of $2 + 3 + 2 = 7$ units. A network of
this type realizes a disjunctive expression of a switching function. Since it consists

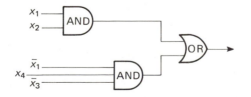

Figure 5.10.1

of AND gates whose outputs feed the inputs of a single OR gate, this type of network is frequently referred to as an AND-TO-OR network. The criterion of simplicity for this class of realizations, calls for finding an expression containing the least number of product terms and of literals. To find such a *minimal* expression, one starts from an expression of the function *f* in disjunctive normal (or canonical) form. Then one uses systematically the identity

$$x + \bar{x} = 1 \qquad \text{(for an arbitrary literal } x\text{)}$$

in order to find all the shortest product terms (i.e., with the least number of literals) such that, for any one of them, *p*, we have *pf* = *p*. This property means that if *p* = 1 then *f* = 1; that is, *p* implies *f* or is an "implicant" of *f*. Since there is no shorter product which implies *p* and *f*, *p* is said to be a *disjunctive prime implicant* of *f* (conjunctive prime implicants are defined dually).[3] It is easy to show that a minimal disjunctive expression of a switching function is a disjunction of prime implicants, and interesting methods exist for the selection of an optimum set of prime implicants of *f* whose disjunction is an expression of *f*.

In some cases, however, a minimal AND-TO-OR expression of a switching function *f* requires the use of gates with a number of inputs larger than that of available manufactured gates; this is effectively referred to by saying that the *fan-in limitation* of the logic blocks has been exceeded (as if the input lines were "fanning into" the gate). One must then look for alternative realizations, as shown by the following example.

In figure 5.10.2 (a) we have a minimal AND-TO-OR realization of a function $f(x_1, x_2, x_3, x_4)$, and in figure 5.10.2 (b) we have an alternative realization of the

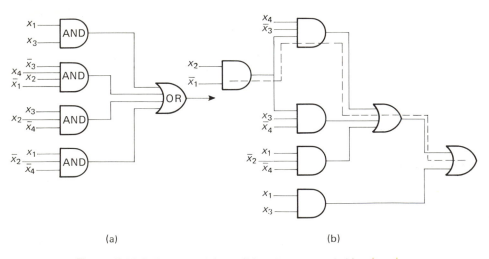

(a)   (b)

Figure 5.10.2. Two networks realizing the same switching function

3. The procedure for finding all the prime implicants of *f*, disjunctive or conjunctive, is known as the Quine-McCluskey method. A detailed description of the Quine-McCluskey method is beyond the scope of this book.

same function subject to a fan-in limitation of 3. Note that the signals in the network in (a) propagate from input to output through a cascade of two logical blocks, or two *levels of logic*; in the network in (b), the longest path of propagation (dotted line) traverses four levels of logic. Since traversing a level of logic requires time (roughly of the order of 10^{-8} seconds or less with 1970 technology), the network in (b) is "slower" than that in (a) if the same types of electronic components are used for both. This fact points to another problem in the design of switching networks: the signal propagation delay. In some instances, the reduction of this delay may be the main design criterion.

Although the traditional criterion of simplicity has been the cost of the components used, several other practical criteria have become increasingly important. In the past decade the price of electronic components has been going down constantly. This fact has given increasing weight to the cost of interconnecting the components, usually by printed ("etched") metallic connections on an epoxy board, and has fostered interest in networks which require a simple interconnection. In graph-theoretic terms, this may be expressed by saying that the network graph—obtained by replacing each gate with a vertex—either is planar (definition 2.6.1) or has a small number of edges, etc.

The decrease in the cost of components is essentially due to an expanding market and a revolutionary technological innovation called *large-scale integration* (LSI). This technology makes it possible to manufacture on a single semiconductor chip a very large number of individual gates (tens or hundreds). Although the cost per manufactured unit becomes very low, the cost of "tooling" for manufacturing a new type of chip is extremely high. Therefore, LSI calls for functional standardization and modularization, that is, the realization of digital systems as the interconnection of relatively few types of rather powerful subsystems ("modules"). Answers to this call have been, for example, the studies of cellular logic and of universal logical modules (we shall return to the latter in Chapter 6).

Finally, physical components may be defective when built, or they become defective in operation. It is therefore essential to be able to test whether a digital network contains faults and, in some instances, to locate them (fault diagnosis). The determination of fault-diagnostic procedures and the design of networks for which diagnostic procedures are more easily obtained are parts of a thriving area of logical design.

Note 5.2. The Design of Sequential Networks

Finite state machines were introduced in Chapter 1 (Note 1.1) and restudied later in the context of applications of formal notions. In this note we shall outline the important problem of implementing (*realizing*) finite state machines by means of hardware.

The physical device that realizes a finite state machine is called a *sequential network*. The relation between a finite state machine and a sequential network that realizes it is conceptually analogous to that between a switching function and a switching network that implements it (see Note 5.1).

The sequential networks we shall consider are called *synchronous* because transitions between states occur synchronously with a clock's marking of the instants of separation between consecutive time units. Omitting the engineering details, we note that synchronous operation means that internal states are stable between successive clock beats ("clock pulses"). Networks for which internal changes are triggered by input changes alone, without clock synchronization, are called *asynchronous,* and their study presents several additional complications. An important property of synchronous sequential networks is that the "next state" $\delta(s_i, i_k)$ is an arbitrary member of the state set S; that is, the finite state machines so far considered are realizable by synchronous sequential networks. Henceforth, we shall omit the word "synchronous" when referring to sequential networks.

The crucial point in the design of a sequential network is the physical representation of the elements of the sets $I \equiv \{i_1, \ldots, i_m\}$, $S \equiv \{s_1, \ldots, s_n\}$, and $Z \equiv \{z_1, \ldots, z_r\}$. Each element (whether an input, a state, or an output) is represented by the physical state of a set of wires (*lines*). Such a representation is called a *coding* of the element, and the lines are usually susceptible of being in one of two conditions (*binary* lines).

In most cases the codings of the input and output symbols are not within the designer's control; they are supplied by the environment with which the network is to interact. The coding of the internal states, however, is always of the designer's choice. Physically, the coding of the network state is stored in binary storage devices, called *flip-flops.* Various types of flip-flops exist. For the sake of simplicity, we shall restrict ourselves to delay flip-flops (*D*-flip-flops, illustrated in figure

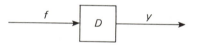

Figure 5.10.3. A delay flip-flop

5.10.3), which have one binary input line, f, and one binary output line, y. Hereafter we shall denote by $w^{(t)}$ and $w^{(t+1)}$ the conditions of a binary line w at the present time unit and at the next time unit, respectively. With this notation, a D-flip-flop is governed by the rule

$$y^{(t+1)} = f^{(t)}; \qquad\qquad (5.10.1)$$

that is, during the next time unit y will have the condition that f has during the present time unit.

Assume now that the input set has been encoded with p binary lines $x_1, x_2, \ldots, x_p$; that is, there is a function $\psi : I \to \{0,1\}^p$. Assume also that q D-flip-flops $D_1, D_2, \ldots, D_q$ are sufficient for encoding the states (this means that $2^q \geqslant n$, since there are n states and 2^q different conditions for the set of flip-flops). The task of the designer is reduced to finding a function $\varphi : S \to \{0,1\}^q$ which assigns a code to each state $s_i \in S$ (*state assignment*). Once a state assignment has been made, the rest is routine design of switching networks, as we shall now see.

For simplicity, we shall consider only the implementation of the transition function $\delta : S \times I \to S$. Let f_j and y_j be input and output lines, respectively, of flip-flop D_j. Then $\varphi : S \to \{0,1\}^q$ is equivalent to q functions $y_j : S \to \{0,1\}$.

Since $\delta(s_i, i_k)$ is the next state, given that s_i and i_k are the present state and input, we have, by (5.10.1),

$$y_j^{(t+1)}[\delta(s_i, i_k)] = f_j^{(t)}[\delta(s_i, i_k)], \qquad j = 1, 2, \ldots, q;$$

that is, f_j is a function from $\psi(I) \times \varphi(S)$ to $\{0, 1\}$, where $\psi(I)$ is the set of the binary codings of the input elements and $\varphi(S)$ is the set of the binary codings of the states. Note that if, for example, the number n of states is less than 2^q, we have $\varphi(S) \subset \{0, 1\}^q$; that is, not all sequences of q binary symbols are needed to code the states. Thus the function f_j is, in general, a partial function $\{0, 1\}^{p+q} \to \{0, 1\}$ (cf. Section 1.3). It can always be transformed into a total function f_j^* by arbitrarily assigning the function values for the unused binary sequences. In this manner we obtain q functions:

$$f_j^* : \{0, 1\}^{p+q} \to \{0, 1\}, \qquad j = 1, 2, \ldots, q.$$

Thus f_j^* is a switching function, and by theorem 5.5.1, we may view it as a boolean function $f(x_1, \ldots, x_p; y_1, \ldots, y_q)$ of the variables $x_1, \ldots, x_p, y_1, \ldots, y_q$. The implementation of the transition function reduces to the design of q switching networks (cf. Note 5.1), and the sequential network has the general structure shown in figure 5.10.4.

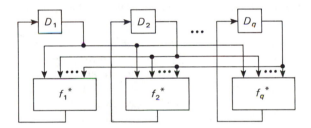

Figure 5.10.4. Implementation of the transition function of a sequential network

Example 5.10.1. Let the transition function of a finite state machine $\mathcal{M}$ be described by the table in figure 5.10.5(a). Since there are three states, two flip-flops are sufficient for coding them. Let their outputs be y_2 and y_1. If we make the

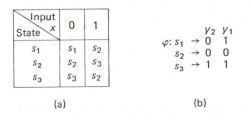

Input State $\backslash x$	0	1
s_1	s_1	s_2
s_2	s_2	s_3
s_3	s_3	s_2

	y_2	y_1
$\varphi: s_1 \to$	0	1
$s_2 \to$	0	0
$s_3 \to$	1	1

(a)   (b)

Figure 5.10.5. Transition function and state assignment of a finite state machine $\mathcal{M}$

state assignment φ shown in figure 5.10.5(b), we can replace the uncoded table of figure 5.10.5(a) with the coded table shown in figure 5.10.6(a). From this table, expressing the functions f_2 and f_1 in the familiar form shown in figure 5.10.6(b) is straightforward. Note that functions are not specified for either $(x, y_2, y_1) = (0, 1, 0)$ or $(x, y_2, y_1) = (1, 1, 0)$; these specifications can be made arbitrarily to obtain two switching functions.

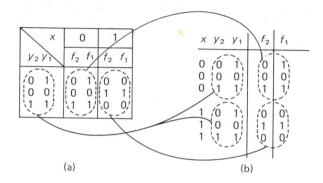

(a)   (b)

Figure 5.10.6. The transition function after the state assignment

As the preceding discussion shows, the complexity of the sequential network depends critically on the state assignment. Unfortunately, there is no known exact procedure, short of exhaustion, which results in an assignment having reasonably defined minimum complexity. Therefore one must usually resort to some heuristic method for finding good state assignments, since the number of possible assignments, even in simple cases, is so large that exhaustion is completely impractical.

References

F. J. Hill and G. R. Peterson, *Switching Theory and Logical Design*, Wiley, New York, 1968.
Z. Kohavi, *Switching and Finite Automata Theory*, McGraw-Hill, New York, 1970.

Note 5.3. An Independent Set of Axioms for Boolean Algebras

The manipulative rules of boolean algebra or, alternatively, the statement of definition 5.4.1 represents an adequate repertoire for a calculus of boolean algebras. When we say "adequate," we mean that they are sufficiently simple to be memorized and yet sufficiently powerful to allow efficient manipulation of boolean expressions. Clearly, one may use the given identities (5.4.1) through (5.4.11) to synthesize more complex identities that may be useful in calculation; for example, we could prove the identity, for $x, y, z \in \mathscr{B}$,

$$xy + \bar{x}z + yz = xy + \bar{x}z.$$

As a rule of thumb, the more complex an identity, the more powerful it is and the more difficult to memorize; and clearly there is no ceiling to the number of valid identities, involving an increasing number of arguments, which can be obtained.

These considerations naturally pose the reverse question. How many of the given identities are deducible, that is, are indeed theorems provable on the basis of an irreducible or independent subset of them? An irreducible set of properties will have to be considered as a set of axioms of boolean algebras. The identification of one such subset is the objective of this note.

As a general pattern of discussion, we shall remove the redundant properties one at a time from a "surviving set" of properties, using only the remaining properties as the basis for the deduction of the property being removed.

We begin by showing that identities (5.4.2), (5.4.5), (5.4.6), and (5.4.9) imply that, for $a \in \mathscr{B}$, the complement $\bar{a}$ of a is unique. Indeed, assume that a has two complements, $\bar{a}_1$ and $\bar{a}_2$. We then have the following.

$$
\begin{array}{lll}
\bar{a}_1 = 1 \cdot \bar{a}_1 & \bar{a}_2 = 1 \cdot \bar{a}_2 & \text{by (5.4.6)} \\
\quad = (\bar{a}_2 + a)\,\bar{a}_1 & \quad = (\bar{a}_1 + a)\,\bar{a}_2 & \text{by (5.4.9)} \\
\quad = \bar{a}_2\,\bar{a}_1 + a\bar{a}_1 & \quad = \bar{a}_1\,\bar{a}_2 + a\bar{a}_2 & \text{by (5.4.5)} \\
\quad = \bar{a}_2\,\bar{a}_1 & \quad = \bar{a}_1\,\bar{a}_2 & \text{by (5.4.9)}
\end{array}
$$

That is, from $\bar{a}_1 = \bar{a}_2\bar{a}_1 = \bar{a}_1\bar{a}_2 = \bar{a}_2$ [by (5.4.2)] we have $a_1 = a_2$; in other words, the complement is unique.

With this result we can prove the following lemma.

Lemma 5.10.1. De Morgan's laws (5.4.11) are deducible from distributivity, commutativity, associativity, complementarity, idempotency, and the properties of the bounds.

Proof. We consider the two expressions $(a + b)$ and $\bar{a}\bar{b}$, and we form their disjunction.

$$
\begin{array}{ll}
(a + b) + \bar{a}\bar{b} = [(a + b) + \bar{a}][(a + b) + \bar{b}] & \text{by (5.4.5)} \\
\quad = [(a + \bar{a}) + b][a + (b + \bar{b})] & \text{by (5.4.2) and (5.4.3)} \\
\quad = (1 + b)(a + 1) & \text{by (5.4.9)} \\
\quad = 1 \cdot 1 & \text{by (5.4.7)} \\
\quad = 1 & \text{by (5.4.1)}
\end{array}
$$

This shows that the disjunction of $(a + b)$ and $\bar{a}\bar{b}$ is the bound 1. We now form the conjunction of $(a + b)$ and $\bar{a}\bar{b}$.

$$
\begin{array}{ll}
(a + b)\,\bar{a}\bar{b} = a\bar{a}\bar{b} + b\bar{a}\bar{b} & \text{by (5.4.5)} \\
\quad = (a\bar{a})\,\bar{b} + (b\bar{b})\,\bar{a} & \text{by (5.4.2) and (5.4.3)} \\
\quad = 0 \cdot \bar{b} + 0 \cdot \bar{a} & \text{by (5.4.9)} \\
\quad = 0 + 0 & \text{by (5.4.7)} \\
\quad = 0 & \text{by (5.4.1)}
\end{array}
$$

We obtain that $(a + b)$ satisfies the requirement of the complement of $\bar{a}\bar{b}$ (identities 5.4.9), and because the complement is unique,

$$
\overline{a + b} = \bar{a}\bar{b}.
$$

Similarly, we can prove the dependence of the dual law $\overline{ab} = \bar{a} + \bar{b}$. ‖

Lemma 5.10.2. Associativity (5.4.3) is deducible from distributivity and absorption.

Proof. As usual, we limit our proof to one of the two dual associativity laws. Consider the conjunction of $[(a + b) + c]$ and $[a + (b + c)]$.

$$
\begin{aligned}
[(a + b) + c][a + (b + c)] &= [(a + b) + c]\, a + [(a + b) + c](b + c) \quad &\text{by (5.4.5)} \\
&= [(a + b)\, a + ca] + \{[(a + b) + c]\, b \\
&\quad + [(a + b) + c]\, c\} \quad &\text{by (5.4.5)} \\
&= [a + ca] + \{[(a + b)\, b + cb] + c\} \quad &\text{by (5.4.4)} \\
&= a + \{[b + cb] + c\} \quad &\text{by (5.4.4)} \\
&= a + (b + c) \quad &\text{by (5.4.4)}
\end{aligned}
$$

The same conjunction can also be handled in the following way.

$$
\begin{aligned}
[(a + b) + c][a + (b + c)] &= (a + b)[a + (b + c)] + c[a + (b + c)] \quad &\text{by (5.4.5)} \\
&= \{a[a + (b + c)] + b[a + (b + c)]\} \\
&\quad + [ca + c(b + c)] \quad &\text{by (5.4.5)} \\
&= \{a + [ba + b(b + c)]\} + [ca + c] \quad &\text{by (5.4.4)} \\
&= \{a + [ba + b]\} + c \quad &\text{by (5.4.4)} \\
&= (a + b) + c \quad &\text{by (5.4.4)}
\end{aligned}
$$

Thus, by the transitive property of equality, we conclude that $(a + b) + c = a + (b + c)$.$\|$

Lemma 5.10.3. Involution (5.4.10) is deducible from complementarity and the uniqueness of the complement.

Proof. Let $\bar{a} \in \mathscr{B}_n$ and let $b = \bar{\bar{a}}$ (the complement of $\bar{a}$). Then we have, by definition of complement (5.4.9).

$$\bar{a} + b = 1, \qquad \bar{a}b = 0;$$

but we know that a satisfies the same identities

$$\bar{a} + a = 1, \qquad \bar{a}a = 0,$$

whence, by the uniqueness of the complement, $a = b = \bar{\bar{a}}$.$\|$

Lemma 5.10.4. Absorption (5.4.4) is deducible from distributivity and the properties of the universal bounds.

Proof. Consider the expression $a + ab$.

$$
\begin{aligned}
a + ab &= a \cdot 1 + ab \quad &\text{by (5.4.6)} \\
&= a(1 + b) \quad &\text{by (5.4.5)} \\
&= a \cdot 1 \quad &\text{by (5.4.7)} \\
&= a \quad &\text{by (5.4.6)}\|
\end{aligned}
$$

Lemma 5.10.5. The universal bound properties (5.4.7) $a0 = 0$ and $a + 1 = 1$ are deducible from distributivity, complementarity, and the universal bound properties $a + 0 = a$, $a \cdot 1 = a$ (5.4.6).

Proof. We need to prove only $a0 = 0$; the other identity will follow by duality.

$$
\begin{aligned}
a0 &= 0 + a0 && \text{by (5.4.6)}\\
&= a\bar{a} + a0 && \text{by (5.4.9)}\\
&= a(\bar{a} + 0) && \text{by (5.4.5)}\\
&= a\bar{a} && \text{by (5.4.6)}\\
&= 0 && \text{by (5.4.9)}\|
\end{aligned}
$$

Lemma 5.10.6. Idempotency (5.4.1) is deducible from distributivity, complementarity, and the universal bound properties (5.4.6).

Proof. Consider the expression $a + a$.

$$
\begin{aligned}
a + a &= (a + a) \cdot 1 && \text{by (5.4.6)}\\
&= (a + a)(a + \bar{a}) && \text{by (5.4.9)}\\
&= a + a\bar{a} && \text{by (5.4.5)}\\
&= a + 0 && \text{by (5.4.9)}\\
&= a && \text{by (5.4.6)}\|
\end{aligned}
$$

At this point we note that we have successively eliminated six of the identities (5.4.1) through (5.4.11), which form the manipulative repertoire of the calculus of boolean algebras. These six identities are in fact formally deducible from the five surviving ones. Thus we have obtained what can be considered an irredundant axiomatic definition of a boolean algebra.

Definition 5.10.1. (Alternative of definitions 5.2.3 and 5.4.1.) A boolean algebra $\mathscr{B}_n$ is a set S of at least two elements with the following properties.

1. $\mathscr{B}_n$ has two binary operations, $+$ and $\cdot$, which are commutative and mutually distributive.
2. $\mathscr{B}_n$ has a unary operation, which obeys complementarity.
3. $\mathscr{B}_n$ contains two mutually complementary bounds, 0 and 1, such that, for $a \in S$, $a + 0 = a \cdot 1 = a$.

In summary, we have given an example of eliminating redundant axioms from a set of axioms of a class of formal systems (boolean algebras). There is indeed considerable logical interest in determining the "least" number of independent statements which completely define a given structure. The lemmas proved above, though slightly tedious, constitute an interesting example of the problem. The "surviving" five dual statements, (5.4.2), (5.4.5), (5.4.6), (5.4.8), (5.4.9) may be taken as a *set of axioms* of boolean algebras. They are usually referred to as the "Huntington postulates." In order to show that the Huntington postulates are indeed independent, we would need to construct mathematical models that would satisfy only four of the five axioms. That task, however, is beyond the scope of this book.

CHAPTER 6

A TASTE OF COMBINATORICS

6.1 INTRODUCTION

Weekly sports lotteries, popular in several countries, distribute fabulous sums of money to very small numbers of winners on the basis of the outcomes of certain combinations of two-team games. A lottery of this type, which we are going to describe, is best suited to low-scoring games, such as soccer or water polo, for reasons which will be immediately apparent. The individual bet consists of the completion of a form containing a list of two-team games (for example, 12 major-league soccer games), each presented as an ordered pair of teams, as suggested in figure 6.1.1.

Figure 6.1.1

The bettor places one of three symbols (1, 2, ×) in each of the 12 boxes of the forecast column, according to the following convention: 1 to denote a team 1 victory, 2 to denote a team 2 victory, and × to denote a tie (very likely to occur in low-scoring games). The bettor may complete as many such forms as he wants, he merely pays a fixed fee for the validation of each. He wins if the outcomes perfectly match his forecasts, and his payoff is a fixed fraction of the lottery revenue divided by the number of winners. Clearly, he increases his chances by increasing the number of his bets. How many bets will he have to enter in order to be mathematically sure of being a winner? To count all the distinct possible bets is very simple. Each of the 12 entries can be chosen independently of the others, in three ways. Thus the top entry can be chosen in 3 ways, the top two entries in 3^2 ways, ..., the 12 entries in $3^{12} = 531,441$ ways, a number large enough to discourage exhaustive betting (since only a fraction of the total bet revenue is distributed to the winners).

The example above, in its extreme simplicity, contains the essential ingredients of combinatorial analysis, or combinatorics, as it is frequently known. There are three sets: a set C, consisting of the distinct boxes of the forecast column; a set S, consisting of the three symbols $\{1, 2, \times\}$; and a set F of mappings $f: C \to S$, each expressing a forecast for the outcomes of the 12 games. Usually the set C presents itself in the form of a geometric configuration of its positions (such as a column with 12 boxes). Therefore we refer to it, for convenience, as the *configuration* set and to the elements of C as *cells*. The set S, on the other hand, is usually a collection of objects to be placed in the cells of the given configuration; for this reason, we call it the *object* set. In the preceding example, the mapping f is arbitrary; that is, the set F is the collection of *all* functions $f: C \to S$. Normally, however, there are some additional properties that f must satisfy beside being a mapping from C to S. Clearly, since f is a function, each cell contains *exactly one* object!

The main objective of combinatorics is the *enumeration* of the members of F, where enumeration must be thought of in its double connotation of "listing" and "counting." In the first sense, we refer to exhibiting the actual members of F by providing some sort of description thereof. In some cases, the generation of

1	3	2	4
2	4	1	3
3	1	4	2
4	2	3	1

$C =$

Figure 6.1.2

members of F may be trivial (as in the preceding example), but great care must be used to ensure that *exactly all* the members of F are generated (that is, no member is omitted and none is generated more than once). In other cases, the generation of even a single member of F may be very difficult, and listing corresponds to a proof of "existence." For example, it is by no means a trivial task to map the 16 cells of the square array configuration C illustrated in figure 6.1.2 to the set $S \equiv \{1, 2, 3, 4\}$ so that no member of S appears more than once in each row and each column of C. In the second sense (counting), we refer to calculating the cardinality of F (census problems). That is why combinatorics is referred to as the "art of counting the ways."

6.2 DISTRIBUTIONS, PERMUTATIONS, COMBINATIONS

Given the sets $C \equiv \{c_1, c_2, \ldots, c_n\}$ and $S \equiv \{s_1, s_2, \ldots, s_m\}$, we begin by considering the case in which $f: C \to S$ is an arbitrary function. Each element $s_i \in S$ will have as its pre-image $f^{-1}(s_i)$, a subset of the elements of C. To visualize it, we may think of having available a number of replicas of each member s_i of S large enough to be able to place s_i in each cell of $f^{-1}(s_i)$. It follows that any such mapping f may be interpreted as a distribution of m distinct types of objects (each type being an element of S) into n distinct cells (the elements of C), with one object per cell. This was, in fact, the classical way of describing the type of function just introduced, and we retain the term "distribution" for its obvious mnemonic convenience.

Definition 6.2.1. A *distribution* of m distinct types of objects into n distinct cells is an arbitrary function $f: C \to S$, where $C \equiv \{c_1, \ldots, c_n\}$ and $S \equiv \{s_1, \ldots, s_m\}$.

We now determine the number $D(n, m)$ of distributions of m distinct objects into n distinct cells. Since for each cell $c_j \in C$, cell c_j will contain object $f(c_j) \in S$ independently of the distribution of objects in the other cells of C, each $f(c_j)$ can be chosen in as many ways as there are elements in S. Thus the object $f(c_1)$ can be chosen in m ways. If we assume that the sequence $(f(c_1), f(c_2), \ldots, f(c_{j-1}))$ can be chosen in m^{j-1} ways, for each such choice there are m choices of the sequences $(f(c_1), f(c_2), \ldots, f(c_{j-1}), f(c_j))$ for a total of $m^{j-1} \, m = m^j$ choices. Therefore, by induction on j, we have shown that

$$D(n, m) = m^n. \tag{6.2.1}$$

Distributions are frequently encountered in ordinary experience, as the following examples indicate.

Example 6.2.1. Consider in how many ways we can distribute the arabic numerals $S \equiv \{0, 1, 2, \ldots, 9\}$ into three decimal positions $C \equiv \{\text{hundreds, tens, units}\}$. If we follow the usual convention of representing the configuration C as

hundreds	tens	units

and ignore the zeros to the left of the leftmost nonzero digit (insignificant zeros), then the distributions we obtain are representations of exactly all the integers from 0 through 999. By formula (6.2.1) they are 1000 in number, $D(3, 10) = 10^3$. Similarly, the distributions of the binary symbols $\{0, 1\}$ into n positions are 2^n in number.

Example 6.2.2. In Section 5.5 we obtained the number of the distinct switching functions of n arguments using a specialized argument. That number will now be obtained again by using the general notions just developed.

A switching function $\varphi:\{0,1\}^n \to \{0,1\}$ is specified when for every binary n-tuple $(a_1, a_2, \ldots, a_n)$—that is, $a_i \in \{0,1\}$—the value of $\varphi(a_1, a_2, \ldots, a_n)$ is selected in $\{0,1\}$. As indicated in example 6.2.1, there are 2^n binary n-tuples. We now form a linear array C, each position of which corresponds to a distinct n-tuple, and φ is specified by distributing the two symbols $\{0,1\}$ into the 2^n positions of C. Clearly this can be done in 2^{2^n} ways—the number of the switching functions of n arguments.

We now start placing some restrictions on the nature of the function $f: C \to S$. Suppose that we require that f be one-to-one, i.e., that $c_i \in C$ be the only pre-image of $f(c_i) \in S$ (definition 1.3.3). Then, since each element of C has a distinct image in S, we must have $n = \#(C) \leqslant \#(S) = m$. We may interpret a function of this class as the selection of n distinct objects $s_{j_1}, s_{j_2}, \ldots, s_{j_n}$ of S (i.e., a subset of S) and their distribution into the n cells of C; clearly, among the objects $s_{j_1}, s_{j_2}, \ldots, s_{j_n}$ there is no repetition, a fact which explains the classical denomination of the mapping just described.

Definition 6.2.2. For $C \equiv \{c_1, \ldots, c_n\}$ and $S \equiv \{s_1, \ldots, s_m\}$, $m \geqslant n$, a *distribution without repetition* of m distinct objects into n distinct cells is a one-to-one mapping $f: C \to S$.

The number $R(n,m)$ of the distributions without repetition of m objects into n cells is readily obtained. In this case the choices of the objects $f(c_1), \ldots, f(c_n)$ are no longer independent. However, object $f(c_1)$ can be chosen in as many ways as there are elements in S, that is, in m ways. Subsequently, all the elements of S except $f(c_1)$ are legitimate choices for $f(c_2)$; that is, there are $(m-1)$ such choices. It follows that the pair $f(c_1), f(c_2)$ can be chosen in $m(m-1)$ ways. A straightforward argument by induction establishes

$$R(n,m) = \underbrace{m(m-1) \cdots (m-n+2)(m-n+1)}_{n \text{ factors}}. \qquad (6.2.2)$$

Introducing, for a positive integer r, the notation $r! = r(r-1)(r-2) \cdots 3 \cdot 2 \cdot 1$ (referred to as "r factorial"), we obtain

$$R(n,m) = m(m-1) \cdots (m-n+1)\frac{(m-n) \cdots 2 \cdot 1}{(m-n) \cdots 2 \cdot 1} = \frac{m!}{(m-n)!}. \qquad (6.2.3)$$

When we introduce the further specialization that $f: C \to S$ is one-to-one and that $m = n$, f becomes also onto (definition 1.3.2); i.e., by theorem 1.3.1, f is invertible. Thus, if we compare an arbitrary invertible function $f: C \to S$ with the invertible function $e: C \to S$ such that $e(c_i) = s_i$ (referred to here as the identity function), we recognize that f realizes a rearrangement of the configuration realized by e; that is, f is a permutation. Since the mapping e is taken as reference, frequently S is identified with C and a permutation is described as an invertible function $f: S \to S$.

Definition 6.2.3. For $C \equiv \{c_1, \ldots, c_n\}$ and $S \equiv \{s_1, \ldots, s_n\}$ a *permutation* of the n objects $s_1, s_2, \ldots, s_n$ is an invertible function $f: C \to S$ (or equivalently, $f: S \to S$).

The number $P(n)$ of the permutations of n objects is easily obtained from (6.2.2). In fact, the argument used to characterize the permutations shows that $P(n) = R(n,n)$; that is

$$P(n) = n \cdot (n-1) \cdots 2 \cdot 1 = n!. \tag{6.2.4}$$

Now, substituting n for m in formula (6.2.3), we have $R(n,n) = n!/(n-n)! = n!/0!$. Comparing this result with (6.2.4), we conclude that the conventional value of $0!$ is 1. Thus the factorial is defined for all *nonnegative* integers.

Example 6.2.3. Consider the problem of determining in how many distinct ways m persons can be seated at a round table. We know that the total number of distinct arrangements of m persons in linear sequence is $P(m) = m!$. However, since only the relative position is important in the seating at a round table, rotating a fixed seating arrangement would not give a new arrangement. Hence there are

$$\frac{P(m)}{m} = (m-1)!$$

distinct ways to seat m persons at a round table.

Finally we consider a very important class of functions $f: C \to S$. We assume that $\#(S) \geqslant \#(C)$ and that f is one-to-one, but we consider two such functions f_1 and f_2 to be indistinguishable if their ranges coincide, i.e., if $f_1(C) = f_2(C)$. Clearly, this corresponds to defining an equivalence relation ε on F, the set of all functions from C to S. We interpret an equivalence class of this relation ε as the choice of the codomain $f(C)$ of a one-to-one function $f: C \to S$, that is, the selection of a subset of S of cardinality n [since both C and $f(C)$ contain n elements]. Traditionally, such selection has been referred to as a "combination" of the members of $f(C)$; this is reflected in the following definition.

Definition 6.2.4. A *combination* of m objects taken n at a time $(m \geqslant n)$—or equivalently, of n out of m objects—is a subset A of cardinality n of the set $S \equiv \{s_1, \ldots, s_m\}$. [For $C \equiv \{c_1, \ldots, c_n\}$, it can also be viewed as the set of all one-to-one functions $f: C \to S$, for which $f(C) = A$.]

The number $C(n,m)$ of the combinations of n out of m objects is now calculated. As in Section 3.3, we represent a function $f: C \to S$ as two rows

$$f = \begin{pmatrix} c_1 & c_2 & \cdots & c_n \\ f(c_1) & f(c_2) & \cdots & f(c_n) \end{pmatrix}$$

or, more simply, since the first row is the same for all such mappings, as a single sequence $(f(c_1), f(c_2), \ldots, f(c_n))$. Suppose now that we compile a vertical list of all the distinct subsets of S which contain n elements (each such subset corresponds to a combination of n out of m objects, according to definition 6.2.4.). For each of these sequences we list—horizontally to the right—all the $n!$ permutations of its elements. Since each row contains the same number $n!$ of sequences, we have generated a rectangular array. This is illustrated below, where $m = 4$ and $n = 3$.

$$\{a,b,c\}:abc\ acb\ bac\ bca\ cab\ cba$$
$$\{a,b,d\}:abd\ adb\ bad\ bda\ dab\ dba$$
$$\{a,c,d\}:acd\ adc\ cad\ cda\ dac\ dca$$
$$\{b,c,d\}:bcd\ bdc\ cbd\ cdb\ dbc\ dcb$$

We claim that the set of sequences thus obtained comprises all the distinct distributions without repetition of n elements of S. In fact, an arbitrary sequence $(s_{j_1}, s_{j_2}, \ldots, s_{j_n})$ appears exactly once in the constructed rectangular array, because there is only one row whose leftmost sequence contains (a permutation of) $s_{j_1}, \ldots, s_{j_n}$, and within this row, there is only one sequence which coincides with the given one (since all permutations are distinct). We conclude that the array contains $R(n,m)$ entries. Being rectangular and consisting of $n!$ columns, it contains

$$C(n,m) = \frac{R(n,m)}{n!} = \frac{m!}{(m-n)!\,n!}$$

rows [use has been made of formula (6.2.3)]. This number, $C(n,m)$, is the desired cardinality. It is customary to use the notation

$$C(n,m) = \frac{m!}{(m-n)!\,n!} = \binom{m}{n}, \tag{6.2.5}$$

where the symbol $\binom{m}{n}$ is the *binomial number* or *binomial coefficient* and is read "n out of m," "m over n," "m choose n."

We conclude this section with applications of the preceding concepts.

Example 6.2.4. We consider the number of distinct bridge and poker hands, where "hand" means a set of cards that may be dealt to an individual player. Since there are 52 cards in total, the number of different bridge hands is equal to the number of possible selections of subsets of thirteen cards and hence is equal to

$$\binom{52}{13} = 635{,}013{,}559{,}600.$$

Similarly, there are $\binom{52}{5} = 2{,}598{,}960$ hands at poker.

Example 6.2.5. A universal logical module for n variables is a switching circuit with m input terminals, which realizes a switching function $U(z_1, z_2, \ldots, z_m)$ of the m variables $z_1, z_2, \ldots, z_m$ with the following property. For *any* given switching function $f(x_1, x_2, \ldots, x_n)$ of the n binary variables $x_1, x_2, \ldots, x_n$ it is possible to find a mapping $\varphi:\{z_1, z_2, \ldots, z_m\} \to \{0, 1; x_1, x_2, \ldots, x_n; \bar{x}_1, \bar{x}_2, \ldots, \bar{x}_n\}$—where 0 and 1 are the two constants of switching algebra (Section 5.8) and $\bar{x}_j$ is the complement of x_j—so that $U(\varphi(z_1), \varphi(z_2), \ldots, \varphi(z_m)) = f(x_1, x_2, \ldots, x_n)$. See figure 6.2.1, where, for example, $\varphi(z_1) = 1$, $\varphi(z_2) = x_n$, $\varphi(z_m) = \bar{x}_1$, etc. Concisely we

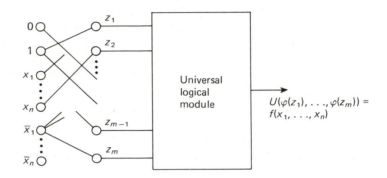

Figure 6.2.1. A universal logical module and a connection of its input terminals

say, "U realizes f." We now want to determine the lowest value of m such that there may exist a universal logical module for n variables having m input terminals. Clearly, there must exist at least as many distinct connections of the input terminals of the module (more precisely, distinct mappings φ) as there are switching functions to be realized. The number v of mappings is obtained by noting that each terminal z_1 can be connected to one of $(2n + 2)$ terminals, that is, in $(2n + 2)$ ways. Thus v is the number of the distributions of the $(2n + 2)$ symbols $\{0, 1, x_1, \ldots, x_n, \bar{x}_1, \ldots, \bar{x}_n\}$ into m positions; that is, $v = (2n + 2)^m$. Since there are 2^{2^n} distinct switching functions of n arguments (example 6.2.2), we have the inequality

$$(2n + 2)^m \geqslant 2^{2^n}$$

or, equivalently,

$$m \geqslant \frac{2^n}{\log(2n + 2)}.$$

n	3	4	5	6
m	4	5	9	17

It sometimes helps our understanding if we reverse the convention about the meanings of sets C and S. As in the following example, it is sometimes useful to interpret C as a set of objects and S as a set of cells or boxes. In this interpretation, each box may contain more than one object.

Example 6.2.6. We want to determine the number v of the integers between 0 and 9999 in which the four digits of their decimal representations form a nondecreasing sequence when read from right to left (e.g., 7442 is one such integer, but 2411 is not). Note that, if we were to require that the four digits form an increasing sequence, all four digits would have to be distinct; that is, v would be the number of combinations of ten digits (the arabic numerals 0, ..., 9) taken four at a time. However, since we are allowing digits to be repeated, v will be the number of "combinations with repetitions of ten digits taken four at a time."

To determine v, we proceed as follows: We have a set $C \equiv \{c_0, c_1, c_2, c_3\}$ of four objects (the digits of a four-digit integer N—c_0 for the units, c_1 for the tens, etc.) and a set $S \equiv \{0, 1, ..., 9\}$ of ten boxes (a box for each of the ten arabic numerals). Each function $f: C \to S$ identifies an integer N in the interval $[0, 9999]$. For example, the function described by the object-box assignment of figure

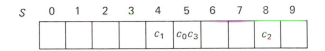

Figure 6.2.2. An illustration of $f: C \to S$, identifying the integer $N = 5845$

6.2.2 identifies the integer N for which $c_1 = 4$, $c_0 = c_3 = 5$ and $c_2 = 8$; that is, $N = 5845$. Figure 6.2.2 may also be viewed as a sequence of 13 symbols: the four symbols c_0, c_1, c_2, c_3 and nine repetitions of the symbol "|", indicating the separations between adjacent boxes. In other words, figure 6.2.2 is equivalent to the sequence $||||c_1|c_0c_3|||c_2|$. The number of distinct 13-symbol sequences is $13!/9!$ since there are $13!$ permutations of 13 symbols, but the 9 separation symbols "|" are indistinguishable. To any such 13-symbol sequence there corresponds one and only one standard sequence in which c_0, c_1, c_2, c_3 appear in this order from left to right; for example, $||||c_0|c_1c_2|||c_3|$ is the standard sequence corresponding to $||||c_1|c_0c_3|||c_2|$. Note that the set of any of the standard sequences is in a one-to-one correspondence with the set of four-digit numbers whose digits form a nondecreasing sequence; since each standard 13-symbol standard sequence is obtained by choosing the 4 positions for the sequence c_0, c_1, c_2, c_3 among 13 positions, v is given by

$$v = c(4, 13) = \binom{13}{4} = 715.$$

Generalizing the argument, one could show that the number of combinations with repetition of m objects taken n at a time is

$$\binom{n+m-1}{n}.$$

_____

EXERCISES 6.2

1. Answer the following questions.
   a) How many words of six letters can potentially be formed with the letters of the English alphabet? (By "words" we simply mean sequences of letters.)
   b) An $m \times n$ binary array has m rows and n columns, and its entries are either 0 or 1. How many $m \times n$ binary arrays can be formed?
   c) How many different outcomes can a throw of four dice have?
   d) How many different outcomes can we have by throwing a die k times?
   e) How many possible sets of initials can be formed if every person has one surname and at most two given names?

2. In how many ways can three distinct numbers be chosen from the set of numbers $1, 2, \ldots, 50$ such that their sum is divisible by 5?

3. A switching function $f(x_1, x_2, \ldots, x_n)$ is said to be *self-dual* if $f(x_1, x_2, \ldots, x_n) = \overline{f(\bar{x}_1, \bar{x}_2, \ldots, \bar{x}_n)}$ for any choice of the binary n-tuple $(x_1, x_2, \ldots, x_n)$.
   a) How many are the self-dual functions of 4 variables?
   b) How many are the self-dual functions of n variables?

4. a) How many binary sequences of m digits contain an even number of 1's?
   b) Imagine writing the integers from 1 through $(10^{10} - 1)$ in decimal form. How many of these expansions contain the digit 0?
   c) How many sequences of 10 letters can be formed from three a's, four b's, two c's, and one d?

5. How many binary operations are there on a set of four elements? How many of them are commutative?

6. Let S be a set of n objects such that k_1 objects are of the first kind, k_2 objects are of the second kind, $\ldots$, and k_p objects are of pth kind. Show that the number of distinct permutations of these n elements is equal to the number of distributions of these n objects into n distinct cells and is equal to

$$\frac{n!}{k_1! k_2! \cdots k_p!}$$

7. Prove the *binomial theorem* embodying the following algebraic identity.

$$(x + y)^n = \binom{n}{0}x^n + \binom{n}{1}x^{n-1}y + \cdots + \binom{n}{n}y^n.$$

8. Prove the following identities and inequalities.

a) $\dbinom{n}{r} = \dbinom{n-1}{r} + \dbinom{n-1}{r-1}$

b) $\dbinom{n}{r} = \dbinom{n}{n-r}$

c) $\dbinom{2n}{0} < \dbinom{2n}{1} < \cdots < \dbinom{2n}{n}$

d) $\dbinom{2n-1}{0} < \dbinom{2n-1}{1} < \cdots < \dbinom{2n-1}{n-1} = \dbinom{2n-1}{n}$

*9. Prove the following identity using combinatorial arguments.

$$\sum_{i=0}^{n} \binom{n}{i} = 2^n.$$

(*Hint*: See exercise 6.2.8.)

6.3 ENUMERATION BY RECURSION (RECURRENCE EQUATIONS)

In the previous chapters (for example, see Section 4.6) we mentioned certain enumeration problems and deferred their solution to a more advanced stage. The time has now come to reconsider those problems. Specifically, we consider the problems of enumerating the partitions of a set.

For convenience we recall (definition 1.5.7) that a partition of a set C is a collection of subsets of C such that the intersection of any two sets in the collection is $\varnothing$, and each $c \in C$ belongs to exactly one of these subsets. Assume that C is subdivided into m nonempty subsets $S_1, S_2, \ldots, S_m$ and denote by S their collection $\{S_1, S_2, \ldots, S_m\}$. Thus a partition is a function $f: C \to S$. (Here again it is preferable to interpret C as a set of objects and S as a set of cells or boxes.) Since each S_i is nonempty, the mapping f is onto (definition 1.3.2). We must also note that each subset of C is identified by the elements it contains and not by the specific label S_i attached to it. Thus each partition into m subsets will be described by as many distinct onto functions $f: C \to S$ as there are permutations of the m labels, that is, $m!$ times. Since we want to enumerate the distinct partitions into m subsets, we must divide the number of onto functions by $m!$. If we now try to calculate this number using arguments analogous to those used in the preceding section, we experience immediate difficulty. First, the cardinality n_i of each S_i must be chosen, subject to the conditions that $n_1 + n_2 + \cdots + n_m = n$ and $n_j > 0$ for every j (in other words, that S be a partition of C into m nonempty subsets). Second, we assign successively elements of C to $S_1, S_2, \ldots$; that is, the n_i elements of S_i must be chosen from those elements of C not yet assigned to $S_1, S_2, \ldots, S_{i-1}$. After obtaining the number of the onto functions for a fixed choice of

$n_1, n_2, \ldots, n_m$, we must sum over all possible such choices to obtain the desired result. The task is indeed quite involved.

However, there is an elegant way around this difficulty. Let $S(n,m)$ denote the number of the partitions of n elements into m nonempty subsets. We assume that we have been informed by a knowledgeable and friendly genie about the number of partitions of $(n-1)$ objects; that is, we know $S(n-1,1)$, $S(n-1,2), \ldots, S(n-1,n-1)$. If we now discover the mechanism by which we can "generate" the partitions of n objects by modifying those of $(n-1)$ objects, then the information given to us can be used to compute $S(n,m)$. Indeed, that mechanism is quite apparent. Each partition of n objects into m classes can be generated either

1) by adding the nth object to one of the classes of an arbitrary partition of $(n-1)$ objects into m classes, or

2) by adding a new class (consisting of the nth element) to an arbitrary partition of $(n-1)$ objects into $(m-1)$ classes,

and in no other way. Generation (1) can be effected in m ways (as many as there are classes) for each of the $S(n-1,m)$ partitions; generation (2) can be effected in one way for each of the $S(n-1,m-1)$ partitions. This is summarized by the equation

$$S(n,m) = mS(n-1,m) + S(n-1,m-1). \tag{6.3.1}$$

This equation exemplifies *recurrence equations* (or *recurrence relations*, as they are frequently called), which play a very important role in combinatorics. The reader has certainly recognized a rather familiar train of thought. We consider the n-tuple $\sum_n = (S(n,1), S(n,2), \ldots, S(n,n))$ as the nth term of a sequence $\sum_1, \sum_2, \ldots, \sum_n, \ldots$. Knowing the initial term (or terms) of this sequence and assuming known $\sum_{n-1}$, we have a procedure for computing $\sum_n$. Thus the method has all the basic features of induction (base and inductive step) (cf. Section 0.4).

Solving a recurrence equation to obtain a closed-form expression for the nth term of a sequence is beyond the scope of this text. However, a recurrence equation is an extremely valuable tool, since it allows the successive, straightforward computation of all the terms desired.

For example, let us return to consider the numbers $S(n,m)$. We note that $S(n,1) = 1$ and $S(n,n) = 1$ (that is, there is only one partition with a single class, and there is only one partition of n objects into n classes). Now we can proceed to calculate the terms $S(n,m)$, which we display in an array (figure 6.3.1) whose rows and columns are identified by the integers n and m, respectively. The rows of this array are the successive terms of the sequence $\sum_1, \sum_2, \ldots$; the 1's in the leftmost column are the terms $S(n,1) = 1$, whereas those on the diagonal are the terms $S(n,n) = 1$. As an example of the application of equation (6.3.1), the term $S(4,2)$ (encircled in figure 6.3.1) is given by $S(4,2) = 2S(3,2) + S(3,1)$; that is, from $S(3,1) = 1$ and $S(3,2) = 3$ we obtain $S(4,2) = 2 \times 3 + 1 = 7$. The illustrated

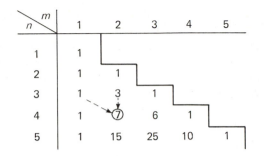

Figure 6.3.1. The array of $S(n,m)$ (Stirling triangle)

array is called the *Stirling triangle*, and the $S(n,m)$ are called *Stirling numbers of the second kind.*

As another example, consider the binomial coefficient $C(n,m)$ given in (6.2.5). This number can also be computed with the aid of the following recurrence relation.

$$C(n,m) = C(n-1, m-1) + C(n, m-1). \qquad (6.3.2)$$

Equation (6.3.2) can be proved by means of the following arguments. Let one of the m objects be marked as special. Then the number of ways of selecting n objects from m objects is equal to the sum of the number of ways of selecting n objects so that the special element is always included and the number of ways of selecting n objects so that the special object is always excluded. There are $C(n-1, m-1)$ ways to make the former selection and $C(n, m-1)$ ways to make the latter selection. The computation of the binomial coefficients $C(n,m)$ is illustrated in the array of figure 6.3.2, known as the *Pascal triangle*.

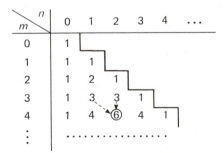

Figure 6.3.2. The array of $C(n,m)$ (Pascal triangle)

EXERCISES 6.3

1. The Pascal triangle (figure 6.3.2) is drawn as a circuit-free directed graph in figure 6.3.3. We refer to the unique vertex with no incoming arc as the root. Show that the binomial coefficient in each vertex corresponds to the number of paths from the root to this vertex.

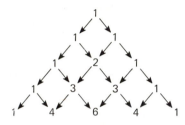

Figure 6.3.3. Pascal triangle

2. Compute the terms a_1, a_2, a_3, a_4, a_5, using the following recurrence relations.
   a) Fibonacci numbers

   $$a_0 = a_1 = 1$$

   $$a_n = a_{n-1} + a_{n-2}, \text{ for } n \geqslant 2$$

   b) Bernoulli numbers

   $$a_0 = 1$$

   $$a_n = \sum_{k=0}^{n} \binom{n}{k} a_{n-k}, \; n > 1$$

3. Define the following sequences of numbers, using recurrence relations.
   a) $1, 2, 2^2, 2^3, 2^4, \ldots, 2^n, \ldots$
   b) $0, 1^2, 2^2, 3^2, 4^2, \ldots, k^2, \ldots$
   c) $1, 1, 2, 3, 5, 8, 13, 21, 34, 55, \ldots$

*4. The *Stirling numbers of the first kind*, denoted by $S_1(m, n)$, are defined by the equation

$$\sum_{n=0}^{m} S_1(m, n) x^n = x(x - 1) \cdots (x - m + 1).$$

Show that $S_1(m, n)$ satisfies the recurrence relation (with $S_1(0, 0) = 1$, $S_1(k, 0) = S_1(0, k) = 0$ for $k > 0$)

$$S_1(m + 1, n) = S_1(m, n - 1) - m S_1(m, n).$$

*5. If n straight lines meet inside a circle and no three lines meet at one point, into how many regions is the circle divided by the lines?

6.4 THE CARDINALITY OF A UNION OF SETS (PRINCIPLE OF INCLUSION AND EXCLUSION)

Suppose now that we want to determine the number of integers from 1 through 85 which are not divisible either by 3 or by 4. A brute-force approach would be to list the 85 integers, delete those which are divisible by either of the two integers, and finally count the survivors. We prefer a more elegant approach.

To this end, we define the following sets: $A \equiv \{n|\ n$ an integer, $1 \leqslant n \leqslant 85\}$; $A_j \equiv \{n|n \in A$ and n is divisible by the integer $j\}$ for $j = 3,4$. The sets A_3 and A_4 are not disjoint, since, for example, the integer 12 belongs to both A_3 and A_4. Denoting by $\bar{A}_j$ the complement of A_j in A—that is, $\bar{A}_j \equiv \{n|\ n \in A$ and n is *not* divisible by $j\}$—we must determine $\#(\bar{A}_3 \cap \bar{A}_4)$—that is, the cardinality of the set of integers which are not divisible by 3 or 4. The situation is visualized in the Venn diagram of figure 6.4.1, where the domains A_3 and A_4

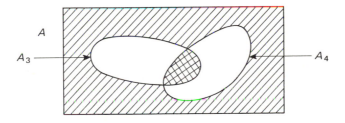

Figure 6.4.1. A Venn diagram

are indicated. We must determine the size of the cross-hatched domain. The sizes of the A_j's, that is, $\#(A_j)$, are readily obtained. With reference to A_3, the largest multiple of 3 contained in A is 84. Thus there are $84/3 = 28$ multiples of 3 in A; that is, $\#(A_3) = [85/3] = 28$, where the notation "$[x]$" means "integer part of x." Similarly, $\#(A_4) = [85/4] = 21$. Now, if $A_3 \cap A_4 = \varnothing$, we would simply subtract $(\#(A_3) + \#(A_4))$ from $\#(A)$ to obtain the result. By so doing in our case, however, we would subtract twice the number of the integers belonging to $A_3 \cap A_4$. Thus we must determine $\#(A_3 \cap A_4)$ and add it to $\#(A) - \#(A_3) - \#(A_4)$ to obtain the result $(\bar{A}_3 \cap \bar{A}_4)$. Since $\#(A_3 \cap A_4) = [85/12] = 7$, we conclude that $\#(\bar{A}_3 \cap \bar{A}_4) = 85 - 28 - 21 + 7 = 43$.

The preceding formula, by straightforward manipulation, can be rewritten as

$$\#(A) - \#(\bar{A}_3 \cap \bar{A}_4) = \#(A_3) + \#(A_4) - \#(A_3 \cap A_4), \qquad (6.4.1)$$

which has a very interesting interpretation. We first note that, for any set $B \subset A$, B and $\bar{B}$ form a partition of A. (Here A plays the role of the universal set, that is, of the universal upper bound of $\mathscr{P}(A)$, Section 5.1.) Therefore we have $\#(B) + \#(\bar{B}) = \#(A)$ or, equivalently, $\#(A) - \#(\bar{B}) = \#(B)$. The expression on the left side of (6.4.1) gives the cardinality of the set $\overline{A_3} \cap \overline{A_4}$, which, by dualization (theorem 5.2.3), coincides with the set $A_3 \cup A_4$. It follows that

$$\#(A_3 \cup A_4) = \#(A_3) + \#(A_4) - \#(A_3 \cap A_4); \qquad (6.4.2)$$

that is, we have reduced our problem to the calculation of the cardinality of a *union of sets*. This formula has an immediate interpretation in the Venn diagram of figure 6.4.1 : The cardinality of the union of A_3 and A_4 is equal to the sum of the cardinalities of A_3 and A_4 less the cardinality of their intersection.

Formula (6.4.2) can be readily generalized to an arbitrary number of subsets $A_1, A_2, \ldots, A_n$ of A. The following argument is an induction on the number n. For $n = 1$, $\#(A_1) = \#(A_1)$, trivially. Suppose now that the cardinality of the union of $(n-1)$ sets $A_1, A_2, \ldots, A_{n-1}$ is given by the formula

$$\#(A_1 \cup \cdots \cup A_{n-1})$$

$$= \sum_{j=1}^{n-1} \#(A_j) - \sum_{\substack{i, j=1 \\ i<j}}^{n-1} \#(A_j \cap A_i) + \cdots + (-1)^{n-2} \#(A_1 \cap \cdots \cap A_{n-1}).$$

We now apply this formula to the subsets $A_1, A_2, \ldots, A_{n-2}, A_{n-1} \cup A_n$ and obtain

$$\#(A_1 \cup \cdots \cup A_{n-2} \cup (A_{n-1} \cup A_n))$$

$$= \sum_{j=1}^{n-2} \#(A_j) + \#(A_{n-1} \cup A_n)$$

$$- \sum_{\substack{i, j=1 \\ i<j}}^{n-2} \#(A_j \cap A_i) - \sum_{j=1}^{n-2} \#((A_{n-1} \cup A_n) \cap A_j) + \cdots$$

$$+ (-1)^{n-2} \#(A_1 \cap \cdots \cap A_{n-2} \cap (A_{n-1} \cup A_n)).$$

We make the following remark:

1. By the associative property of set union, $A_1 \cup \cdots \cup (A_{n-1} \cup A_n) = A_1 \cup \cdots \cup A_{n-1} \cup A_n$.
2. By (6.4.2), $\#(A_{n-1} \cup A_n) = \#(A_{n-1}) + \#(A_n) - \#(A_{n-1} \cap A_n)$.
3. Similarly, since $(A_{n-1} \cup A_n) \cap A_j = (A_{n-1} \cap A_j) \cup (A_n \cap A_j)$, by distributivity and by virtue of (6.4.2), we have $\#((A_{n-1} \cup A_n) \cap A_j) = \#(A_{n-1} \cap A_j) + \#(A_n \cap A_j) - \#(A_{n-1} \cap A_n \cap A_j)$. The same arguments can be applied to all other terms, thus proving that

$$\#(A_1 \cup \cdots \cup A_n)$$

$$= \sum_{j=1}^{n} \#(A_j) - \sum_{\substack{i, j=1 \\ i<j}}^{n} \#(A_j \cap A_i) + \cdots + (-1)^{n-1} \#(A_1 \cap \cdots \cap A_n), \qquad (6.4.3)$$

which is known as the *principle of inclusion and exclusion* (actually a theorem). The cardinality of a union of sets can be expressed in terms of the cardinalities of the individual sets and of the intersections of two, three, ..., n sets at a time as stated by formula (6.4.3).

Example 6.4.1. Twelve boxes are distinguishable in the following way.

Two are painted red.
Three are painted white.
One is not painted.
One is painted black.
Two are painted red and white.
Three are painted white and black.

How many boxes are painted either white or red or are unpainted? Let $P_1, P_2 P_3, P_4$ be the properties that a box is painted red, white, black, and unpainted, respectively, and let A_i be the set of boxes having property P_i, $1 \leqslant i \leqslant 4$. Then by formula (6.4.3) we have

$$\#(A_1 \cup A_2 \cup A_4) = (\#A_1 + \#A_2 + \#A_4) - \#(A_1 \cap A_2) - \#(A_1 \cap A_4)$$
$$- \#(A_2 \cap A_4) + \#(A_1 \cap A_2 \cap A_4)$$
$$= (4 + 8 + 1) - 2 - 0 - 0 + 0 = 11.$$

_____

EXERCISES 6.4

1. In the set of integers $\{1, 2, ..., 300\}$, how many are divisible by 3 or 5 but not by 7?

2. In how many ways can the walls of a rectangular room be painted in three distinct colors so that color changes only at the corners?

3. In a battle at least 70% of the combatants lost an eye, at least 60% lost an arm, at least 50% lost a leg, and at least 40% lost an ear. At least what percentage of the combatants lost all four?

4. If B and C are subsets of a set A, under what conditions will the following identities hold?
   a) $\#(B \cup \bar{C}) = \#(B) + \#(\bar{C})$
   b) $\#(\bar{B} \cup C) = \#(A)$

5. Let A, B, C be subsets of a set U. Show that
   $$\#(A \cap \bar{B} \cap \bar{C}) = \#(A) - \#(A \cap B) - \#(A \cap C) + \#(A \cap B \cap C).$$

6. Let B and C be subsets of a set A, with $\#(B) = \#(\bar{B})$ and $\#(C) = \#(\bar{C})$. What are the relationships among the following numbers?
   $$\#(B \cap C), \qquad \#(B \cap \bar{C}), \qquad \#(\bar{B} \cap C), \qquad \#(\bar{B} \cap \bar{C})$$

*7. A permutation of the integers $1, 2, ..., n$ is said to be a *derangement* of these n integers if no integer appears in its natural position, e.g., if k does not appear in kth position for $1 < k \leqslant n$. Show that the number of derangements of n objects is
   $$n! \left(1 - \frac{1}{1!} + \frac{1}{2!} - \cdots + (-1)^n \frac{1}{n!} \right).$$

_____

*6.5 ENUMERATION OF A SET RELATIVE TO A PERMUTATION GROUP (ELEMENTS OF PÓLYA'S THEORY)

We now consider a more subtle problem. We are given the tetrahedron shown in figure 6.5.1 (with faces labeled with the numbers 1, 2, 3, 4), and we are asked to

Clockwise rotation

Figure 6.5.1. A tetrahedron to be painted

state how many ways we can paint the four faces of it black and white, each face being entirely painted either one or the other. The problem would be very simple if the four faces were geometrically distinguishable. The ways would then be exactly the distributions with repetition of two objects {black and white} into four cells {1,2,3,4}, and we know by (6.2.1) that their number is 2^4. However, in the given tetrahedron faces 1, 2, and 3 are identical; i.e., the tetrahedron has a symmetry for a 120° rotation around the vertical axis, with the following consequences. If we have two identical tetrahedrons, labeled as in figure 6.5.1, and we paint faces 1 and 2 of the first black and faces 3 and 4 white, and if we paint faces 1 and 3 of the second black and faces 2 and 4 white, the two painted tetrahedrons are identical. Thus different assignment of colors results in painted objects that are indistinguishable.

A coloration of a tetrahedron may be viewed as a function $f: C \to S$, where $C \equiv \{1,2,3,4\}$ is the set of the faces and $S \equiv \{b, w\}$ is the set of the colors (b for black, w for white). Clearly, the two functions

$$\begin{pmatrix} 1\,2\,3\,4 \\ b\,b\,w\,w \end{pmatrix} \quad \text{and} \quad \begin{pmatrix} 1\,2\,3\,4 \\ b\,w\,b\,w \end{pmatrix}$$

result in two identically painted tetrahedrons, if the second is rotated 120° clockwise after it is painted (see figure 6.5.1).

This simple example is typical of a variety of enumeration problems in which the configuration set $C \equiv \{c_1, \ldots, c_n\}$—to be thought of as a rigid body—possesses symmetries such that there are rotations which map C onto itself. Each of these rotations is clearly a permutation of the cells of C; for example, a 120° clockwise rotation of the tetrahedron of figure 6.5.1 around its vertical axis is represented by the following permutation of its faces (cells):

$$\begin{pmatrix} 1234 \\ 2314 \end{pmatrix}.$$

A rotation which maps C onto itself is denoted as an *applicable rotation* of C. Similarly, a permutation representing an applicable rotation of C is denoted as an *applicable permutation* of C.

It is simple to show that, in general, the set of applicable permutations of a configuration set C is a group G (definitions 3.5.1, 3.2.3, 3.2.1). In fact, if $\pi_1 \in G$ and $\pi_2 \in G$ are two applicable rotations of C, we may apply them in sequence (composition); that is, $\pi_1\pi_2$ is also applicable, or $\pi_1\pi_2 \in G$ (closure). Moreover, this composition is associative; G contains the identity, i.e., the null rotation which leaves C unaltered. Finally, for every rotation $\pi \in G$, there exists the inverse rotation π^{-1} also in G. In our example, G consists of the three permutations

$$\pi_0 = \begin{pmatrix} 1234 \\ 1234 \end{pmatrix}, \qquad \pi_1 = \begin{pmatrix} 1234 \\ 3124 \end{pmatrix}, \qquad \pi_2 = \begin{pmatrix} 1234 \\ 2314 \end{pmatrix}.$$

The intrinsic symmetries of C—completely described by the permutation group G acting on the elements of C—determine the indistinguishability of some functions $f: C \to S$, whose domain is C. We shall now make this notion precise. Consider the set of functions $F \equiv \{f \mid f: C \to S\}$. Each permutation π of the elements of C induces a transformation of a given function $f_1: C \to S$; in our example

$$\pi_1 = \begin{pmatrix} 1234 \\ 3124 \end{pmatrix}$$

transforms

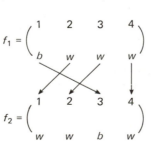

into

Thus each $\pi \in G$ induces a function $\varphi_\pi : F \to F$ on the set F, described by: $\varphi_\pi(f_1) = f_2$ means that for every $c_i \in C$, $f_1(c_i) = f_2(\pi(c_i))$. This function φ_π is one-to-one because, if there are two functions $f_1, f_2 \in F$ such that $\varphi_\pi(f_1) = \varphi_\pi(f_2) = f_3$, then for every $c_i \in C$, $f_1(c_i) = f_3(\pi(c_i))$ and $f_2(c_i) = f_3(\pi(c_i))$ and, by transitivity, $f_1(c_i) = f_2(c_i)$; that is, the two functions coincide. Clearly, a one-to-one function on a set is also onto. We conclude that φ_π is invertible; that is, φ_π is a permutation of the elements of F. For example, suppose that we list the $2^4 = 16$ functions $f : C \to S$ of our current example, as in figure 6.5.2. Then permutation π_2 of C induces a permutation $\varphi_{\pi_2} \in F$, whose diagram is also illustrated.[1] It is a simple matter to show that the set $\Phi \equiv \{\varphi_\pi | \pi \in G\}$ is a group and that Φ and G are isomorphic; these details, however, are omitted. A function $f \in F$ such that $\varphi_\pi(f) = f$ is said to be left *invariant* by the permutation φ_π of F.

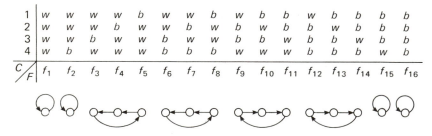

Figure 6.5.2. The permutation φ_{π_2} induced by π_2 (120° rotation) on the 16 colorings of the tetrahedron

In accordance with our previous discussion, two functions f_1 and f_2 (two ways of painting the tetrahedron) are indistinguishable if there is some $\pi \in G$ such that $f_2 = \varphi_\pi(f_1)$. In other words φ_π determines the relation of indistinguishability, which is clearly an equivalence relation, since the following relationships hold.

1. Function f_1 is related to itself, since for the identity $\pi_0 \in G$, $f_1 = \varphi_{\pi_0}(f_1)$. (reflexivity)

2. If f_1 is related to f_2, there is a $\pi \in G$ such that $f_2 = \varphi_\pi(f_1)$, whence $f_1 = \varphi_{\pi^{-1}}(f_2)$; that is, f_2 is related to f_1.   (symmetry)

3. If f_1 is related to f_2, and f_2 to f_3, there exist π_1 and π_2 in G such that $f_2 = \varphi_{\pi_1}(f_1)$ and $f_3 = \varphi_{\pi_2}(f_2)$. But, by the isomorphism of G and Φ, $\varphi_{\pi_1 \pi_2} = \varphi_{\pi_1} \varphi_{\pi_2}$, whence $f_3 = \varphi_{\pi_2}[\varphi_{\pi_1}(f_1)] = \varphi_{\pi_1 \pi_2}(f_1)$; that is, f_3 is related to f_1.   (transitivity)

Therefore G, through Φ, determines a partition of the set F of functions into equivalence classes. Our problem reduces exactly to the enumeration of these classes.

---

1. The reader should recall that the graph (or diagram) of a permutation, considered as a function $p : S \to S$, consists of disjoint cycles (see Note 3.2).

Before proceeding further, we define a very convenient descriptor of a function $f: C \to S$. The tabular representation of f consists of a sequence of symbols $s_1, s_2, \ldots, s_m$ of S; that is, s_1 appears e_1 times, s_2 appears e_2 times, and so on. Thus we could take as a descriptor of f the sequence $(e_1, e_2, \ldots, e_m)$, but for reasons to be apparent later, we prefer the product $s_1^{e_1} s_2^{e_2} \cdots s_m^{e_m}$. This product, called the *weight* $W(f)$ *of the function* f, is a monomial in the "variables" $s_1, s_2 \ldots, s_m$—the elements of the codomain S of f.

Note that functions in the same equivalence class have the same weight. For example, in figure 6.5.2, functions f_3, f_4, and f_5 are equivalent since they belong to the same cycle in the graph of φ_{π_1}. Their common weight is $w^3 b$. Note also that, although an equivalence class uniquely identifies its weight, there may be distinct classes having the same weight (for example, we will see that $\{f_6, f_7, f_8\}$ and $\{f_9, f_{10}, f_{11}\}$ are two distinct classes although their weights, $w^2 b^2$, are identical).

We now have all the premises to prove the following fundamental lemma.

Lemma 6.5.1. (Burnside.) The number p of equivalence classes into which F is partitioned by the equivalence relation induced by G through Φ is

$$p = \frac{1}{\#(G)} \sum_{\pi \in G} q(\varphi_\pi), \qquad (6.5.1)$$

where $q(\varphi_\pi)$ is the number of elements of F left invariant by the permutation φ_π of F.

Proof. Let $F \equiv \{f_1, \ldots, f_r\}$ and $G \equiv \{\pi_1, \ldots, \pi_s\}$. We form a rectangular array (the $\Phi \times F$ array, figure 6.5.3), whose columns and rows correspond to the elements of F and Φ, respectively. For a given φ_{π_i} we place a cross ($\times$) at the intersection of the ith row and the jth column if and only if f_j is left invariant by φ_{π_i} (that is, if $\varphi_{\pi_i}(f_j) = f_j$).

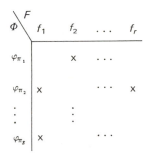

Figure 6.5.3

We want to show that each equivalence class (i.e., the columns corresponding to the functions of each such class) contributes the same number of crosses. Assume that $\{f_1, f_2, \ldots, f_\nu\}$ is an equivalence class, and let $\Phi_i \subseteq \Phi$ be the set of the permutations which leave f_i invariant. We claim that Φ_i is a subgroup of Φ. In fact, if $\varphi_\pi \in \Phi_i$, the inverse φ_π^{-1} of φ_π also leaves f_i invariant, whence $\varphi_\pi^{-1} \in \Phi_i$. Similarly, if φ_{π_1} and φ_{π_2} belong to Φ_i, then $\varphi_{\pi_2} \varphi_{\pi_1}^{-1}$ also leave f_i invariant, whence $\varphi_{\pi_2} \varphi_{\pi_1}^{-1} \in \Phi_i$. By theorem 3.5.4 we conclude that Φ_i is a subgroup of Φ. Now, by lemma 3.5.5, the family of all the subsets of permutations of the form $\Phi_i \varphi_\pi$, for $\varphi_\pi \in \Phi$, is a partition of Φ (the right cosets of Φ, lemma 3.5.5). Thus, if φ_π maps f_i into f_j, there are exactly $\#(\Phi_i)$ permutations—the elements of $\Phi_i \varphi_\pi$—which do so. Therefore Φ is partitioned into ν families (Φ_i and its right cosets), of which one maps f_i into itself, and the remaining $(\nu - 1)$ map f_i into the other members of $\{f_1, \ldots, f_\nu\}$. The same argument can be repeated for any other member f_j of the class; i.e., all the columns corresponding to $f_1, f_2, \ldots, f_\nu$ have *identical* numbers of crosses. Since each column contains $\#(\Phi)/\nu$ crosses and there are ν columns in the equivalence class, the latter contributes exactly $\#(\Phi)$ crosses, independently of its cardinality ν. It follows that the number of equivalence classes is given by the total number N of crosses in the array divided by $\#(\Phi) = \#(G)$. Counting the crosses *by rows* we have

$$N = \sum_{\pi \in G} q(\varphi_\pi),$$

and the lemma is proved. $\|$

Burnside's lemma is effectively illustrated by our current example. We consider the functions $\{f_1, \ldots, f_{16}\}$ of figure 6.5.2 and the permutation group $\Phi \equiv \{\varphi_{\pi_0}, \varphi_{\pi_1}, \varphi_{\pi_2}\}$ (where π_0, π_1 and π_2 are as defined above) and construct the corresponding array (figure 6.5.4). In this case, $q(\varphi_{\pi_0}) = 16$, $q(\varphi_{\pi_1}) = q(\varphi_{\pi_2}) = 4$; that is, $\sum q(\varphi_\pi) = 24$. It follows that $\nu = 24/3 = 8$, and the eight classes are indicated by the vertical separations.

Φ \ F	f_1	f_2	f_3	f_4	f_5	f_6	f_7	f_8	f_9	f_{10}	f_{11}	f_{12}	f_{13}	f_{14}	f_{15}	f_{16}
φ_{π_0}	x	x	x	x	x	x	x	x	x	x	x	x	x	x	x	x
φ_{π_1}	x	x													x	x
φ_{π_2}	x	x													x	x

Figure 6.5.4

The task of counting the classes is accomplished by means of Burnside's lemma. However, this method is not satisfactory in two respects.

1. To compute p we must know the numbers $q(\varphi_\pi)$ for every $\varphi_\pi \in \Phi$; that is, we must construct the group Φ (permutations of F) from the group G (permutations of C). Although $\#(\Phi) = \#(G)$, the fact that $\#(F)$ is usually much larger than $\#(C)$ greatly complicates the task. A method based directly on G, rather than on Φ, would be more appealing.

2. Beyond simple counting, we may want to get some additional insight into the structure of the equivalence classes by obtaining their weights.

These objectives are both achieved by Pólya's theory, which we are about to illustrate. First, however, we need some additional nomenclature.

Let $W_1, W_2, \ldots, W_n$ be the distinct weights of the functions $f \in F$, and let F_i be the set of the functions whose weight is W_i. Since functions in the same equivalence class have the same weight, F_i is a collection of equivalence classes. If we consider the $(\Phi \times F_i)$ array (that is, only the columns of the $(\Phi \times F)$ array corresponding to functions in F_i), the total number of crosses in $(\Phi \times F_i)$ is given by $\sum_{\pi \in G} q_i(\varphi_\pi)$, where $q_i(\varphi_\pi)$ is the number of elements of F_i left invariant by φ_π. Since the proof of Burnside's lemma contains the result that each equivalence class contributes the *same* number of crosses to the $(\Phi \times F)$ array, we have immediately the following lemma.

Lemma 6.5.2. The number p_i of equivalence classes into which the set $F_i \equiv \{f \mid f \in F$ and $W(f) = W_i\}$ is partitioned by the equivalence relation induced by G through Φ is

$$p_i = \frac{1}{\#(G)} \sum_{\pi \in G} q_i(\varphi_\pi), \tag{6.5.2}$$

where $q_i(\varphi_\pi)$ is the number of elements of F_i left invariant by φ_π.

A description of the weight structure of the set of functions F is given by the polynomial

$$\mathscr{I} = p_1 W_1 + p_2 W_2 + \cdots + p_n W_n, \tag{6.5.3}$$

usually called the *inventory* (indeed, it tells us which "types" of classes are in F and how many of each type there are). We can now express $\mathscr{I}$ in terms of the expression (6.5.2) of p_i. We have

$$\mathscr{I} = \sum_{i=1}^{n} p_i W_i = \sum_{i=1}^{n} W_i \frac{1}{\#(G)} \sum_{\pi \in G} q_i(\varphi_\pi) = \frac{1}{\#(G)} \sum_{\pi \in G} \sum_{i=1}^{n} q_i(\varphi_\pi) W_i, \tag{6.5.4}$$

where we have exchanged the order of summation over i and π. We concentrate on the expression

$$\sum_{i=1}^{n} q_i(\varphi_\pi) W_i.$$

Note that this expression, which is the sum of the distinct weights W_i, each multiplied by its cardinality $q_i(\varphi_\pi)$, depends on φ_π, that is, on a single row of the

$(\Phi \times F)$ array (the row corresponding to φ_π). Its reinterpretation is now the crucial part of the proof.

Let us consider the row in the $(\Phi \times F)$ array corresponding to φ_π. This row contains a certain number of crosses.

If f_j is left invariant by φ_π, this row contains a cross in correspondence to f_j. If we replace each such cross by the weight $W(f_j)$, then clearly

$$\sum_{i=1}^{n} q_i(\varphi_\pi) W_i = \sum_{\{j\,|\,f_j \text{ has a cross in the row of } \varphi_\pi\}} W(f_j). \tag{6.5.5}$$

The fact that f_j is left invariant by φ_π means, by definition, that $f_j(c) = f_j(\pi(c))$ for every $c \in C$. In turn, $f_j(c) = f_j(\pi(c))$ expresses the fact that c and $\pi(c)$, which belong to the same cycle of the graph of π, have the same value under f_j. It follows that if the diagram of π has a cycle of length l and the codomain of f_j is $S \equiv \{s_1, \ldots, s_m\}$, since all the elements on this cycle have the same value under f_j, this cycle contributes to $W(f_j)$ a factor of the form s_i^l, for some $s_i \in S$. In general, if the diagram of π consists of n_1 cycles of length 1, n_2 cycles of length 2, ..., n_k cycles of length k, the weight $W(f_j)$ of an f_j left invariant by φ_π is a product of the form

$$\underbrace{s_{i_1} \cdots s_{i_{n_1}}}_{n_1 \text{ terms}} \quad \underbrace{s_{j_1}^2 \cdots s_{j_{n_2}}^2}_{n_2 \text{ terms}} \quad \cdots \quad \underbrace{s_{m_1}^k \cdots s_{m_{n_k}}^k}_{n_k \text{ terms}} \tag{6.5.6.}$$

Conversely, since each function $f: C \to S$ corresponds to a column of the $(\Phi \times F)$ array, an arbitrary product of the type (6.5.6) identifies a function left invariant by φ_π. This proves that f_j is left invariant by φ_π if and only if $W(f_j)$ is a term like (6.5.6). But any such term is a product term obtained by applying the distributive property to the product

$$(s_1 + \cdots + s_m)^{n_1} (s_1^2 + \cdots + s_m^2)^{n_2} \cdots (s_1^k + \cdots + s_m^k)^{n_k},$$

whence, from 6.5.5,

$$(s_1 + \cdots + s_m)^{n_1} (s_1^2 + \cdots + s_m^2)^{n_2} \cdots (s_1^k + \cdots + s_m^k)^{n_k} = \sum_{\substack{\{f_j\,|\,f_j \text{ left} \\ \text{invariant by } \varphi_\pi\}}} W(f_j) = \sum_{i=1}^{n} q_i(\varphi_\pi) W_i.$$

This is the basic equality we wanted to obtain; we now derive some simple consequences.

Consider the following representation of π. We define a function μ_π from the cycle set of π to the codomain $X \equiv \{x_1, x_2, \ldots, x_k\}$, so that a cycle of length j of

π is mapped to x_j. It follows that the weight of μ_π is the monomial $x_1^{n_1} x_2^{n_2}, \ldots, x_k^{n_k}$ and that the inventory of $M \equiv \{\mu_\pi | \pi \in G\}$ is the polynomial

$$\sum_{\pi \in G} x_1^{n_1} x_2^{n_2} \cdots x_k^{n_k}.$$

This polynomial divided by $\#(G)$ is traditionally called the *cycle index of G*, $P(G; x_1, \ldots, x_k)$:

$$P(G; x_1, \ldots, x_k) = \frac{1}{\#(G)} \sum_{\pi \in G} x_1^{n_1} x_2^{n_2} \cdots x_k^{n_k}. \tag{6.5.7}$$

Combining (6.5.4), (6.5.6) and (6.5.7), we have

$$\mathscr{I} = \frac{1}{\#(G)} \sum_{\pi \in G} \left[\sum_{i=1}^{n} q_i(\varphi_\pi) W_i \right]$$

$$= P(G; (s_1 + \cdots + s_m), (s_1^2 + \cdots + s_m^2), \ldots, (s_1^k + \cdots + s_m^k)), \tag{6.5.8}$$

which embodies the following beautiful theorem.

Theorem 6.5.3. (Pólya). The inventory of the equivalence classes of the functions $f : C \to S$ under a permutation group G acting on C is given by (6.5.8), where $P(G; x_1, x_2, \ldots, x_k)$ is the cycle index of G.

Returning to our current example we note that π_0 is represented by x_1^4, π_1 and π_2 by $x_3 x_1$, whence $P(G; x_1, x_2, x_3) = \frac{1}{3}(x_1^4 + 2x_3 x_1)$. Since $S \equiv \{b, w\}$, it follows that the inventory is given by

$$\mathscr{I} = \tfrac{1}{3}[(b + w)^4 + 2(b^3 + w^3)(b + w)] = b^4 + w^4 + 2b^3 w + 2b^2 w^2 + 2bw^3.$$

We now consider an additional application of Pólya's theorem.

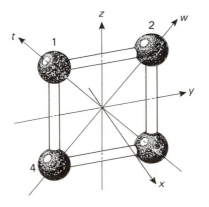

Figure 6.5.5

Example 6.5.1. Consider the arrangement of four spheres and four rods illustrated in figure 6.5.5. We want to count the distinct ways to paint the spheres with two colors, black (b) and white (w). The configuration C consists of the four cells $\{1,2,3,4\}$ and has the following symmetries.

1. rotations around the axis x in multiples of $90°$ corresponding to the permutations

$$\pi_1 = \begin{pmatrix} 1\,2\,3\,4 \\ 1\,2\,3\,4 \end{pmatrix}, \; \pi_2 = \begin{pmatrix} 1\,2\,3\,4 \\ 4\,1\,2\,3 \end{pmatrix}, \; \pi_3 = \begin{pmatrix} 1\,2\,3\,4 \\ 3\,4\,1\,2 \end{pmatrix}, \; \pi_4 = \begin{pmatrix} 1\,2\,3\,4 \\ 2\,3\,4\,1 \end{pmatrix}.$$

2. $180°$ rotations around the axes y and z (reflections) corresponding to the permutations

$$\pi_5 = \begin{pmatrix} 1\,2\,3\,4 \\ 4\,3\,2\,1 \end{pmatrix}, \; \pi_6 = \begin{pmatrix} 1\,2\,3\,4 \\ 2\,1\,4\,3 \end{pmatrix}.$$

3. $180°$ rotations around the axes w and t (reflections), corresponding to the permutations

$$\pi_7 = \begin{pmatrix} 1\,2\,3\,4 \\ 3\,2\,1\,4 \end{pmatrix}, \; \pi_8 = \begin{pmatrix} 1\,2\,3\,4 \\ 1\,4\,3\,2 \end{pmatrix}.$$

It is easily verified that $\{\pi_1, \pi_2, \ldots, \pi_8\}$ form a group G—indeed, a very famous group, known as the *group of the square*. We now compute the cycle index of G. We have the following weights:

Permutation	π_1	π_2	π_3	π_4	π_5	π_6	π_7	π_8
Weight	x_1^4	x_4	x_2^2	x_4	x_2^2	x_2^2	$x_1^2 x_2$	$x_1^2 x_2$

whence

$$P(G; x_1, x_2, x_3, x_4) = \tfrac{1}{8}(x_1^4 + 2x_4 + 3x_2^2 + 2x_1^2\, x_2);$$

thus, by Pólya's theorem we have

$$\mathscr{I} = \tfrac{1}{8}[(b+w)^4 + 2(b^4 + w^4) + 3(b^2 + w^2)^2 + 2(b+w)^2(b^2 + w^2)]$$
$$= b^4 + w^4 + b^3\, w + bw^3 + 2b^2\, w^2;$$

i.e., there are six equivalence classes with the indicated weights. We leave as a nontrivial exercise for the reader to show that we have just enumerated the equivalence classes of the boolean functions of two variables, where two boolean functions are considered equivalent if one can be obtained from the other by permuting and/or complementing their arguments.

Pólya's theorem is very useful even when we simply want to count the number of the equivalence classes of F, as shown by the following corollary.

Corollary 6.5.4. The number of the equivalence classes of the functions $f: C \rightarrow S$ under a permutation group G acting on C is given by

$$p = P(G; \#(S), \#(S), \ldots, \#(S)), \tag{6.5.9}$$

where $P(G; x_1, x_2, \ldots, x_k)$ is the cycle index of G.

Proof. If each $s_j \in S$ is replaced by the integer 1, the weight of each class in (6.5.3) also becomes 1; that is, $\mathscr{I} = \sum p_i = p$. This means that in (6.5.8) the sum $(s_1^j + \cdots + s_m^j)(j = 1, 2, \ldots, k)$ is replaced by $(1 + \cdots + 1) = m = \#(S)$, thus proving (6.5.9).||

_____

EXERCISES 6.5

1. We have defined as applicable a permutation which maps a configuration C into itself. Give the applicable permutations of the following configurations.

   a) The ends of a rod
   b) The edges of an equilateral triangle
   c) The edges of a rectangle
   d) The edges of a rhombus
   e) The edges of a square
   f) The faces of a cube

2. The edges C of a rectangle are to be painted with two different colors S (so that each edge is painted with only one color).

   a) Determine the applicable permutations π for the rectangle.
   b) For each π, determine the corresponding induced permutation φ_π of the colorings $F: C \rightarrow S$.
   c) Repeat parts (a) and (b) for a square.

3. Compute the cycle index for each of the permutation groups specified in exercise 6.5.1.

4. Bracelets are made with five beads of three colors: red, yellow, and blue.

   a) Find the number of distinguishable bracelets, assuming that the bracelets may be not only rotated but also flipped over. For example, if r, y, and b in figure 6.5.6 stand for red, yellow and blue, the bracelet given in (a) is rotated one bead position to the right in (b), and in (c) it is flipped over around the dotted line.

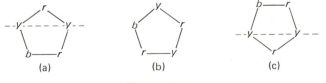

(a)   (b)   (c)

Figure 6.5.6

   b) With the same conventions as in (a), find the number of distinguishable bracelets if one bead is red, two are yellow, and two are blue.

5. Find the number of black and white 2×3 configurations made up of 3 black squares and 3 white squares.

6. In some applications, two binary sequences of length n are considered equivalent if one can be obtained from the other by applying the cyclic permutation

$$\begin{pmatrix} 1 & 2 & 3 & \cdots & n \\ n & 1 & 2 & \cdots & n-1 \end{pmatrix}$$

an appropriate number of times.

   a) Compute the inventory of the equivalence classes of binary sequences of length 6 and 7, using z to denote 0 and u to denote 1.
   b) Give the number of equivalence classes of binary sequences of length 6 and 7.

7. Two boolean functions of n variables are considered equivalent if one can be obtained from the other by permuting and/or complementing the variables. Prove that the inventory of the equivalence classes of the boolean functions of two variables is as obtained in example 6.5.1.

6.6 BIBLIOGRAPHY AND HISTORICAL NOTES

The combinatorial viewpoint can be discerned in several puzzles, arithmetic and geometric results developed by ancient civilizations (Greek, Chinese).

Only in modern times has combinatorics emerged as a mature discipline, principally because of the work of Euler, Laplace, Pascal, and Fermat. Its modern origin is closely intertwined with that of the theory of graphs. In recent times, combinatorial analysis has evolved into a flourishing branch of discrete mathematics.

The literature in combinatorics is very extensive; besides textbooks and research monographs, there are several scholarly journals devoted to the subject. A clearly written, extensive and transparent textbook is

C. L. Liu, *Introduction to Combinatorial Mathematics*, McGraw-Hill, New York, 1968.

The following book is a fascinating guided tour of the subject, rich with very illuminating remarks:

C. Berge, *Principle of Combinatorics*, Academic Press, New York, 1971.

Of the many more compact books, we simply cite, with no intention to slight other valuable works, the classical reference

H. J. Ryser, *Combinatorial Mathematics*, Carus Monograph No. 14, Wiley, New York, 1963.

6.7 NOTES AND APPLICATIONS

Note 6.1. Bounds for Error-Correcting Codes

In Note 3.3 we introduced the notion of error-correcting codes. We recall now, for the reader's convenience, that a code is a subset of sequences, consisting of n symbols, from a finite set called the alphabet. When the alphabet consists of the two symbols 0 and 1, the code is called a *binary code* of length n. Each code has specific properties, depending on some relationships existing among its member sequences. Suppose that some kind of disturbance is acting on code sequences by changing 0's to 1's, and vice versa; each such change we call an *error*. Since it is reasonable to assume that errors are rare events, one error is more likely than two errors, two errors are more likely than three errors, and so on. With this assumption, we intuitively conclude (and this could be justified more formally) that our best policy is to interpret (to *decode*) a sequence affected by errors as the code sequence which differs from the altered sequence in the smallest number of places. For example, 01101 differs from 01011 in two places, the third and the fourth from the left. Note that with this criterion, decoding is a mapping from the set of all binary sequences of length n onto the sequences of the binary code. Therefore two sequences which map to the same code sequence are equivalent from the viewpoint of decoding. This means that decoding determines a partition of the set of the binary sequences into equivalence classes, called *decoding subsets*, each of which is identified (or associated) with a code sequence.

We now ask a question: Which is the maximum number k of members of a code of length n such that at least all single errors can be corrected? This is tantamount to determining the maximum number k of decoding sets of a code with the property of correcting single errors. If we can determine the smallest cardinality of decoding sets compatible with the correction of single errors, then k cannot exceed the ratio

$$\frac{\text{number of binary sequences of length } n}{\text{minimum cardinality of decoding sets}}.$$

The binary sequences of length n are exactly the distributions of the two symbols 0,1 into n positions, and their number was found to be 2^n in example 6.2.1. The minimum cardinality of decoding sets is determined by observing that, if at least single errors must be corrected, a decoding subset must contain at least the following elements.

1. The code sequence associated with it.

2. All the sequences (*not* belonging to the code) which differ from the code sequence in exactly one position.

Since (1) contributes 1 sequence and (2) contributes n sequences, each decoding set contains at least $(n + 1)$ sequences. Therefore, for a single-error-correcting binary code of length n,

$$k \leqslant \frac{2^n}{n+1}.$$

More generally, if we consider codes which guarantee correction of up to and including t errors, each decoding set must contain at least the following elements.

1. A code sequence w.
2. All the sequences differing from w in 1 position, whose number is n.
3. All the sequences differing from w in 2 positions. They are obtained by selecting 2 out of n positions, and their number is given by $\binom{n}{2}$ [formula (6.2.5)].
    .
    .
    .
$t+1$. All the sequences differing from w in t positions. They are obtained by selecting t out of n positions, and their number is given by $\binom{n}{t}$ [formula (6.2.5)].

We conclude, therefore, that a t-error-correcting binary code of length n may have at most k codewords, where

$$k \leqslant \frac{2^n}{1 + n + \binom{n}{2} + \cdots + \binom{n}{t}}.$$

This important inequality for error-correcting codes is the Hamming bound, named for its inventor, R. W. Hamming. For example, for $n = 7$ and $t = 1$ we have

$$k \leqslant \frac{2^7}{1 + 7} = \frac{128}{8} = 16;$$

that is, a single-error-correcting code of length 7 can have at most 16 sequences. Such a code indeed exists, but exhibiting it is beyond the scope of this book. Coding theorists, however, became rapidly aware of their inability to construct codes which meet the Hamming bound for $t \geqslant 2$ (with very few exceptions). Thus it appears that the Hamming bound promises much more than can usually be achieved. It is therefore of interest to obtain, for a given code length n, the number k of code sequences which *can* be constructively selected if the code is to correct up to t errors. Such a bound has been obtained by Varshamov, Gilbert, and Sacks, but its derivation goes beyond the scope of this note.

References

R. B. Ash, *Information Theory*, Interscience, New York, 1965 (Section 4.5).

W. W. Peterson, *Error Correcting Codes*, M.I.T. Press, Cambridge, Mass., and Wiley, New York, 1961 (Chapter 4).

Note 6.2. On the Complexity of the Traveling Salesman Problem

The traveling salesman problem, introduced in Section 2.4, is now restated for convenience. A salesman must visit a number of cities, and given the distance between pairs of cities, he must find the shortest route by which he can visit each city exactly once and return to the point of departure. We remarked in Section 2.4 that there is no efficient general algorithm for this problem; i.e., the solution can be obtained only by exhaustive inspection of all cases. The fact that this note appears here instead of in Chapter 2 should suggest that the difficulty arises when the number of cases to be exhaustively examined is large.

Suppose that there are N cities the salesman must visit. Let us label the cities with integers $1, 2, \ldots, N$. Without loss of generality, we may assume that the salesman starts at 1. Since there are $(N-1)$ places the salesman must visit, there are a total of $(N-1)!$ possible paths. We can reduce this number by observing that if the salesman has already visited k places beside 1, only the remaining $N-1-k$ places need be considered.

Let $f(i_1, i_2, \ldots, i_k)$ be a function whose value is the minimum remaining distance for the salesman to travel, beginning at i_k, with the knowledge that he has already visited cities $i_1, i_2, \ldots, i_k$, where $2 \leqslant i_1, i_2, \ldots, i_k \leqslant N$. If we let $d(i,j)$ denote the distance between cities i and j, then the function $f(i_1, i_2, \ldots, i_k)$ is a recurrence relation since, for $2 \leqslant i_j \leqslant N$,

$$f(i_1, i_2, \ldots, i_{j-1}, i_{j+1}, \ldots, i_{N-1}) = d(i_{N-1}, i_j) + d(i_j, 1),$$

and

$$f(i_1, \ldots, i_k) = \min_j [d(i_k, j) + f(i_1, i_2, \ldots, i_k, j)].$$

The recurrence relation introduced above gives us a systematic procedure for calculating the minimal route for the salesman.

Clearly, the complexity grows with N. The question now is: How large an N can be handled, even by a computer? This is equivalent to asking in how many ways can we choose k numbers $(i_1, i_2, \ldots, i_k)$ from $N-1$ numbers for $k = 1, 2, \ldots, N-1$. This quantity is the binomial coefficient

$$\binom{N-1}{k} = \frac{(N-1)!}{k!(N-1-k)!}$$

It is a property of binomial coefficients that if $N-1$ is odd,

$$\binom{N-1}{N/2} \text{ is the largest among the } \binom{N-1}{j}.$$

If $N-1$ is even,

$$\binom{N-1}{(N-1)/2}$$

is the largest. If, for example, $N-1$ is even—that is, $N-1 = 2m$—then we must search over at least

$$\frac{(2m)!}{m!\,m!}$$

paths in order to determine a shortest path. If $N = 21$, the quantity above is 923,780.

Reference
R. Bellman, K. L. Cooke, and J. A. Lockett, *Algorithms, Graphs and Computers*, Academic Press, New York, 1970.

CHAPTER 7

ALGORITHMS AND TURING MACHINES

7.1 THE CONCEPT OF ALGORITHM

A wife wants to please her husband by trying out a fancy new recipe but ends up in total disappointment. We say she is not an experienced cook. Yet her failure is probably due to the fact that instructions given in recipes are often ambiguous and subject to different interpretations. Such phrases as "add the scallops, stir, and fry until white but not too firm," "add a dash of salt," and "cook on top of range over low heat till tender" are a few samples from cookbooks. On the other hand, in the preceding chapters we have introduced a number of "algorithms" consisting of a set of instructions for solving various classes of problems. Although we have not given an explicit definition of an algorithm, the underlying assumption was that the reader would know exactly how to carry out a given instruction. In other words,

an *algorithm*, or *effective procedure*, is a sequence of instructions, each of which can be executed in a precise manner.[1] The execution of an *instruction* is called a *step*.

There are several questions concerning the concept of algorithm. For example, will the execution of a given algorithm ever terminate? Is any sequence of instructions an algorithm? Does there exist an algorithm for solving any class of problems? Indeed, the investigation of these and several other problems associated with the concept of algorithm has drawn a major effort from many twentieth-century mathematicians.

In order to gain some insight into the concept of algorithm, let us now analyze a well-known algorithm, named for the Greek mathematician Euclid. The

---

1. Observe that the notion of algorithm as stated here is not a precisely defined mathematical concept.

euclidean algorithm consists of five instructions and will give as answer the greatest common divisor k of two positive integers m and n.

Instruction 1. Select two numbers m and n. Proceed to the next instruction.

Instruction 2. Subtract n from m, and let r be the difference. Proceed to the next instruction.

Instruction 3. If $r = 0$, the algorithm stops; k equals m. If $r \neq 0$, proceed to instruction 4.

Instruction 4. If $r > 0$, replace m by n and n by r, and proceed to instruction 2. If $r < 0$, proceed to instruction 5.

Instruction 5. Interchange m and n and proceed to instruction 2.

Example 7.1.1. The euclidean algorithm as applied to integers $m = 15$ and $n = 6$ is illustrated in figure 7.1.1. Although there are only five instructions, 17 steps are needed in this example to execute the algorithm.

Sequence of steps	Active instruction	m	n	r	k
1	1	15	6		
2	2	15	6	9	
3	3			$9 \neq 0$	
4	4	6	9	$9 > 0$	
5	2	6	9	-3	
6	3			$-3 \neq 0$	
7	4			$-3 < 0$	
8	5	9	6		
9	2	9	6	3	
10	3			$3 \neq 0$	
11	4	6	3	$3 > 0$	
12	2	6	3	3	
13	3			$3 \neq 0$	
14	4	3	3	$3 > 0$	
15	2	3	3	0	
16	3			$0 = 0$	3
17	Stop				

Figure 7.1.1. An illustration of the euclidean algorithm

A characteristic of the euclidean algorithm is its *deterministic* nature, in the sense that anyone who understands the rules of arithmetic can carry out its instructions without stretching his imagination. In other words, the algorithm specifies a deterministic process which is independent of the "calculator" that carries out the algorithm. Determinism should, in fact, be a common characteristic of all algorithms; that is, all the operations specified by the instructions of an algorithm must be sufficiently elementary to be carried out "effectively." A non-effective algorithm follows.

Let r be equal to 2 if Fermat's last theorem is true.[2]

Clearly, this instruction cannot be carried out effectively until either someone proves Fermat's last theorem or someone comes up with a counterexample to it.

A word of caution is needed at this point. In order to convey exactly what to do at any time when executing the instructions of an algorithm, we must describe the algorithm in language that is relatively rich but unambiguous. For example, if we apply the euclidean algorithm to real numbers, what is the "result of the subtraction" of the number $\sqrt{2}$ from the number π? The ambiguous nature of natural languages is evidenced by the phrases from recipes at the beginning of this section. And bear in mind that English is not the only offender.

Another characteristic of the euclidean algorithm is that when it is applied to any given pair of positive integers, it will always stop in a finite number of steps because, after each cycle of execution of the algorithm, one of the given integers is reduced, as illustrated in example 7.1.1. This "finiteness" condition is clearly desirable for any algorithm used to systematically produce solutions to a particular class of problems. On the other hand, not all effective procedures satisfy the "finiteness" condition, as illustrated by the following example.

Example 7.1.2. Consider the following sequence of instructions.

Instruction 1. Compare the two integers m and n. Proceed to the next instruction.

Instruction 2. If $m \geqslant n$, proceed to instruction 4. Otherwise, proceed to the next instruction.

Instruction 3. Interchange m and n. Proceed to instruction 1.

Instruction 4. Replace n by $n + 1$. Proceed to instruction 1.

By trying out a few examples, the reader will discover that, although the four instructions are effective, the execution of the algorithm requires an infinite number of steps.

- -

EXERCISES 7.1

1. Carry out the euclidean algorithm for the integers 49 and 28.
2. Design an algorithm as a list of instructions for each of the following problems.
   a) To solve the quadratic equation $ax^2 + bx + c = 0$ for x, where a, b, and c are integers.
   b) For the first player to win or draw in the game of tic-tac-toe.

---

2. Fermat's last theorem states that there exist integers a, b, c, and $(d > 2)$ such that $a^d + b^d = c^d$. No one knows whether this is true or not.

c) To color planar graphs with five colors (see Section 2.6 and Note 2.7).
d) To check if a sequence of 0's and 1's contains the sequence 0010 as a subsequence.
e) To order alphabetically a set of words.

_____

7.2 FLOWCHART REPRESENTATION OF ALGORITHMS

Since an algorithm consists of a finite list of instructions specifying elementary operations to be executed in a certain order, it is sometimes convenient to represent it pictorially as a directed graph indicating the "flow" or "sequencing"

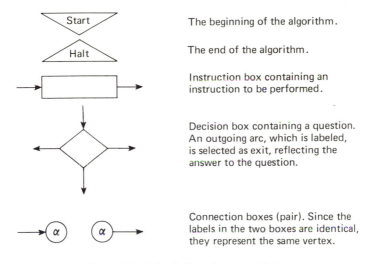

The beginning of the algorithm.

The end of the algorithm.

Instruction box containing an instruction to be performed.

Decision box containing a question. An outgoing arc, which is labeled, is selected as exit, reflecting the answer to the question.

Connection boxes (pair). Since the labels in the two boxes are identical, they represent the same vertex.

Figure 7.2.1. Basic flowchart notations

of the elementary operations (recall example 2.1.1). Each vertex of the graph is a box with an instruction written in it, and the order in which the instructions are to be executed is given by directions of the arcs. Such a graph is called a *flowchart* of the algorithm. Since there are different types of instructions, we shall use boxes of different shapes to indicate the instructions contained in them. Some flowchart notations were introduced in example 2.1.1. A more adequate repertoire is given in figure 7.2.1. In this context we will also adopt the notation $a \leftarrow b$ to mean that the symbol a is replaced by the symbol b.

Example 7.2.1. The flowchart representation of the euclidean algorithm is given in figure 7.2.2.

When an algorithm can be meaningfully decomposed into simpler algorithms (called subalgorithms or component algorithms), it is advantageous to represent

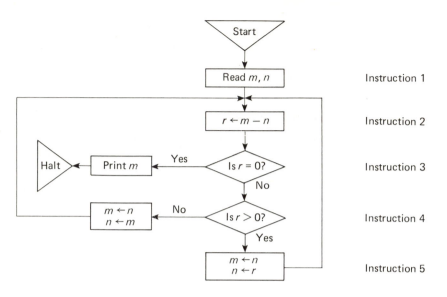

Figure 7.2.2. Flowchart representation of the euclidean algorithm

it by means of several flowcharts, each representing a component algorithm. The connection box is then useful, as illustrated in figure 7.2.3, where an algorithm A consists of three component algorithms A_1, A_2, and A_3. When executing algorithm A_1, we will reach either α or β, that is, the starting point of algorithm A_2 or A_3, respectively. In flowchart jargon, this event is referred to as "transfer of control" from A_1 to A_i ($i = 2, 3$), and A_i is said to be "called in" by A_1. Note that figure 7.2.3(b) indicates that algorithm A_2, after being called in by algorithm A_1, will return control to A_1 through γ.

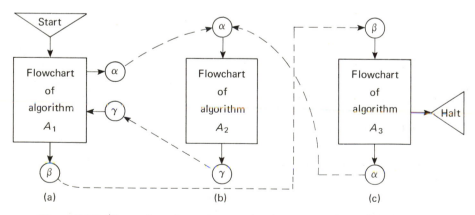

Figure 7.2.3. Illustration of use of connection box in flowchart of an algorithm

EXERCISES 7.2

1. Draw a flowchart of each of the algorithms for the problems in exercise 7.1.2.

2. Let the vertices of oriented rooted trees (definition 2.5.4) be labeled with lower-case letters of the English alphabet. Design an algorithm for detecting a specific labeled subtree in an oriented rooted tree and for relabeling its vertices in a prescribed manner. The algorithm must consist of the following three component algorithms.

    a) An algorithm A_1 for traversing all vertices of the oriented rooted tree.

    b) An algorithm A_2 for detecting the labeled subtree shown in figure 7.2.4.

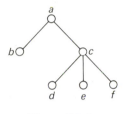

Figure 7.2.4

    c) An algorithm A_3 for relabeling the subtree detected by A_2 as shown in figure 7.2.5.

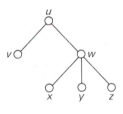

Figure 7.2.5

Design and interconnect the component algorithms A_1, A_2, and A_3 (see figure 7.2.3).

7.3 REQUIREMENTS OF A FORMAL DEFINITION OF ALGORITHM

Consider the instruction "Go to see Paul's uncle, who wears a red hat, and ask for the bicycle he owes me." Although this instruction would be perfectly meaningful to someone who knew the person—Paul's uncle, who wears a red hat—it certainly could not be carried out by any person who had no knowledge of Paul's uncle. Thus, although we intend an algorithm to specify a deterministic procedure independent of the "calculator," the rules or instructions of the algorithm are subject to the interpretation of that individual. As we know, a person's interpretation of instructions depends on his background and intellectual ability. How can we avoid the problem of interpretation? Can we distill the intuitive notion of algorithm into some precisely defined mathematical concept?

Human beings are well known for their ability to construct tools or machines to help in or to do their routine work. Let us consider one such tool, the adding machine. If you enter two numbers into an adding machine and push the button labeled ADD, after a short time the adding machine will print out the sum of the two numbers. A moment's reflection will convince the reader that adding two multidigit numbers represented in the familiar decimal notation involves no more than the following types of simple operations.

1. Consider a pair of digits and a carry (zero or one) from the right.
2. Look up the addition table and write the sum digit.
3. Mark a carry (zero or one) to the left.

Thus we may say that the algorithm for adding two numbers is *realized* by an adding machine in the sense that applying the algorithm to any pair of numbers will obtain the same result as entering the two numbers into an adding machine and pushing the ADD button.

This illustration of an adding machine which realizes an algorithm suggests that a precise formulation of the concept of algorithm may be one in which we set up

a) a language in which instructions of algorithms are to be expressed, and

b) a machine which can "understand" statements of the given language and therefore execute the operations specified by the instructions.

In other words, we need to design an abstract machine which can understand a set of instructions, expressed in some language, and do what the instructions command. Conceptually, it is difficult to imagine that a *single* machine is powerful enough to understand and execute instructions for all algorithms. The fact that such a machine can be built is the formal basis of modern computers.

We shall motivate the definition of the instruction-obeying machine by first analyzing how we carry out a computation. For illustrative purposes, let us analyze how we add two numbers, say 14 and 28, step by step. First, we write the two numbers on paper, as shown in figure 7.3.1.[3] We next decide to compute

```
14   Two given numbers
28
──
42   Sum
1    Carry from the sum of two digits 4 and 8
```

Figure 7.3.1. An illustration of the addition of two numbers

---

3. We have not forgotten that some people can add strings of numbers mentally. The point to be stressed here is that information needs to be stored somewhere, whether on paper or in a person's brain.

the sum of the pair of digits on the right by looking at an addition table or recalling the rules of addition. The carry to the unit digit position is always 0. Since the sum of 4 and 8 is 12, we write down 2 under 8 and either write a carry of 1 somewhere on the paper or remember it. We then repeat the process for the next two digits and the carry.

From this illustration, we see that there are three ingredients that are essential to computation.

1. A memory medium for storing information. In the illustration above the medium is a piece of paper. The information stored includes not only data to be processed (the numbers 14 and 28) and standard reference data (the addition table), but also the instructions for solving the problem and intermediate results of the computation. Thus the piece of paper must be viewed as a generic medium for storing information.

2. Some means for carrying out the elementary operations specified by the algorithm. In the illustration the device could be, for example, an adding machine. Each step of the computation consists of taking some information from the memory, performing a specific operation on it, and recording the result somewhere in the memory.

3. Some means of controlling the process, that is, a device that can determine which operation is to be carried out next. In the illustration, the human being is the controlling device.

The relationship of the three factors of human computation is illustrated in figure 7.3.2.

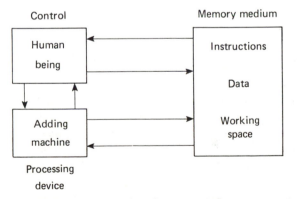

Figure 7.3.2. Illustration of information flow among the components of human computation

7.4 TURING MACHINES

Having analyzed how human beings perform computations, we are now ready to describe an abstract machine which can carry out instructions specified by any algorithm. First of all, we need a language in which instructions can be expressed. For our purpose, we shall simply consider a language as a set of finite strings of symbols (or words) from a finite set which we call an *alphabet*. We note that this assumption of a language does not restrict the machine's ability to carry out algorithms written in the English language since we can transform English sentences into strings of the machine's language, as is done in telegraph transmission or in the writing of programs for computers.

The machine will have three main components: a *memory unit* for storing information; an *arithmetic unit* for performing the elementary operations; and a *control unit* for determining what the machine should do next.

The memory unit will be a one-dimensional tape of unbounded length, divided into squares such that a single symbol from the alphabet can be written on each square. We shall use a special symbol, say *, to denote the fact that the square with * printed on it is blank. The use of a special symbol to denote a blank square is merely for convenience in describing the machine. Our assumption of a one-dimensional tape also does not create any problem, since we may translate information written on any two-dimensional medium, say, a piece of paper, into a one-dimensional tape. For example, the information exhibited in figure 7.3.1 can be expressed on a one-dimensional tape as follows:

$$\ldots\ *\ 1\ 4\ *\ 2\ 8\ *\ 4\ 2\ *\ 1\ *\ \ldots$$

Finally, the unboundedness of the memory tape is not such an unreasonable assumption as it may appear to be. The initial data occupies a finite portion of the tape, and after a finite amount of time, still only a finite amount of tape is used. One may look on this assumption as meaning that initially only a finite tape is given, but whenever the given tape is filled, one can always attach another finite blank tape to it for use. Essentially, this amounts simply to having a generous supply of scratch paper so that for performing computations there is no need to worry about running out of paper.

We now turn our attention to the control unit and the arithmetic unit. Since we are describing an abstract machine, we will not discuss the structure of these two units but will simply describe their capabilities. Thus the important aspects of our abstract machine are: (1) what kinds of elementary operations it can perform; (2) what the instructions are for it; and (3) how communication takes place between its various components.

Again, let us refer to human computation. When we perform simple arithmetic operations on a piece of paper, the following elementary operations are involved.

1. Move the pencil on the paper.
2. Read symbols (extract information) written on the paper.
3. Erase and write symbols on the paper.

In order to simulate these operations, we assume that in our abstract machine there is a read-write head (RWH) which moves along the tape and reads and writes symbols on the tape. To make the operation elementary, we assume that the RWH can move at most one square (to the right or left) at a time. Finally, since in the course of carrying out the operations specified by an algorithm, the instruction to be executed next depends on the given algorithm, the control unit of the abstract machine must be capable of going from one instruction to another instruction; that is, it must have access to a list of instructions (the *program*) sequentially numbered. These numbers are called the *labels* of the instructions.

With these requirements in mind, we now give a list of simple but necessary instructions that our machine is capable of executing.

1. R: Move the RWH one square to the right.

2. L: Move the RWH one square to the left.

3. N: No movement of the RWH.

4. a: Erase the symbol presently being scanned by the RWH and write the symbol a on the square under the RWH.

5. J k: Unconditional jump—go to the kth instruction.

6. J a k: Conditional jump—go to the kth instruction if the symbol being scanned is a.

7. H: Halt.

We are now ready to describe a version of the machine which bears the name of its inventor, the mathematician Alan Turing. A *Turing machine* consists of an unbounded one-dimensional tape, an alphabet having a finite number of symbols, a program, which is a sequence of instructions from the list above, and a control unit with an RWH. Initially the RWH is positioned to scan the leftmost non-blank symbol, and the control unit begins to execute the first instruction of the program. For each instruction of the program we should specify which instruction

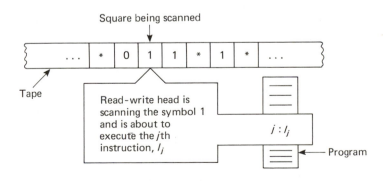

Figure 7.4.1. Illustration of a Turing machine

must be executed next. However, in order to economize in the instruction representation, we establish that normally the instruction to be executed next in time is the
next instruction in the program, except when the current instruction is a jump
(either J k or J a k). Thus, if the current instruction is one of the first four types
and is the j-th of the program, control is transferred to the $(j + 1)$-st instruction
of the program. When control is transferred to H or to an instruction label
outside the program, the machine halts. A pictorial illustration of a Turing
machine is given in figure 7.4.1. Note that instructions J k and J a k are program
jumps and not moves on the tape.

Example 7.4.1. We now consider an application of the Turing machine for solving
arithmetic problems. We adopt the notation that any positive integer, say k, will
be represented by a sequence of k 1's on the tape (unary notation). For example,
3 is represented as *111*, where * denotes a blank space. The special problem
under consideration is the addition of two numbers. We would like to design a
Turing machine such that, given the representations of two positive integers on
the tape, separated by the special symbol ¢, it will stop with the representation of
the sum on the tape. The program of the Turing machine and its flowchart are
given in figure 7.4.2, and the operation of the machine is illustrated in figure 7.4.3,
in which an up-pointed arrow is used to indicate the symbol presently scanned
by the RWH.

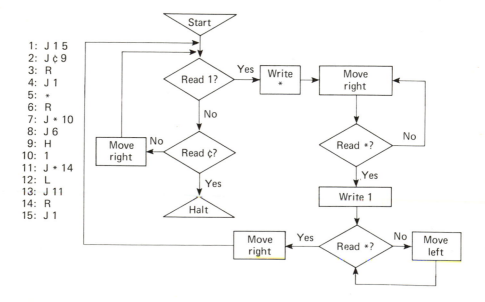

```
 1:  J 1 5
 2:  J ¢ 9
 3:  R
 4:  J 1
 5:  *
 6:  R
 7:  J * 10
 8:  J 6
 9:  H
10:  1
11:  J * 14
12:  L
13:  J 11
14:  R
15:  J 1
```

Figure 7.4.2. (a) A program of a Turing machine and (b) its flowchart representation

Active instruction	Tape before	Tape after
1	*11¢1* ↑	*11¢1* ↑
5	*11¢1* ↑	**1¢1* ↑
6	**1¢1* ↑	**1¢1* ↑
.		
.		
.		

(7, 8)6	**1¢1*
	  ↑
		**1¢1*
		  ↑

10	**1¢1*
	  ↑
		**1¢11*
		   ↑

(11)12	**1¢11*
	  ↑
		**1¢11*
		  ↑

.		
.		
.		
(13, 11)12	**1¢11* ↑	**1¢11* ↑
(13, 11)14	**1¢11* ↑	**1¢11* ↑
(15, 1)5	**1¢11* ↑	***¢11* ↑
6	***¢11* ↑	***¢11* ↑
.		
.		
.		

(7, 8)6	***¢11*
	  ↑
		***¢11*
		  ↑

10	***¢11*
	   ↑
		***¢111*
		   ↑

(11)12	***¢111*
	   ↑
		***¢111*
		   ↑

.		
.		
.		
(13, 11)12	***¢111* ↑	***¢111* ↑
(13, 11)14	***¢111* ↑	***¢111* ↑
(15, 1, 2)9	***¢111* ↑	***¢111* ↑

Figure 7.4.3. Illustration of the operation of the Turing machine of figure 7.4.2. (The instructions in parentheses are executed by the machine, but do not determine modifications of the tape)

Example 7.4.2. Consider a Turing machine whose tape symbols are 1, 0, and * (the blank symbol). The program of this Turing machine and its flowchart are given in figure 7.4.4. A *word* is defined as a sequence of 0's and l's between two *'s.

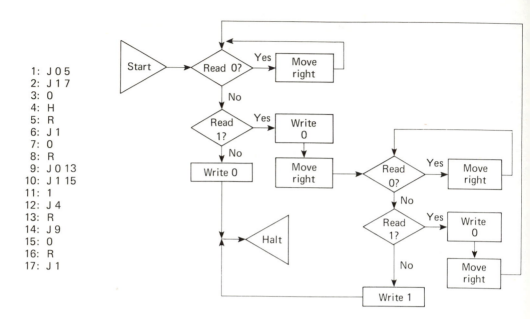

```
 1: J 0 5
 2: J 1 7
 3: 0
 4: H
 5: R
 6: J 1
 7: 0
 8: R
 9: J 0 13
10: J 1 15
11: 1
12: J 4
13: R
14: J 9
15: 0
16: R
17: J 1
```

Figure 7.4.4. (a) Program of a Turing machine and (b) its flowchart representation

This Turing machine acts as a parity counter; i.e. it counts the number of 1's occurring in any given word. The machine rewrites the word as a sequence of 0's, followed by a 1 if the number of 1's in the original word was odd, and by a 0 otherwise. Thereafter the machine stops. This example is an instance of a Turing machine used as a sequence recognizer.

Since any program of a Turing machine consists of a sequence of elementary operations which can be carried out effectively, it specifies an effective, deterministic process, as illustrated by the preceding examples. If the algorithms corresponding to the flowcharts of figures 7.4.2 and 7.4.4 were given first, they certainly could be coded in the language of Turing machines and hence realized by Turing machines. This means that, with respect to any initial data, the result obtained by manually executing the algorithm and the result obtained by applying the Turing machine to the initial data coincide. Thus we have two instances in which the intuitive concept of algorithm is replaced by the precise mathematical concept of a Turing machine. Indeed, these are illustrations of the now famous Turing's thesis:

> Any process which could naturally be called an effective procedure can be realized by a Turing machine.

Since we know how simple the instructions of a Turing machine are, this thesis is certainly a surprise. However, we cannot hope to prove the validity of this proposition since it is a statement about the general concept of effective procedure, which is not precisely defined and hence cannot be a proper subject of mathematical discussion. However, its validity is more or less established, since all known algorithms are realizable by Turing machines. The readers are referred to Turing's original paper for a brilliant intuitive argument in support of his thesis (see Section 7.7).

EXERCISES 7.4

1. For each of the following problems, design a Turing machine that will begin at the given initial configuration and halt at the prescribed final configuration. The letters X and W are words over some alphabet, and $*$ denotes a blank square; the vertical arrow indicates the position of the RWH.

	Initial configuration	Final configuration
a)	$\underset{\uparrow}{*}\,W\,*$	$*\,W\,\underset{\uparrow}{*}$
b)	$*\,W\,\underset{\uparrow}{*}$	$\underset{\uparrow}{*}\,W\,*$
c)	$*\,Wa\,*\,*\cdots\,\underset{\uparrow}{*}$	$*\,W\underset{\uparrow}{a}\,*\,*\cdots\,*$
d)	$\underset{\uparrow}{*}\,W\,*\,*$	$*\,*\,W\,\underset{\uparrow}{*}$
e)	$*\,X\,*\,W\,\underset{\uparrow}{*}$	$*\,W\,\underset{\uparrow}{*}$

2. Design a Turing machine which can decide whether a sequence of left and right parentheses is well formed (see Note 3.7). Draw the corresponding flowchart.

3. Design a Turing machine which will never stop. Draw the corresponding flowchart.

*4. Design a Turing machine which can perform the multiplication of nonnegative integers represented in binary. Draw the corresponding flowchart. (The binary representation of a nonnegative integer N is the sequence $a_n a_{n-1} \cdots a_0$, where $a_i \in \{0, 1\}$ for $i = 0, 1, \ldots, n$, and $N = a_0 2^0 + a_1 2^1 + \cdots + a_n 2^n$.)

*5. Design a Turing machine which can compute the exponent function M^N for two nonnegative integers M and N represented in binary (see exercise 7.4.4). Draw the corresponding flowchart.

7.5 ANOTHER DESCRIPTION OF TURING MACHINES

If we consider again the parity counter machine of example 7.4.2, we see intuitively that the machine can have two different states (or modes of operation): the "even" and "odd" modes. It changes mode whenever it reads a 1. At the same time, it makes decisions as to what to write on the square being scanned and in which direction to move next. Thus, the kind of stimulus-response relation involved can be represented as a quintuple:

(current state, symbol being read, next state, symbol to be written, direction of motion).

If we represent the two states "even" and "odd" of the Turing machine in example 7.4.2 by the symbols s_1 and s_2, respectively, then the machine can be described by the set of quintuples given in figure 7.5.1, in which H denotes "halt" and N denotes "no motion."

Machine in "even" mode
$$\begin{cases} (s_1, 0, s_1, 0, R) \\ (s_1, 1, s_2, 0, R) \\ (s_1, *, H, 0, N) \end{cases}$$
← Change to "odd" mode when encountering a 1.

Machine in "odd" mode
$$\begin{cases} (s_2, 0, s_2, 0, R) \\ (s_2, 1, s_1, 0, R) \\ (s_2, *, H, 1, N) \end{cases}$$
← Change to "even" mode when encountering a 1.

Figure 7.5.1. Quintuple description of a Turing machine

We see that the particular machine discussed in example 7.4.2 can be represented more compactly by a set of six quintuples. However, it should be clear from our discussion that this new representation is not due to any special characteristic of the Turing machine in question; rather, it can be used for any Turing machine. Indeed, the quintuple representation of the program can be obtained directly from the instruction list of the program. Specifically, one can take the labels of instructions as states, and if the program contains k_1: R, k_2: L, k_3: a, and k_4: J k, then the quintuple representation would contain, for each input symbol b, respectively $(k_1, b, k_1 + 1, b, R)$, $(k_2, b, k_2 + 1, b, L)$ $(k_3, b, k_3 + 1, a, N)$, (k_4, b, k, b, N). If the program contains k_5: J a k, then several quintuples are used, namely, (k_5, a, k, a, N) and $(k_5, b, k_5 + 1, b, N)$ for each input symbol $b \neq a$. We note that this transformation would result in a large number of quintuples, many of which are redundant.

Applying the transformation outlined in the preceding paragraph to the program given in example 7.4.2, we would end up with twenty-five quintuples, since there will be three quintuples corresponding to each of the following instructions: 1: J 0 5, 2: J 1 7, 9: J 0 13, and 10: J 1 15; and one quintuple will correspond to each of the other instructions. On the other hand, figure 7.5.1 gives six quintuples which describe the same Turing machine. This clearly illustrates the redundancy of the given transformation procedure. Although there are procedures for eliminating this redundancy, they are beyond our scope here. We note that the quintuple description of a Turing machine must satisfy the condition that distinct quintuples must not begin with identical state-input pairs because a Turing machine in a particular state reading a particular input symbol must behave in a unique way.

If we wish to transform a quintuple description of a Turing machine into a program of the machine, we need a sequence of instructions to realize the effect of a quintuple. In fact, it is necessary to determine the symbol being scanned by the machine, to direct the machine to move, and to transfer control. It is usually easier to transform the quintuple description into a flowchart and then write the program. A flowchart obtained from the set of quintuples in figure 7.5.1 is given

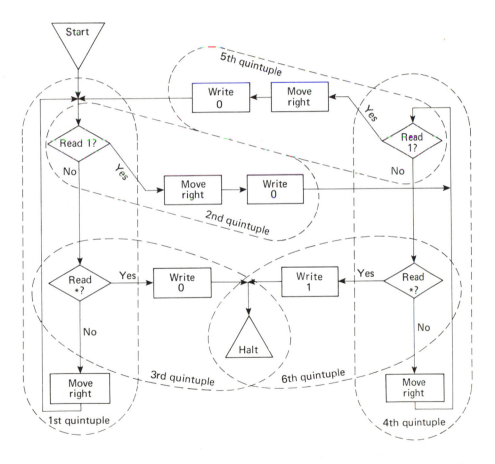

Figure 7.5.2. Flowchart of a program obtained from the quintuple description in figure 7.5.1

in figure 7.5.2. For illustrative purposes, we have shown the instructions corresponding to the quintuples, which are labeled according to their top-to-bottom ordering in figure 7.5.1. We will leave as an exercise the writing of the program.

A Turing machine can thus be viewed as consisting of a finite set of *states*, a finite alphabet, and a finite set of quintuples (defined above) such that no two begin with the same pair of state and tape symbols. Again, the input is written on a one-dimensional tape, and the machine has an RWH.

The fact that no two quintuples of a Turing machine begin with the same pair of state and input symbols means that the operation of the Turing machine can be expressed as a function $f: S \times A \to S \times A \times D$, where S, A, and D are respectively the sets of states, tape symbols, and directions of motion. Thus we may represent a Turing machine by a table such that a quintuple (s_i, a_j, s_k, a_l, d) will appear as shown below.

We shall refer to this table as the *transition table* of the Turing machine in question (cf. the transition table of finite state machines, introduced in Note 1.1).

Example 7.5.1. The transition tables of the Turing machines in examples 7.4.1 and 7.4.2 are given in figure 7.5.3(a) and (b), respectively; the dash symbolizes "no change."

	1	*	¢
s_0	$s_2 * R$	$--R$	$H * N$
s_1	$--L$	$s_0 - R$	$--L$
s_2	$--R$	$s_1 1N$	$--R$

	0	1	*
s_1	$--R$	$s_2 0R$	HON
s_2	$--R$	$s_1 0R$	H1N

Figure 7.5.3. Transition tables of two Turing machines: (a) addition; (b) parity counter

Example 7.5.2. We now illustrate how a Turing machine can be used as a generator of numbers. More specifically, we shall design a Turing machine that will start with an empty tape and print on it, sequentially, all the positive integers, repre-

sented in unary notation and separated by blanks. In other words, the infinite
tape would look like figure 7.5.4.

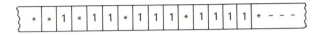

Figure 7.5.4. An infinite tape storing positive integers

The desired Turing machine will write a 1 on the empty tape, copy the last
number, and then add 1 to the copy. Thus, one may view the machine to be
designed as composed of two simpler machines: one having the function of
writing a 1 and the other having the function of copying. The transition table of
the machine is given in figure 7.5.5. The additional symbol $ is used by the
machine to keep track of positions.

	*	1	$
s_0	s_1 1N	s_1 1N	s_1 1N
s_1	s_2 - R	- -L	- -L
s_2	s_6 - R	s_3 $N	- -R
s_3	s_4 - R	- -R	- -R
s_4	s_5 1N	- -R	- -R
s_5	- -L	- -L	s_2 1R
s_6	s_0 -N	- -R	- -R

Figure 7.5.5. Transition table of a Turing machine which can generate all positive
integers in unary notation

An explanation of the function of each state follows.

s_0: Writes 1 and goes to state s_1 (to begin the generating process).

s_1: Moves left until encountering *, and then goes to state s_2 (to begin the copying
process). This * is the left marker of the number to be copied.

s_2: If 1 is encountered, writes $ and goes to state s_3. If * is encountered, goes to
state s_6 and moves right. (The machine uses $ to keep track of which 1 is to
be copied next. If all digits of a number have been copied, the process begins
again with the next number.)

s_3: Moves right until encountering *, and then goes to state s_4. (This * is the
separation marker between the number to be copied and that being written.)

s_4: Moves right until encountering *, then writes 1, and goes to state s_5. (Copies the digit marked by \$ on the first blank to the right of the last separation blank between numbers.)

s_5: Moves left until encountering \$, then writes 1, moves to the right, and goes to state s_2. (Restores the digit being copied to its original place. The machine is now ready to copy the next digit.)

s_6: Moves right until encountering * and then goes to state s_0. (The machine is ready to add a 1 to the last number and begin to copy the modified number.)

A sequence of tape configurations is given in figure 7.5.6 to illustrate how the given Turing machine operates.

```
Initial tape        · · ·  *  *  *  *  *  · · ·
First cycle    {    · · ·  *  1  *  *  *  · · ·
               (    · · ·  *  $  *  *  *  · · ·
Second cycle  {     · · ·  *  $  *  1  *  · · ·
               (    · · ·  *  1  *  1  1  *  · · · ·
               (    · · ·  *  1  *  $  1  *  · · · ·
               (    · · ·  *  1  *  $  1  *  1  *  · · · ·
Third cycle   {     · · ·  *  1  *  1  $  *  1  *  · · · ·
               (    · · ·  *  1  *  1  $  *  1  1  *  · · · ·
               (    · · ·  *  1  *  1  1  *  1  1  1  *  · · · ·
```

Figure 7.5.6. Tape configurations of the Turing machine given in figure 7.5.5, starting with a blank tape

---

EXERCISES 7.5

1. Write a program corresponding to the flowchart given in figure 7.5.2.

2. Describe the Turing machines obtained in exercise 7.4.1 as sets of quintuples.

3. A Turing machine is said to *recognize* a set of sequences L if the Turing machine will eventually halt with a 1 on its tape if the nonblank portion of the initial tape is an element of L. Design a Turing machine to recognize precisely each of the following sets.

   a) $\{0^n 1^n | n \geqslant 1\} = \{01, 0011, 000111, \ldots\}$

   b) $\{a^n b^n c^n | n \geqslant 1\} = \{abc, aabbcc, \ldots\}$.

   c) The set of all binary sequences having exactly the same number of 0's and 1's.

   *d) $\{1^p | p \text{ is a prime}\} = \{11, 111, 11111, \ldots\}$.

4. Design a two-state Turing machine with alphabet $\{0,1\}$ such that, beginning with blank tape, it will halt with four 1's printed on its tape. (This is a special case of the Busy Beaver problem. See exercise 7.6.5.)

5. Design a Turing machine to generate all well-formed strings of parentheses; i.e., beginning with the blank tape, it will produce an infinite tape like

$$¢ (\ )¢ (\ )(\ )¢ (\ (\ ))¢ \cdots$$

The order of the occurrence of different well-formed sequences does not matter. However, no sequence may be generated twice.

*6. Two Turing machines are said to be equivalent if, given the same initial tape, they stop and produce the same terminal tape. Show that every Turing machine Z is equivalent to a Turing machine Z' such that only one direction of motion is associated with each state of Z'. In other words, when Z' is in a particular state, say s', it can move in only one direction (right or left) for all input symbols.

7.6 UNIVERSAL TURING MACHINES AND THE HALTING PROBLEM

The reader by now will already have some experience in tracing the operation of a Turing machine. Indeed, anyone who has a precise understanding of how to interpret the description (or transition table) of a Turing machine can imitate the machine if given the initial data, the description of the machine, and a storage medium, say, pencil and paper. The imitation process really involves no more than looking up the transition table of the machine to find out what symbol to write, which way to move, and what state (or row) to look at next. Thus the very process of imitating a machine can itself be given as an exact sequence of instructions. Such an imitation procedure can be realized by a Turing machine U, called a Universal Turing machine since, given the initial data and the description of the machine, it can imitate the behavior of any other Turing machine. We shall not go into the details of constructing a Universal Turing machine; we merely note here that Universal machines have been constructed and then clarify a couple of points.

     A Universal Turing machine is a fixed machine having a fixed number of tape symbols and states. To enable it to imitate the operations of other Turing machines (some of which will have more states and more tape symbols than the Universal machine in question), we need to code the data as well as the description of the given machine by means of the tape symbols of the Universal machine. A Universal Turing machine U will simulate a given Turing machine T one step at a time, given the description of T and the initial data for T on U's tape. The initial condition of T—that is, the initial state of T and the square initially scanned by T—will also be given. Then U will keep a complete account of what T's tape looks like at each moment. It will remember what state T is supposed to be in, and it can see what T would read on the "simulated" tape. Then U will look at the description of T to see what T is supposed to do next and do it. The reader interested in the details is referred to Minsky (see Section 7.7) for a clear presentation of the construction of a Universal Turing machine.

Having now passed from the vague concept of an algorithm to the precise concept of a Turing machine, we find it desirable to give a classification of problems solvable by algorithmic methods. In other words, we want to know whether a given class of problems can be solved by means of an algorithm. Stated in another way, we want to know whether there exists a Turing machine for solving a given class of problems. In the following discussion, we will present a particular problem which cannot be solved by a Turing machine (or, as we say, which is algorithmically unsolvable).

When a Turing machine starts to operate on a given tape, it may continue for some time before it comes to a halt. For many machine-tape pairs, the machine may never stop. Hence it would be desirable to have a procedure for deciding, given any machine T and a tape t, whether T will eventually halt on it. This problem is usually referred to as the *halting problem*.

We will now demonstrate the algorithmic unsolvability of the halting problem. Let us assume, as a tentative hypothesis, that there is indeed an effective decision procedure for the halting problem. Then there must exist a Turing machine Z_1 such that, given a copy of the tape t of the machine T and the description d_T of T on its tape, it will eventually halt and communicate to us whether T will eventually halt. (Note that both T and t are arbitrary.) The communication can be effected in the following way: If T eventually halts, then Z_1 also halts and prints a 1 on the tape after erasing all other symbols. If T will never halt, then Z_1 halts and prints a 0 on the tape after erasing all other symbols. The transformation of tape configurations performed by Z_1 is illustrated in figure 7.6.1.

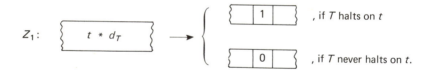

Figure 7.6.1. An illustration of the transformation performed by the Turing machine Z_1

Since t could be any tape, we now select a specific tape for consideration, namely, d_T. We shall now construct a new Turing machine Z_2 from Z_1 such that Z_2 is composed of two machines Z_3 and Z_1, of which Z_3 is a machine that simply duplicates the nonblank portion of the tape. Thus Z_2 is a machine which, when started with a tape

will convert it into

$$\{\ t\ *\ t\ \}$$

and then transfer control to Z_1. The transformation of the tape configuration, as performed by Z_2, is illustrated in figure 7.6.2.

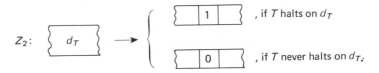

Figure 7.6.2. Transformation performed by the machine Z_2

Since the machine Z_2 must halt eventually and output a 1 or a 0, depending on whether T halts or not, it must have the two quintuples

$$(s_1, a_1, H, 0, N) \qquad \text{and} \qquad (s_2, a_2, H, 1, N) \tag{7.6.1}$$

for some states s_1, s_2 and tape symbols a_1 and a_2. Let s_1' and s_2' be two states not belonging to Z_2. We construct a new machine Z_4 from Z_2 by introducing two new states s_1' and s_2' and by replacing in Z_2 the right-hand quintuple of (7.6.1) with the following set of quintuples

$$
\begin{aligned}
&(s_2, a_2, s_1', 1, N) \\
&(s_1', \ 1, s_2', 1, R) \tag{7.6.2}\\
&(s_2', *, s_1', *, L)
\end{aligned}
$$

The effect of the three quintuples of (7.6.2) is to move back and forth on the tape

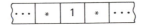

without ever stopping. In other words,

Z_4 halts on d_T if T does not halt on d_T; Z_4 does not halt on d_T if T halts on d_T.

Since T was arbitrary, so is d_T; therefore let us consider the application of Z_4 to d_{Z_4}. With the appropriate substitutions the result is:

Z_4 halts on d_{Z_4} if Z_4 does not halt on d_{Z_4}; Z_4 does not halt on d_{Z_4} if Z_4 halts on d_{Z_4}.

This statement is clearly a contradiction, and we must conclude that Z_4 and hence Z_2 and hence Z_1 cannot exist. This demonstrates the algorithmic unsolvability of the halting problem. The format of the statement above is closely reminiscent of Russell's paradox (see Note 1.3).

--- --- --- --- --- --- --- --- --- --- --- --- --- --- ---

EXERCISES 7.6

1. Does there exist an algorithm for deciding, given a Turing machine Z, a tape t, and the fact that Z starts on t, whether Z will use more than n squares of tape before it halts?

*2. Show that it cannot be proved that the halting problem for a particular Turing machine Z and a particular tape t is unsolvable.

 3. Show that the halting problem for initial blank tape is unsolvable; i.e., there does not exist a Turing machine which will answer, for each machine Z, the question "Does Z halt if started on a blank tape?" (*Hint*: For each Turing machine Z and tape t, construct another Turing machine Z_1, which, when started on a blank tape, will initially print t and then transfer control to Z.)

*4. Show that there is no effective procedure for deciding whether a Turing machine Z will halt on every input tape. (*Hint*: Reduce the problem to that of exercise 7.6.3.)

*5. Consider the problem of constructing an effective procedure to determine whether an n-state Turing machine with alphabet $\{0,1\}$, started on a blank tape, will eventually halt with a maximum number $v(n)$ of 1's printed on its tape. Show that the problem of computing $v(n)$ for arbitrary n is unsolvable. This problem is called the *busy beaver* problem because it is designed to determine how busy a Turing machine can be when the numbers of its states and inputs are restricted. (*Hint*: Suppose that there is a Turing machine Z which can compute $v(n)$. Construct another machine Z' which is a composite of Z and another machine Z'' such that whenever Z halts, Z' transfers control to Z'', which prints an additional 1. In other words, Z' computes the function $v(n) + 1$. Show that the existence of Z' leads to a contradiction.)

7.7 BIBLIOGRAPHY AND HISTORICAL NOTES

Throughout the history of mathematics, the task of developing algorithms for the solution of problems has always been important. However, it was not until the late 1930's that mathematicians finally grasped the essence of the notion of algorithm. Several formalizations of the concept of the algorithm, motivated by widely different considerations, were suggested around 1936. The fact that these different formalizations were shown to be equivalent has finally given the intuitive notion of algorithm a formal, precise interpretation. Today the theory of algorithms has gained increasing importance as the theoretical foundation of the present-day digital computers.

There are several texts valuable for further readings. The second part of M. L. Minsky, *Computation: Finite and Infinite Machines*, Prentice-Hall, Englewood Cliffs, N.J., 1967

is a clear and beautiful treatment of Turing machines and computation with many profound remarks. The more classical and complete references, though more advanced, are

S. C. Kleene, *Introduction to Metamathematics*, Van Nostrand, Princeton, 1952.
H. Hermes, *Enumerability, Decidability, Computability: An Introduction to the Theory of Recursive Functions* (translated by G. T. Hermann and O. Plassmann), Springer-Verlag, New York, 1969.

The following book treats not only Turing machines but also related topics, such as finite automata and formal languages

R. J. Nelson, *Introduction to Automata*, Wiley, New York, 1968.

Finally, advanced readers should read Turing's original paper:

A. M. Turing, "On Computable Numbers, With an Application to the Entscheidungsproblem," *Proceedings, London Mathematical Society*, Ser. 2–42 (1936), 230–265.

7.8 NOTES AND APPLICATIONS

Note 7.1. Finite State Machines as Sequence Recognizers

In Section 7.5, we showed that the quintuples of a Turing machine may be thought of as a function $f: S \times A' \to S \times A' \times D$, where S, A', D are sets of states, tape symbols, and movement symbols, respectively. In fact, we may decompose f into three functions, $f_1: S \times A' \to S$, $f_2: S \times A' \to A'$ and $f_3: S \times A' \to D$. In the latter formulation, we see that every finite state machine (see definition 1.7.1) $\mathcal{M} = [S,A',B',\delta,\lambda]$ is a Turing machine in that $A = A' \cup B'$, $\delta = f_1 \mid S \times A'$, $\lambda = f_2 \mid S \times A'$, and D is a set consisting only of the symbol R; that is, $f_3: S \times A \to \{R\}$. Since for a finite state machine f_3 is a constant function, it may be omitted. Hence every finite state machine is a Turing machine.

Consider now a Turing machine that always moves to the right. The restriction that T must always move right implies that it cannot read the symbols it writes on the tape. Indeed, the quintuples of T must all have the form

$$(s_i, a_j, f_1(s_i, a_j), f_2(s_i, a_j), R). \tag{7.8.1}$$

Since the direction of motion is always to the right, R may be omitted, and the quintuple (7.8.1) becomes

$$(s_i, a_j, f_1(s_i, a_j), f_2(s_i, a_j)) \tag{7.8.2}$$

If we let $\delta = f_1$, and $\lambda = f_2$, we see that T can be represented by the system $[S,A',A',\delta,\lambda]$, which is precisely a finite state machine.

Thus far we have considered finite state machines as sequence transducers. Now we want to consider them, in an equally important application, as *sequence recognizers*. For example, the parity counter machine given in example 7.4.2 may be thought of as a sequence recognizer in the sense that when it stops and a 1 (0) is the rightmost nonblank symbol on the tape, we know that the input tape is a binary sequence containing an odd (even) number of 1's.

Definition 7.8.1. A *finite automaton*[4] M is a quintuple $[S,I,\delta,s_0,F]$, where S is a finite set of states, I the input alphabet, $\delta: S \times I \to S$ is the *transition function* of M, $s_0 \in S$ is the *initial state* of M, and $F \subseteq S$ is the set of *terminal states*.

---

4. In the literature, the terms "finite state machine" and "finite automaton" are used interchangeably. We prefer to use finite state machine as sequence transducer and finite automaton as sequence recognizer.

Intuitively, a finite automaton is a finite state machine whose output alphabet consists of only two symbols, 0 and 1. The dichotomy $(F, S - F)$ of the state set signifies that "recognition" is shown by the last output symbol (which is dependent exclusively on the state of the machine). In other words, a finite automaton $M = [S, I, \delta, s_0, F]$ is a finite state machine $\mathcal{M} = [S, I, \{0,1\}, \delta, \lambda]$ such that, for each input symbol $i \in I$,

$$\lambda(s, i) = \begin{cases} 0, \text{ if } s \notin F, \\ 1, \text{ if } s \in F. \end{cases}$$

Schematically, a finite automaton may be represented as in figure 7.8.1.

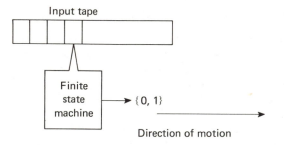

Figure 7.8.1. Schematic representation of a finite automaton

Example 7.8.1. Consider the finite automaton whose state set is $\{s_0, s_1\}$, inputs are $\{0,1\}$, $\{s_1\} = F$, and δ is given in figure 7.8.2. We see that the given finite automaton is the parity counter machine given in example 7.5.1, with state s_1 selected as terminal state.

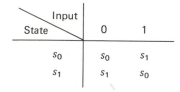

Figure 7.8.2. Transition table of a finite automaton

We note that the structure of the machine as described by the table in figure 7.8.2 can also be described by a *state graph* as can the finite state machine. The initial state is denoted by a vertex with an arrow pointing to it, and terminal states are denoted by encircled vertices. The state graph of the finite automaton in example 7.8.1 is given in figure 7.8.3.

We have mentioned that finite automata may be regarded as sequence recognizers. The automaton given in example 7.8.1 can recognize all sequences of

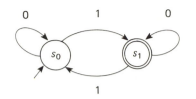

Figure 7.8.3. State graph of the finite automaton given in example 7.8.1

0's and 1's which have an odd number of 1's, in the sense that any such input sequence will take the initial state s_0 into a terminal state in F, that is, will cause the machine to output a 1. It is natural to ask what input sequence will cause a given automaton to output a 1. In other words, how do we characterize those inputs which will take the automaton from the initial state to a terminal state? First, we extend the function δ to the domain $S \times I^*$ as we have done for finite state machines (see Note 3.1, formula 3.10.1).

Definition 7.8.2. Let $M = [S,A,\delta,s_0,F]$ be a finite automaton. A word x over A is said to be *recognizable* by M if

$$\delta(s_0, x) \in F.$$

We will denote by $L(M)$ the set of all words recognizable by M.

We note that a finite automaton whose terminal state set is empty cannot recognize any word. Using definition 7.8.2, we see that if M is the automaton given in figure 7.8.3, then $L(M)$ consists of all words over $\{0,1\}$ containing an odd number of 1's. In general, may we ask for a given M whether $L(M)$ is empty or not? The answer is affirmative, and it is based on the following result.

Lemma 7.8.1. Let $M = [S,I,\delta,s_0,F]$ be a finite automaton with n states. Then M recognizes words over I if and only if it recognizes a word over I of length less than or equal to $n - 1$.

Proof. It suffices to show that if M recognizes any word in I^*, then it recognizes one whose length is less than or equal to $n - 1$. Let $x = i^{(1)}\cdots i^{(r)}$ be a word of $r > n - 1$ symbols which is recognized by M. When the input sequence x is applied, M will traverse the state sequence $s^{(1)}s^{(2)}s^{(3)} \cdots s^{(r+1)}$, where $s^{(1)} = s_0$. This sequence contains at least $(n + 1)$ states, more than there are distinct states in S. Therefore at least two states of $s^{(1)}s^{(2)} \cdots s^{(r+1)}$ must be identical. Let them be $s^{(i)}$ and $s^{(j)}$, and without loss of generality, assume that $j > i$. Then the word

$$x' = i^{(1)}\cdots i^{(i-1)} i^{(j)}\cdots i^{(r)}$$

must also be recognizable by M since $\delta(s_0,x) = s^{(r+1)} \in F$, and

$$\begin{aligned}
\delta(s_0, x) &= \delta(\delta(s_0, i^{(1)}\cdots i^{(i-1)}), i^{(i)}\cdots i^{(r)}) = \delta(s^{(i)}, i^{(i)}\cdots i^{(r)})\\
&= \delta(\delta(s^{(i)}, i^{(i)}\cdots i^{(j-1)}), i^{(j)}\cdots i^{(r)})\\
&= \delta(s^{(j)}, i^{(j)}\cdots i^{(r)}) = \delta(\delta(s_0, i^{(1)}\cdots i^{(i-1)}), i^{(j)}\cdots i^{(r)})\\
&= \delta(s_0, i^{(1)}\cdots i^{(i-1)} i^{(j)}\cdots i^{(r)}) = \delta(s_0, x').
\end{aligned} \tag{7.8.3}$$

Equation (7.8.3) is illustrated as follows:

$$s_0 \xrightarrow{i^{(1)}} s^{(2)} \cdots \xrightarrow{i^{(l-1)}} s^{(i)} \cdots \xrightarrow{i^{(j-1)}} s^{(j)} \xrightarrow{i^{(j)}} s^{(j+1)} \cdots \xrightarrow{\hspace{1cm}} s^{(r+1)} = \delta(s_0, x)$$

$$s_0 \xrightarrow{i^{(1)}} s^{(1)} \xrightarrow{\hspace{1cm}} \cdots \xrightarrow{i^{(l-1)}} \underset{\overset{\shortparallel}{s^{(j)}}}{s^{(i)}} \xrightarrow{i^{(j)}} s^{(j+1)} \xrightarrow{\hspace{1cm}} \cdots \xrightarrow{\hspace{1cm}} s^{(r+1)} = \delta(s_0, x')$$

Now, if the length of x' is less than or equal to $n - 1$, the proof is complete. If not, we may apply the procedure above to x' to obtain a shorter word, x''. Since x is of finite length, we need only to apply the given procedure a finite number of times to obtain a word of length less than or equal to $n - 1$. Thus the lemma holds.‖

Theorem 7.8.2. There is an algorithm to decide, for any finite automaton M, whether $L(M)$ is empty or not.

Proof. By lemma 7.8.1, if n is the number of states of M, the algorithm is simply to test each word of length less than or equal to $n - 1$, by applying it as an input to M. If M does not recognize any of these words, then we know $L(M) = \varnothing$. Since there are only a finite number of words of a given length, the procedure is effective.‖

We will now give a characterization of languages recognizable by finite automata using the formal grammar concepts described in Note 3.7. First, we recall that a *phrase-structure grammar* G is a quadruple $[V_N, V_T, S, P]$ such that V_N and V_T are finite sets called the set of *nonterminals* and *terminals*, respectively, such that $V_N \cap V_T = \varnothing$. Symbol $S \in V_N$ and is referred to as the *sentence symbol*, and P is a set of *productions* of the form $\alpha \to \beta$, where α is a word over $V = V_N \cup V_T$ containing at least one nonterminal, and $\beta \in V^*$. The language $L(G)$ generated by the grammar G is the set of words x over V_T such that, beginning with S, x is obtained by using the production rules $\alpha \to \beta$ of G. We now recall the definition of a special class of phrase-structure grammars (Note 3.7).

Definition 7.8.3. A phrase-structure grammar $G = [V_N, V_T, S, P]$ is said to be *regular* if all productions in P are either of the form $A \to aB$ or $A \to a$, or $A \to \Lambda$ where A, B are in V_N, Λ is the empty word, and $a \in V_T$.

Example 7.8.2. Consider the grammar G with $V_N \equiv \{S, A\}$, $V_T \equiv \{0, 1\}$. Grammar G has the set P of productions:

1.	$S \to 0S$	4.	$A \to 1S$
2.	$S \to 1A$	5.	$A \to 0$
3.	$A \to 0A$	6.	$S \to 1$

We see that $L(G)$ consists of all words over $\{0, 1\}$ containing an odd number of 1's. For instance, the word 0 0 1 1 1 0 1 1 is derived as follows:

$$S \underset{1}{\Rightarrow} 0S \underset{1}{\Rightarrow} 00S \underset{2}{\Rightarrow} 001A \underset{4}{\Rightarrow} 0011S \underset{2}{\Rightarrow} 00111A$$

$$\underset{3}{\Rightarrow} 001110A \underset{4}{\Rightarrow} 0011101S \underset{6}{\Rightarrow} 00111011,$$

where the number under "$\Rightarrow$" indicates which production is used.

Looking at the finite automaton described by figure 7.8.3, we see that if we equate the state s_0 to the symbol S and the state s_1 to the symbol A in example 7.8.2, then the grammar G of example 7.8.2 can be obtained by looking at the state diagram in figure 7.8.3, and vice versa. The exact relation between finite automata and regular grammars is given by the following theorem.

Theorem 7.8.3. A language L is recognizable by a finite automaton M if and only if it is generated by a regular grammar G.

Proof. If $L = L(M)$ for some finite automaton $M = [S,I,\delta,s_0,F]$, then we define a regular grammar $G_M = [V_N,V_T,S,P]$ such that $V_N = S$, $V_T = I$, $S = s_0,$[5] and for each $i \in I$ we define a production $s \to i\delta(s,i)$ in P. If $\delta(s,i) \in F$, then we also define the production $s \to i$ in P. Finally, if $A \in L(M)$, that is, $s_0 \in F$, then we also add the production $S \to A$ in P.

We must show that $L(M) = L(G_M)$. We begin by showing that if x is recognized by M—that is, $x \in L(M)$—then there is a derivation in G_M

$$S \overset{*}{\Rightarrow} x. \tag{7.8.4}$$

Now, $x \in L(M)$ implies that $\delta(s_0,x) = s' \in F$. If either $x = A$ or $x = i \in I$, there exist corresponding productions $S \to A$ or $S \to i$ in P to generate i. If the length of x is greater than 1, then we may write x as yi for some $y \in I^*$ such that the length of y is one less than the length of x. By the extension rule of δ to $S \times I^*$, we have

$$\delta(s_0,x) = \delta(\delta(s_0,y),i).$$

We now make the inductive hypothesis that if $\delta(s_0,y) = s''$, there is a derivation of the form

$$S \overset{*}{\Rightarrow} ys''. \tag{7.8.5}$$

Since $s' \in F$ and $\delta(s'',i) = s'$, we have, by definition of G_M, the production $s'' \to i$. Applying this production to (7.8.5), we have

$$S \overset{*}{\Rightarrow} ys'' \Rightarrow yi = x.$$

Hence $x \in L(G_M)$, and we have shown that $L(M) \subseteq L(G_M)$. Since all the arguments used are reversible, we can also show that $L(M) \supseteq L(G_M)$, whence we conclude that $L(M) = L(G_M)$.

Since the construction and the proof of the converse are exactly analogous to the previous arguments, they will be omitted here.$\|$

Clearly, using the grammar given in example 7.8.2, we can reconstruct the automaton described in figure 7.8.3.

The preceding discussion is a rapid introduction to the theory of finite automata. The study of the relationship between classes of formal languages and classes of automata which recognize them is a fertile area in the theory of computation.

---

5. Note that S denotes the sentence symbol, and S the set of states.

Reference

M. Harrison, *An Introduction to Switching and Automata Theory*, McGraw-Hill, New York, 1965.

Note 7.2. Relations Between Turing Machine and Post Canonical System

This note covers briefly the relationship between Turing machines and a general approach to string manipulation originally due to the logician Emil Post. We first recall that in Chapter 0 we gave an intuitive definition of a formal system as a set of axioms from which statements (theorems) are deduced by using rules of inference. We now formalize that notion.

A formal system $\mathscr{L}$ (also called a *logic* or *logistic system*) is a triple $[A,\Gamma,P]$ where $A \equiv \{a_1,...,a_n\}$ is a finite *alphabet*, $\Gamma \equiv \{A_1,...,A_k\}$ is a finite set of *axioms* which are words over A or, equivalently, members of A^* (see Note 3.6), and P is a finite set of *inference rules* $R_1,...,R_m$ such that each R_i is a function mapping $(A^*)^{n_i+1}$ to the set $\{0,1\}$, for some $n_i \geqslant 1$. If $R_i(\alpha_1,\alpha_2,...,\alpha_{n_i},\alpha) = 1$, we say that "$\alpha$ is directly derivable from $\alpha_1,...,\alpha_{n_i}$ by the rule R_i." For notational convenience, the notation $R_i(\alpha_1,\alpha_2,...,\alpha_{n_i},\alpha) = 1$ is replaced by

$$\alpha_1\alpha_2\cdots\alpha_{n_i} \xrightarrow[R_i]{} \alpha, \tag{7.8.6}$$

which is referred to as a *production* (see Note 3.7). Here $\alpha_1, \alpha_2, ..., \alpha_{n_i}$ are called the *antecedents* of the production, and α is called the *consequent* of the production. Note that the terminology is borrowed from logic. In this vein, a *proof* in $\mathscr{L}$ of a word $\alpha \in A^*$, is a sequence of words $(\alpha_1,\alpha_2,...,\alpha_{k-1},\alpha_k)$ such that $\alpha_k = \alpha$ and each α_i is either an axiom in Γ or is directly derivable via a sequence of productions from a set of words $\alpha_{i_1}, ..., \alpha_{i_n}$, all of which precede α_i in the sequence. Finally, a *theorem* in $\mathscr{L}$ is a word for which there is a proof in $\mathscr{L}$.

Although the terminology is strongly reminiscent of "logic," a logistic system is essentially a device for generation and manipulation of strings of symbols. The "generation" mechanism is explicitly evidenced by the notation (7.8.6), in which a sequence of words $\alpha_1\alpha_2\cdots\alpha_{n_i}$ is transformed to another word α. The reader who is familiar with phrase-structure grammars (see Note 3.7) should have no trouble seeing that every phrase-structure grammar $G = [V_N,V_T,S,P]$ can be interpreted as a logistic system $\mathscr{L}_G = [A,\Gamma,P']$, where $A = V_N \cup V_T$, $\Gamma \equiv \{S\}$, and for each production $\alpha \to \beta$ in P there corresponds an inference rule R in P' such that $R(x\,\alpha\,y, x\,\beta\,y) = 1$ (that is, $x\,\alpha\,y \xrightarrow[R]{} x\,\beta\,y$) for arbitrary words x and y over A. We now give two illustrative examples.

Example 7.8.3. A formal system whose "theorems" are all the strings of 1's (integers in unary notation) is specified as follows:

Alphabet: $\{1\}$
Axiom: $\{1\}$
Production: $\alpha \to \alpha 1$

The production means that if the word α is a theorem in the system, so is the word $\alpha 1$. Thus theorems in this system include 1, 11, 111, etc.

Example 7.8.4. A formal system whose "theorems" are the set of all well-formed strings of parentheses is specified as follows:

   Alphabet : $\{(,)\}$
   Axiom : $\{(\ )\}$
   Production : $AB \rightarrow A(\ )B$

The following steps are used to derive the well-formed string of parentheses $((()) (()))$:

$(\ ) \rightarrow ((\ ))$,	$(\ )$ is an axiom, $A = (\ , B =)$
$((\ )) \rightarrow ((\ )(\ ))$,	$A = ((\ )\quad, B =)$
$((\ )(\ )) \rightarrow (((\ ))(\ ))$,	$A = ((\quad, B =)(\ ))$
$(((\ ))(\ )) \rightarrow (((\ ))((\ )))$,	$A = (((\ ))(\ , B =))$

When the productions of $\mathscr{L}$ are of a specified type, then we have the following class of systems.

Definition 7.8.4. A *Post Canonical System* is a formal system consisting of a finite set A (the alphabet), a finite subset $\Gamma \subset A^*$ (the axioms), and a finite set P (productions) of the form

$$\alpha_0 A_1 \alpha_1 A_2 \alpha_2 \cdots A_n \alpha_n \rightarrow \beta_0 B_1 \beta_1 B_2 \beta_2 \cdots B_m \beta_m, \qquad (7.8.7)$$

where $\alpha_0, \alpha_1, ..., \alpha_n, \beta_0, \beta_1, ..., \beta_m$ are *fixed words* in A^*, $A_1, A_2, ..., A_n$ are the *variable words* (in A^*), and each $B_i \in \{A_1, A_2, ..., A_n\}$.

   The productions given in examples 7.8.3 and 7.8.4 are special cases of (7.8.7). For example, $AB \rightarrow A(\ )B$ can be rewritten as $\alpha_0 A \alpha_1 B \alpha_2 \rightarrow \beta_0 A \beta_1 B \beta_2$ with $\alpha_0 = \alpha_1 = \alpha_2 = \beta_0 = \beta_2 = \Lambda$ (the empty word) and $\beta_1 = (\ )$. The Post canonical systems discussed here are referred to as *single-antecedent* Post canonical systems.
   We shall now illustrate how the computations performed by a Turing machine can be formulated as theorems of a Post canonical system. First of all, at a time unit t, the condition of a Turing machine T may be represented as follows: Let $A_1' A_2'$ (catenation of A_1' and A_2') be the nonblank portion of the tape such that the RWH is positioned on the leftmost symbol of A_2'. Note that both A_1' and A_2' may be empty. We define $A_1 = **A_1'$ and $A_2 = A_2'**$. Then if the current state of T is s, the string

$$I_t = A_1 s A_2$$

is called the *instantaneous description* of T (this notation is slightly different from the usual one, in order to simplify the following discussion). For example, if at time t the state of T is s, and the tape of T looks like figure 7.8.4, the instantaneous description of T is

$$I_t = \underbrace{**1011}_{A_1}\ s\ \underbrace{1\$10010**}_{A_2}. \qquad (7.8.8)$$

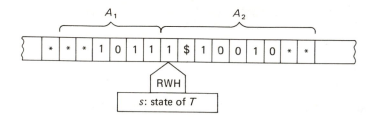

Figure 7.8.4. The condition of T at time t

We now want to explore whether the sequence of the instantaneous descriptions of T can be obtained as theorems of a Post canonical system. For instance, if $(s,1,s',0,R)$ is a quintuple of T and I_t is as given by (7.8.8), then the instantaneous description I_{t+1} of T at time $t+1$ is

$$I_{t+1} = **10110 \; s' \; \$10010**.$$

This observation can be generalized to all cases, and we obtain the correspondences shown in figure 7.8.5. (Note that, in cases (2) and (4), $B = *$; in cases (6) and (8),

	I_t	I_{t+1}	Quintuple of T
1	$Aa''sa\,B$	$Aa''s'\,a'\,B$	(s, a, s', a', N)
2	$Aa''s*B$	$Aa''s'\,a'\,B$	$(s, *, s', a', N)$
3	$Aa''sa\,B$	$Aa''a'\,s'\,B$	(s, a, s', a', R)
4	$Aa''s*B$	$Aa''a'\,s'\,B$	$(s, *, s', a', R)$
5	$Aa''sa\,B$	$As'\,a''a'B$	(s, a, s', a', L)
6	$A*sa\,B$	$As'*a'\,B$	
7	$Aa''s*B$	$As'\,a''a'\,B$	$(s, *, s', a', L)$
8	$A*s*B$	$As'*a'\,B$	

Figure 7.8.5

$A = *$.) We see therefore that for each quintuple of T we may write one (or two) productions of the type

$$I_t \rightarrow I_{t+1}$$

which are exactly of the type (7.8.7). For example,

$$Aa''\;saB \rightarrow Aa''\,a'\,s'\,B$$

is interpreted as

$$\alpha_0(Aa'')\,\alpha_1\,\varLambda\alpha_2\,B\alpha_3 \rightarrow \beta_0(Aa'')\,\beta_1\,\varLambda\beta_2\,B\beta_3,$$

with $\alpha_0 = \alpha_3 = \beta_0 = \beta_3 = \varLambda$, $\alpha_1 = s$, $\alpha_2 = a$, $\beta_1 = a'$, $\beta_2 = s'$, which is in canonical form (7.8.7).

If we denote by Z the set of the quintuples of a Turing machine T, then P_Z denotes the set of canonical productions obtained from Z according to the rules expressed by the table in figure 7.8.5. If the input tape of T is a word A' and the Turing machine starts in state s_0 at $t = 0$ with the RWH on the leftmost symbol of A', the initial instantaneous description of T is

$$I_0 = **s_0 A'.$$

It is then clear, by the preceding construction of the table in figure 7.8.5, that the sequence of instantaneous descriptions I_0, I_1, I_2, ... of T are theorems generated in P_Z. Therefore, given T we can construct P_Z. The converse process—i.e., the construction of a T given P_Z—is also always possible; because it is complex, however, we shall not present it here.

Reference

M. Minsky, *Computation: Finite and Infinite Machines,* Prentice-Hall, Englewood Cliffs, N.J., 1967.

Prerequisite Structure and Teaching Plans

In this appendix we describe the prerequisite structure of the book by means of a Hasse diagram of a partially ordered set, introduced in Section 4.2.

The elements of this set are the sections into which the book is subdivided, and the partial ordering is the relation "to be prerequisite for." This relation is described in figures I and II, where edges must be thought of as being directed from top to bottom on the page, and circles represent sections of the book. The diagram is to be interpreted as follows: Section $X.Y$ is a prerequisite for section $Z.W$ if and only if there is a *down-directed* path from $X.Y$ to $Z.W$. Technically speaking, the Hasse diagram illustrates the immediate-predecessor relation associated with the given partial ordering. In other words, the diagram does not describe a teaching sequence; rather, it illustrates the essential background of any given section of the book. *The order in which sections must be covered is that in which they are presented in the table of contents.* The Hasse diagram should help the instructor extract a self-contained program with no effort. Figures I and II—essentially the same figure—describe more than the prerequisite structure. In fact, by means of heavy lines we have indicated our suggested teaching plans. In figure I we illustrate the essential core of a course in discrete structures: sets, relations, graphs, lattices, boolean algebras, semigroups and groups. In figure II we illustrate four optional supplementary programs: algebraic structures, topics in boolean algebras, combinatorics, and algorithms.

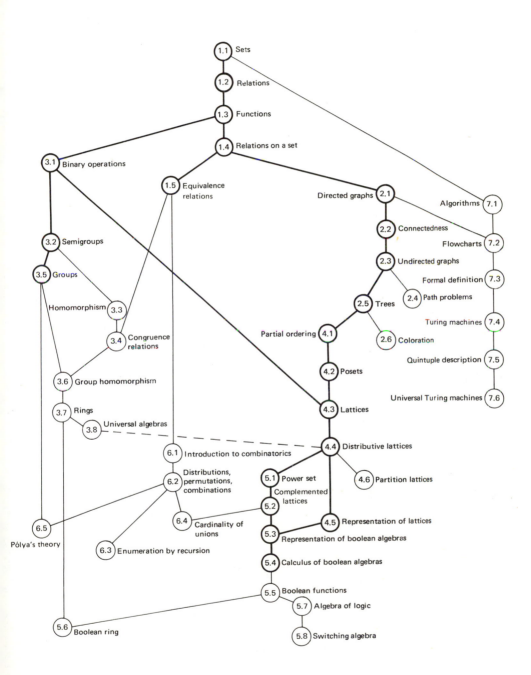

Figure I. Prerequisite structure and core topics

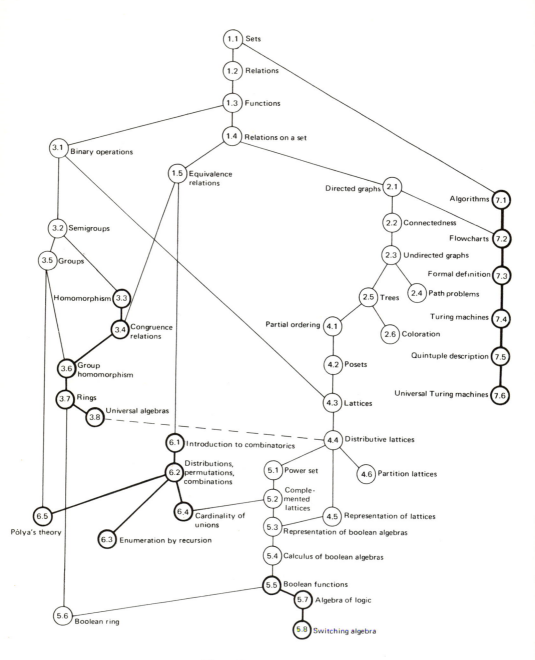

Figure II. Optional topics

INDEX

†Italic numbers refer to pages containing definitions; numbers with an asterisk refer to pages with an Exercise Section.